THE EARTH AND ITS OCEANS

THE EARTH AND ITS OCEANS

ALYN C. DUXBURY
University of Washington

ADDISON-WESLEY PUBLISHING COMPANY
Reading, Massachusetts
Menlo Park, California · London · Amsterdam · Don Mills, Ontario · Sydney

 Printed in the United States of America. Published simultaneously in Canada. Library of Congress Catalog Card No. 73-131202.

ISBN 0-201-01616-1
EFGHIJKLM-HA-798765

To my Parents

FOREWORD

In recent years numerous books have been written that treat the earth, the ocean, and ocean science. Some authors use a broad descriptive approach, whereas others pursue in depth the application to the oceans of a particular single scientific discipline such as chemistry, physics, biology, or geology. Those using the general approach frequently fail to connect in a meaningful way the ever changing ocean to its scientific framework, including the processes that have led to the present state of the ocean and that tend to alter it over both short and long periods of time. Conversely those that stress a particular scientific discipline as applied to the oceans rarely give the desired integration with the other disciplines that also contribute. All too frequently perspective is lost to the extent that the real ocean becomes the handmaiden to the discipline.

In this volume Dr. Duxbury uses the contributing scientific disciplines as tools, not goals. He employs the physical-geophysical approach to better describe the world ocean and its properties as they occur geographically and as they change with time. More significant than the description is the insight he gives into the processes that tend to alter the distribution of properties and the special way these processes apply to the sea. To give an integrated picture he has closely related the ocean to its immediate boundaries, the lithosphere and the atmosphere, and to the sun-moon-earth system of which these bodies are interacting components. The physical and associated chemical environment provide a background against which to view the biological and geological regimes of the oceans.

A balanced presentation includes material ranging from the classical approaches to the newer theories and to the applications of modern technology. He has used a direct but novel approach that should raise many questions in the reader's mind, yet at the same time should either answer or give the pathways through which many of the answers might be approached. The treatment is at an introductory level for science majors, but the combination of description with the scientific *why* should give the general reader a feeling for the ocean and its processes and encourage him to probe more deeply into this interesting area of study. Those having an elementary acquaintance with the basic sciences and mathematics will learn much of how these are applied to the oceans and the atmosphere. Those concerned with a specific scientific discipline as it applies to marine phenomena should find an entry for pursuing selected topics in depth using more sophisticated approaches, and also to gain some feeling of how disciplines outside their own can help them to better understand the sea. The nonscientist should gain an elementary understanding and appreciation of the ocean system as a dynamic entity, and of the rationale of ocean processes extending far beyond that given by any purely qualitative description regardless of its glamour.

Dr. Duxbury has assembled conveniently under one cover much of the information ranging from basic concepts to practical tools and methods needed by anyone undertaking the study of the earth and its oceans. He is to be commended on his careful selection of subject material and on presenting even the more complex concepts clearly and concisely.

Clifford A. Barnes
Professor
Department of Oceanography
University of Washington

PREFACE

The approaches used in organizing and presenting material about the earth and its oceans are as varied as the individuals who undertake the task. Entire books are dedicated to single topics that in themselves form only small parts in the study of the oceanic environment; others are devoted to generalized description of the marine system. The choice of approach is governed in part by the purpose for which the material is assembled, the intended educational level, the degree of understanding that is desired, and also by the particular way in which the author himself considers the material.

The approach used in this book is that of an overview of the oceans as an environment: an environment that is affected by what the earth does, what the earth is, the earth's setting within our solar system, and the properties of the materials that form our earth. The purpose of this approach is to present an integrated picture of the oceans for those who are not scientists but who are interested in increasing their understanding of the marine environment, its whys and wherefores. In some cases topics will be presented and left as open questions without definite conclusions, but such is the state of our knowledge of natural phenomena.

Although you are to a certain extent bound by my approach, you may agree or disagree with what I have felt to be important, the emphases and omissions of material which I have chosen, and the manner in which I have presented it. However, to do so, you must develop informed opinions about the oceans as a dynamic system and the relative importance of processes in the sea and their distribution in time and space. If you do this, then I will have accomplished a major portion of my purpose in writing the book.

The book emphasizes interactions among atmospheric processes, oceanic processes (static, dynamic, physical, chemical, geological, and biological), and processes occurring in and on the solid earth. By necessity it contains details to support its general approach. I have attempted to supply sufficient details about processes and properties of the earth, water, atmosphere, biota, and energy balances and their effects on each other to explain why the sea has its observed features and why they change in time and space.

You may tour the land and observe at first hand its environmental changes and even postulate correctly why such changes occur. You cannot tour the marine environment with equal ease, however; neither can your senses directly register the subtle changes that are present in the marine realm. Thus I have attempted here to develop concepts that are sufficiently general to allow you to use your knowledge of the grosser land changes to gain a fuller understanding of the subtleties of the water-covered portion of the earth.

No book has ever been set between its covers as the result of the author's efforts alone. This book involves many to whom I owe much and can extend only token thanks. Its shortcomings I accept as my responsibility, but credit for its strengths I hope to share. My special thanks go to Professor Richard H. Fleming for his encouragement in this project, the reviewers who commented on the material and assisted me in improving its content, and those who assisted in preparation of the manuscript—Martha Ellis, Jan Jones, Shirley Patterson and Robert Munson. I would also like to thank the production staff at Addison-Wesley Publishing Company for their kind and able assistance. Last, but certainly not least, I wish to thank my wife, Alison, whose assistance, encouragement and patience have allowed me to undertake and complete this task.

Seattle, Washington *A. C. D.*
December 1970

CONTENTS

Chapter 4 The Ocean's Basins and Borderlands

Chapter 5 Sediments of the Deep Sea

Chapter 6 The Chemistry of Seawater

Chapter 7 Physical Properties of Seawater

CHAPTER 1

GEODESY OF THE EARTH AND THE EARTH'S STRUCTURE

INTRODUCTION

The oceans of the world are a feature of the outer surface of the earth. Being a part of the earth, these bodies of water reflect in many ways the structure of, and the processes that occur within, the earth. Thus, to understand the *what*, the *why*, the *where*, and the *when* about the oceans as a system, we must first understand some of the general features of our planet. We could, of course, list the gross properties of the earth and describe its physical configuration, expecting the reader to accept our statements as facts. However, we feel that a far more interesting course is to show how we have gained our knowledge, considering in particular that the earth has never been opened for detailed inspection. We shall therefore discuss in this chapter how the geophysicist proceeds from measurements made on the earth's surface to statements about the interior structure of the earth. His approach is very similar to the familiar process of diagnosis used by a physician to determine the disease from measurable external symptoms. The geophysicist, like the physician, progresses by the process of elimination and the support of cause-and-effect relationships; only the measured parameters change.

We have also included a section on cartography to acquaint the reader with the various methods of representing features of the earth on maps. Projecting the spherical earth's surface on a flat plane involves distortion. The kind of distortion and the properties of the distorted image are important. For example, on a standard Mercator map the area portraying a 1°-square grid segment centered on the equator is only half as large as that portraying a 1°-square segment centered at approximately 59.5° Lat. The distortion of the earth's features created by the expansion of the Mercator grid with latitude results in the portrayal of one square mile of the earth's surface centered on the equator as only half the size of one square mile centered at 45° Lat. In reality, of course, both are the same size on earth.

Time is also important. We live and measure by our local time. On the global scale, however, our time is not the same as that of a person at another location. Thus, since time is related to the longitude system used to denote locations on the earth, we must have a standard system of time relationships in order to place events on the global scale accurately in relative time.

BASIC CHART PROJECTIONS

Sooner or later in the course of studying the earth it becomes necessary to portray the distribution of its topography (properties or features of the world) on a much reduced scale. The charts or maps used in such representations are developed by *cartographers*, who specialize in the techniques of mapmaking and the projecting of the spherical earth on a flat plane. Charts and maps are basically the same. The term map, however, is more appropriate for land areas, while chart is reserved for representations of the sea. To represent the earth accurately, its spherical shape should be scaled down, retaining its form; i.e., ideally the earth should be represented by a globe. Since a globe cannot be folded into a book or carried about in one's pocket, two-dimensional charts are used to portray the three-dimensional earth. Each two-dimensional chart, however, is a distorted representation of the earth's curved surface.

The two-dimensional charts displaying the latitude and longitude grid of the earth are called *maps*. They are obtained by projecting the grid system of the earth onto a two-dimensional plane. Three basic surfaces are used for projections: cylindrical, conical, and a tangent plane. In the first two projection types, the earth is projected onto a cone or cylinder, which is then cut and unrolled to form a plane. There are many chart or map projections, but most of these are merely slight modifications of the three basic types.

The basic cylindrical chart is obtained by assuming an imaginary light source at the center of the earth which casts the earth's features as a shadow on a cylinder placed around the earth. If the cylinder's circle of tangency is the earth's equator, the result is an *equatorial cylindrical projection*. Moving the cylinder so its circle of tangency passes through the poles creates a *transverse cylindrical projection*. A circle of tangency at any other great-circle position produces an *oblique cylindrical projection*.

The well-known *Mercator* projection frequently used in navigational charts at mid- and low-latitudes and found in many books is an adjusted version of the equatorial cylindrical projection (Fig. 1.1). In this projection, the earth's features are portrayed without distortion along the equator, the line of tangency, while the distortions in area become large as latitude increases; the poles cannot be shown.

A cone placed hatlike over the earth can also be used as a projection surface. The cone can be assumed to be tangent to the earth's surface at one small circle or it can be made to intersect the earth's surface along two small circles. The earth's features are again cast onto the surface by a centrally located light source to produce the *conical projection*. Conical projections are typically oriented in such a way that the cone's axis coincides with the earth's rotational axis and hence are called *polar conics*. The circle of tangency of a polar conic is therefore a latitude circle, called the *standard* parallel of the projection. This standard parallel marks the zone of least distortion. The location of the standard parallel is determined by the apex angle of the cone; the standard parallel produced by a sharply pointed cone will be at a lower latitude than that produced by a flattened cone (Fig. 1.2). In some instances, several tangent cones of differing apex angles are used. In this case, each cone has its own standard latitude. A composite projection from the multiple conics can then be made, resulting in a wider latitude range of low distortion. This

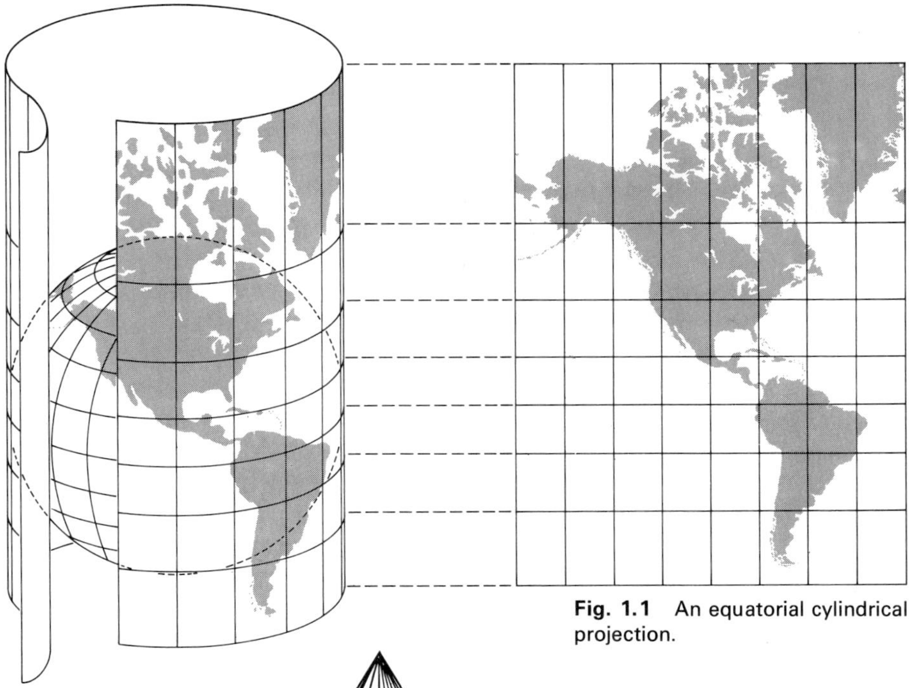

Fig. 1.1 An equatorial cylindrical projection.

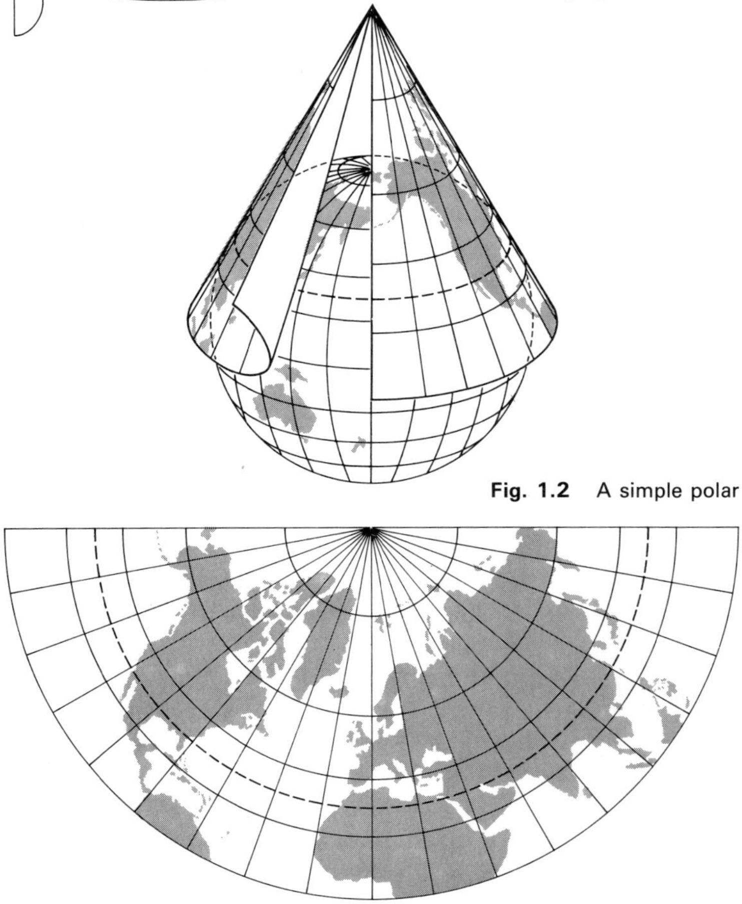

Fig. 1.2 A simple polar conic projection.

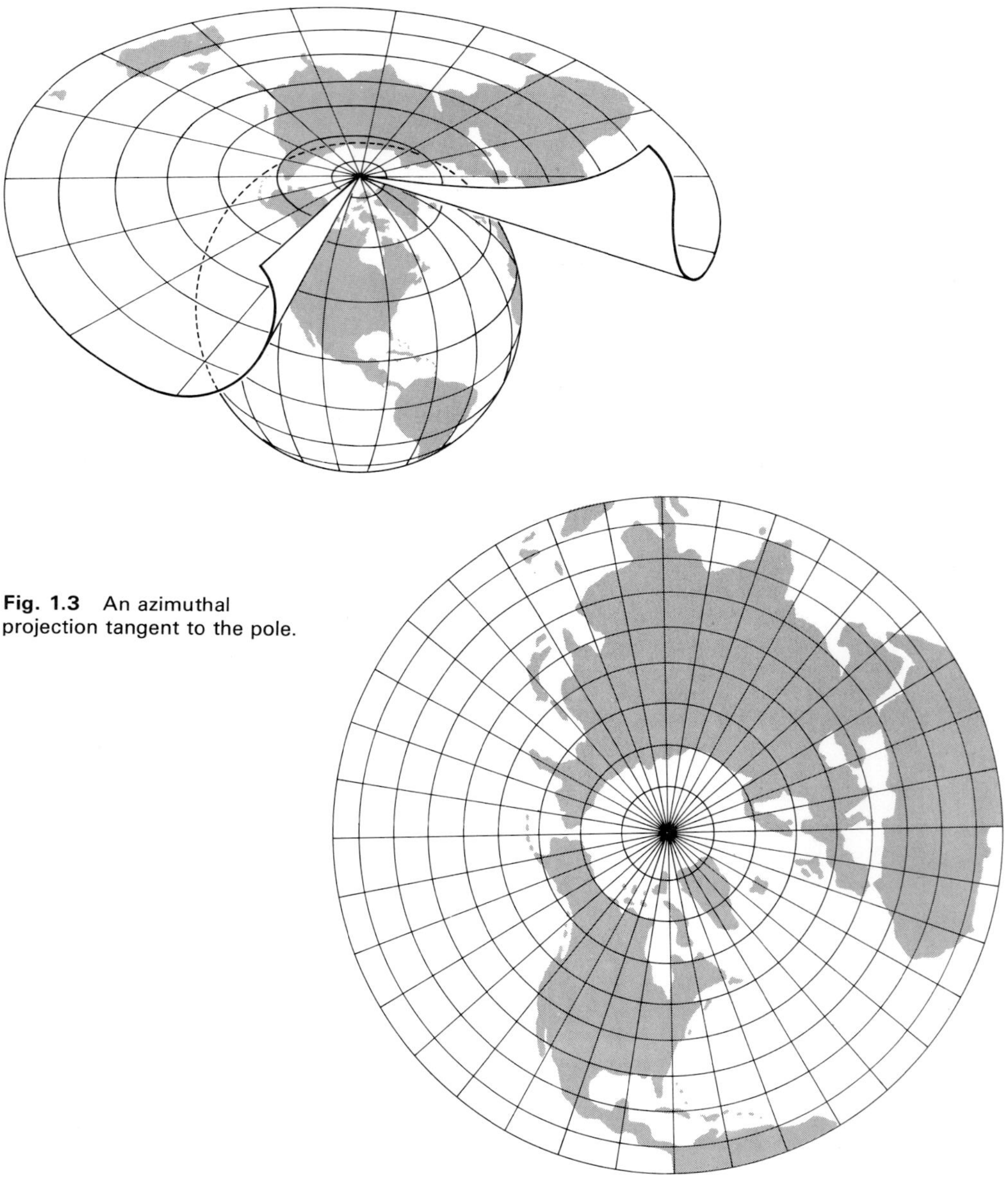

Fig. 1.3 An azimuthal projection tangent to the pole.

projection is called *polar polyconic projection*. Many mid- to high-latitude charts are produced in this manner.

If the polar cone intersects the earth rather than being tangent to its surface, there are two parallels of latitude at which the conical surface enters and exits the earth. This type of cone is called a *secant cone*, and the two parallels of intersection are the standard parallels. The chart obtained in this manner is a *conformal conic* projection.

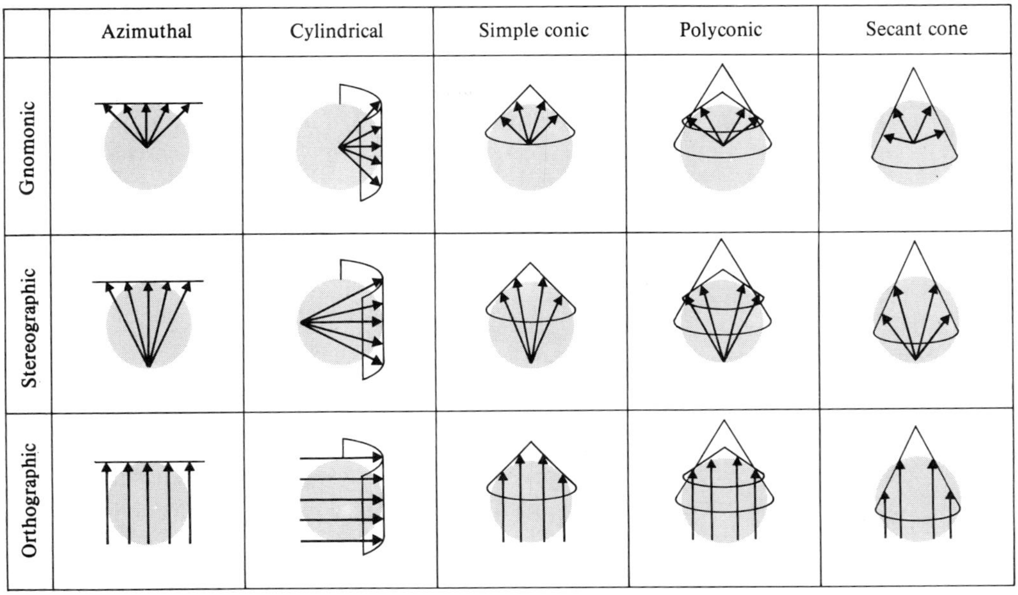

Fig. 1.4 Principal chart projection types.

The third type of projection is the tangent-plane or *azimuthal projection* (Fig. 1.3). The plane is placed tangent to a point on the equator or on either pole. The area of least distortion is at this single point of tangency. Azimuthal projections can also be made obliquely to portray small regions about any point on the earth.

Thus far we have mentioned only the basic shapes of projection surfaces and the types of projections obtained when the light source is assumed to be at the earth's center. However, the projecting source can also be located at a point on the earth antipodal to the point of tangency of the projection surface, or at infinity. If the projection source is at the earth's center, the projection is called *gnomonic*: if the source is antipodal, it is *stereographic*, and if it is at infinity so that parallel projecting rays pass through the earth onto a plane tangent to the sphere, it is *orthographic* (Fig 1.4). Thus the terms *oblique*, *transverse*, *equatorical*, and *polar* describe the attitude of the projection surface to the earth, while *gnomonic*, *stereographic*, and *orthographic* identify the location of the projecting point.

There are also many special-purpose charts that are produced mathematically and are unrelated to the simple projection techniques described above. Each chart type has its own characteristics and is used because of these. Since the conversion of the three-dimensional (spherical) surface to a two-dimensional plane necessarily leads to distortion, the user should select the chart type with the least distortion which is best suited for his purposes. When choosing a projection, we must consider several questions: is equality of area preserved throughout the chart? Do positions have the correct angular relationship with each other? Is the projection conformal (do the features on the chart appear as they do on the globe)? Are *rhumb lines* (lines that cross all meridians at a constant angle) or *great circle* paths portrayed as straight lines? The properties of each projection and their effect on the usefulness of the chart determine the type to be selected. It is impossible to discuss all facets of the numerous possible projections. The reader is referred to a basic text on cartographic techniques for further details.

LATITUDE, LONGITUDE, AND LOCAL TIME

The graticule, or grid of lines, that makes up the latitude and longitude system of the earth is not produced from linear measurements over the earth's exterior, but rather by angular measurements from the earth's interior. *Latitude*, expressed in angular degrees, is the angle subtended by a point on the earth's surface, the center of the earth, and the equatorial plane. Latitude is given in degrees north or south of the equatorial plane, ranging from 0° at the equator to 90° at the poles. *Longitude*, expressed in angular degrees, measures the angular distance of a given location, along a great circle,* from an arbitrary reference point. Unlike latitude, longitude has no obvious reference plane or position from which angular measurement should start (Fig. 1.5).

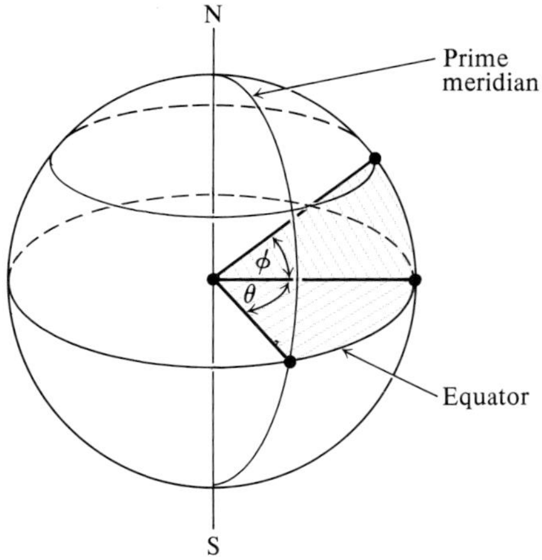

Fig. 1.5 Latitude (ϕ) and longitude (θ) shown as interior angles of a sphere and measured relative to the equatorial plane and the prime meridian, respectively.

To alleviate this problem, an arbitrary zero longitude was chosen that passes through Greenwich, England, and extends from pole to pole. This zero

* A *great circle*, or *geodesic*, is the intersection of a plane passing through the center of the earth with the earth's surface. The shortest distance between any two points on the surface of a sphere is along the great circle between them.

longitude line is called the *prime meridian.* From this reference meridian, angular measurements can be made either in only one direction for a full 360 degrees, or one can measure up to 180 degrees each west or east of the zero line. It has been customary for quite some time to measure longitude angles east and west of the prime meridian. However, on old charts of the British Admiralty longitude is expressed as eastward angles from 0° to 360°. Even today the grid system superimposed on satellite pictures uses a 360-degree longitude angle for the sake of convenience.

The linear measure on earth is related to the angular measure of latitude. If the earth's deviation from a perfect sphere is neglected, equal angles of latitude divide the earth into equal segments of circumference, whereas equal angles of longitude do not necessarily produce equal linear segments. At high latitude, a longitude angle describes a segment of a smaller circle than it does at low latitude. As a measure of surface length, longitude therefore is latitude dependent. On a sphere, the distance described by one degree of longitude equals the length of one degree of latitude times the cosine of the latitude at which the measurement is taken.

The nautical unit of linear measure is the *International Standard Nautical Mile* (6076 ft) defined as approximately one minute of latitude. Hence the length of one degree of latitude along a meridian is about 60 nautical miles. Since the earth is not a true sphere, this linear distance-to-angle relationship varies slightly with latitude. Therefore, distances between points on a nautical chart are always measured against the latitude-degree scale using the local latitude. The standard meter was originally specified as a fraction of the distance from the equator to the pole measured along a meridian. However, with advances in measuring techniques it became obvious that the value established for this distance was subject to change and hence it was abandoned as reference. From 1889 to 1960 the meter was defined as the distance between two lines on a platinum-irridium bar stored in Sèvres near Paris. It is now defined as 1,650,-763.73 wavelengths of the orange-red radiation of krypton-86 under specific conditions.

The time relationships between different locations on the earth are also related to the graticule system. Since the earth rotates 360° approximately every 24 hours relative to the sun, it turns through 15° in one hour; or, conversely, the sun traversing the sky relative to the earth will pass through 15 degrees of longitude in one hour of its westward trek. For example, when the sun is directly overhead at 40° W Long, there is local noon at this meridian. At the same moment, 15 degrees farther west, at 55° W Long, sun time is 11 a.m. since the sun has not yet reached its zenith at this position. Similarly, it is 1 p.m., or one hour past noon, at 35° W Long.

To establish a universal time system, the prime meridian at Greenwich was chosen as the starting point. The solar time changes by one hour every 15 degrees of longitude. The starting time zone, 15 degrees of longitude wide, is centered on the prime meridian and is called the *Greenwich mean time* (GMT)

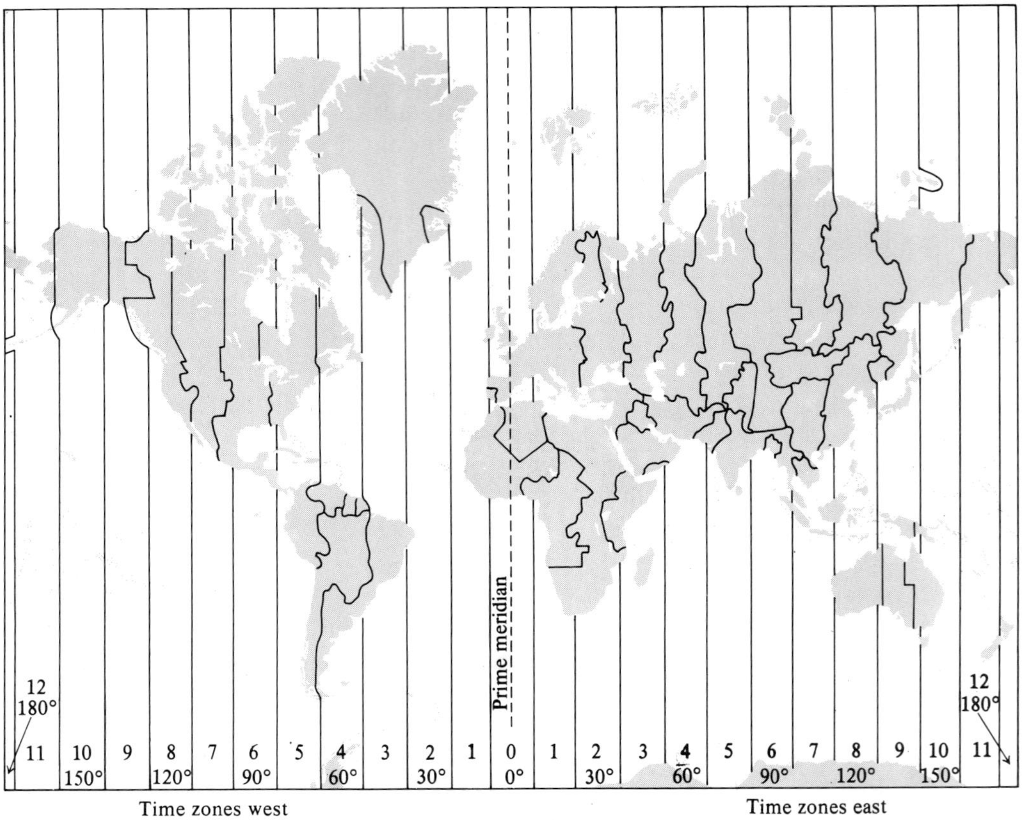

Fig. 1.6 Distribution of time zones on the earth. Each zone is about 15° of longitude wide, with irregularities present on the continental land masses.

time zone. Relative to the sun, the eastern boundary of a time zone is always an hour later than the western boundary (Fig. 1.6).

Adherence to the rigid geometry governing the time zones (15 degrees of longitude wide) would give rise to insurmountable problems in practice. To prevent the division of cities, population groups, and states by rigid zone boundaries, the boundaries of the time zones have been allowed to meander considerably (Fig. 1.7). This manipulation of the time boundaries makes it easier for local populations to keep common time, but in no way alters the actual time of local noon when the sun is at its zenith.

Consider a 15° Long zone centered on a particular meridian: all people within this zone keep the same clock time, which registers 12 noon when the sun is directly over the zone's central longitude. According to the sun, however, at the eastern edge of the boundary, it is 30 min later than local sun noon, and at the western edge, it is 30 min earlier.

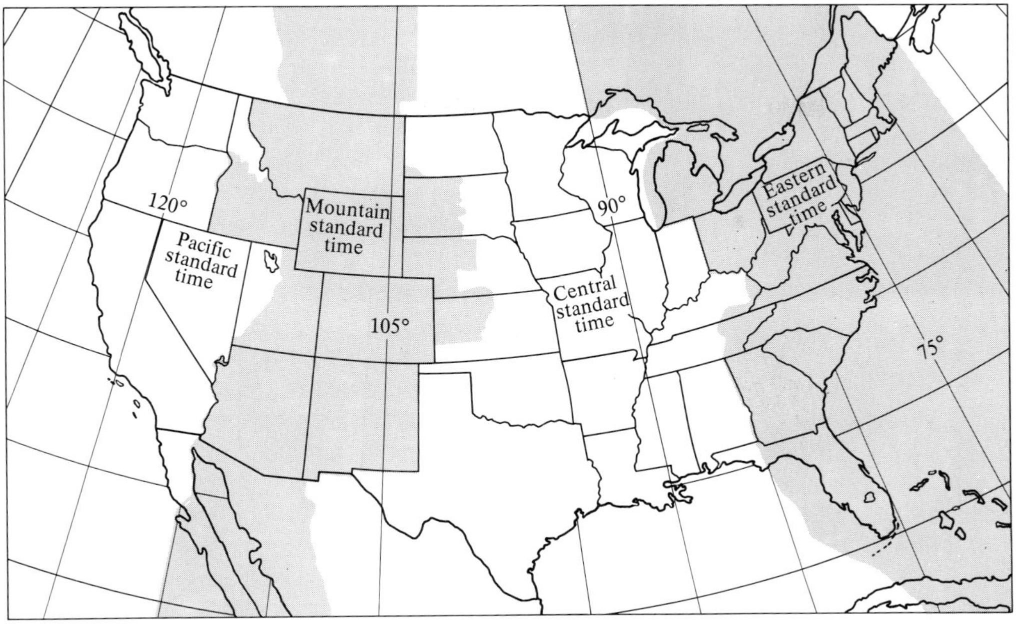

Fig. 1.7 Time zones of the continental United States.

The time zones, like the meridians of longitude, are numbered eastward and westward from the Greenwich meridian: plus zones to the west, and minus zones to the east.

Since time increases—it is later—as one moves east, and decreases—it is earlier—as one moves west, the two time zones meet at +12 and −12, the 180-degree meridian. To avoid confusion, 180° Long has been established as the international date line, on which the day begins (Fig. 1.8). Thus one must set the calendar back by one day when crossing the international date line from the east longitude zone (minus zone) to the west longitude zone (plus zone) and ahead one day when moving in the opposite direction. The hour remains the same during the crossing. When it is precisely noon at Greenwich, the calendar date is the same on either side of the date line. However, it is 0000 hours on the west longitude side of the date line and 2400 hours on the east longitude side. Therefore 24 hours still separate the two sides of the line.

It seems that the consequences of crossing this line were first noted by Magellan's crew. Having completed their circumnavigation of the world, they found that their daily ship's log was in error by one day even though each day had been carefully accounted for.

Fixing one's position accurately on the earth's surface and then transferring this location accurately to a chart is just as important as determining the

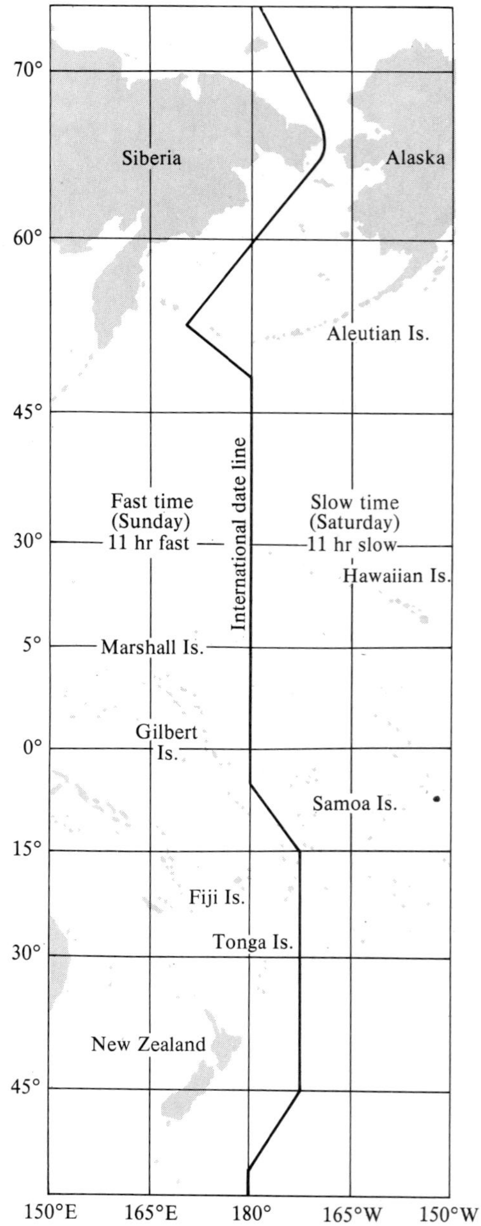

Fig. 1.8 The international date line.

physical characteristics of the earth. Accurate measurements of the earth's parameters are useless if both the time and place at which the measurements were made are not known accurately.

The classical method of determining position on the trackless ocean and even on land is celestial navigation. We can determine latitude and longitude if we are able to find the angular bearing to celestial bodies and if we know the time. Latitude is determined more easily than longitude, and can be obtained, for instance, by measuring the elevation of the north star, Polaris, above the horizon. Longitude, however, is very sensitive to time since the earth rotates on its axis quite rapidly, causing celestial bodies to move swiftly from east to west. If an accurate clock is set to Greenwich time and carried with the traveler, the clock time at which the sun reaches its maximum zenith angle or the relative position of the navigator to the stars can be used to establish the longitude at which the observation is made.

If the time of the zenith of the sun is determined to be 1430 hours GMT, then the longitude is $2\frac{1}{2} \times 15^\circ$ (one hour equals 15° Long), or 37.5° W of Greenwich. Clocks capable of keeping accurate time aboard ships were not available until 1735, when John Harrison completed his first marine chronometer. This historic timepiece still ticks away in the National Maritime Museum in Greenwich. A version of this chronometer (Model No. 4) was used by Captain Cook on his 1772 voyage in search of Antarctica. In the days before longitude determination was made relatively simple, latitude sailing was standard practice. A ship would sail north or south until it arrived at the latitude of its destination. It would then proceed due east or west along this constant latitude.

For slow-moving vessels, accurate positioning in midocean was not very important unless the ship was carrying out survey work. Today, because vessels are faster, ships' courses are changed to avoid storms and obstructions at sea, and detailed survey work is being performed in the oceans, position control must be performed hourly or even more often. In the cloudy areas of the world, celestial fixes are nearly impossible, and dead reckoning must be used to determine position. (Dead reckoning is the method of estimating one's position using relative speed, time, and direction of travel in the interval between more accurate fixes.) To reduce the errors inherent in this method and to improve the accuracy of position control, ships today use electronic navigational aids.

A relatively crude electronic device is the radio direction finder used to determine position from bearings taken from the ship to several broadcast stations. Sonic depth sounders are used with compass headings to determine depth distribution, which is then compared with a bathymetric chart to yield information on position. Where land is within the line of sight of electronic beams, radar aids in position control. Loran networks are used farther out at sea; they enable the navigator to determine his position from the time delay between the radio signals sent from two pairs of known stations. One of the most accurate and newest devices is the satellite navigator which broadcasts its position as a function of time. From the Doppler shift in the frequency of the satellite's signal as it traverses overhead, the surface ship is able to determine its distance to the satellite and the time at which this distance is shortest.

Distance and time are then used by a special-duty computer to calculate the position of the ship relative to the known position of the satellite.

Of the many electronic aids available for time and position control, only a few have been mentioned here. However, we wish to emphasize that it is man's ability to determine position accurately, day or night, in any weather, and at a high rate of repetition that has made it possible to make detailed measurements of the physical properties of the earth and their distribution in time and space. This ability is a relatively recent achievement.

THE SHAPE OF THE EARTH

The earth, though large by man's standards, is relatively small in comparison to other celestial bodies. We stare in wonderment at the magnitude of the high mountains and deep canyons on the earth's surface and can scarcely visualize the width and depth of the oceans. However, if we view these surface features in relation to the earth as a whole, they shrink to near insignificance. The question of how big is big and how small is small cannot be answered in the absolute but requires an appreciation of size on the scales of both man and earth.

Our earth is an oblate spheroid, larger in diameter at the equatorial plane than along the axis of rotation. Because of its plasticity and rotation, the earth has assumed the shape of a ball flattened slightly at the poles. Any deformable round plastic body, when set spinning, tends to elongate in a direction perpendicular to its rotational axis. The earth is no exception.

The force which acts on the plastic material, tending to make a spheroid of rotation oblate is called the *centrifugal force.* It causes the material to creep outward and away from the axis of rotation. The greater the rate of rotation, the greater the centrifugal force, and the more oblate a spinning plastic spheroid becomes. If the plastic sphere is not uniform in composition, or if its mass is not distributed uniformly about its center, or if the cohesive forces that hold it together are not uniform, irregularities in shape will occur and the bulging will not be symmetrical.

The earth is plastic and suffers deformation due to rotation. It is also slightly inhomogeneous in its structure and therefore slightly asymmetrical in its shape. The polar radius (the distance from the center of the earth to the pole) is 6356.9 km, while the equatorial radius is 6378.4 km, demonstrating the effect of the centrifugal force. Recent satellite measurements (Fig. 1.9) have shown that the earth is not symmetrical about the equatorial plane and deviates from the shape of a perfect spheroid of revolution. For reference purposes, scientists have postulated a theoretical rotational shape. This is the *geoid,* which represents a level surface, i.e., one that is everywhere perpendicular to the direction of the force of gravity. The geoid is assumed to have the same overall dimensions as the earth. Satellite measurements show the following differences between the earth's surface and the geoid: The earth's surface is depressed 15 m below the geoid at the south pole and

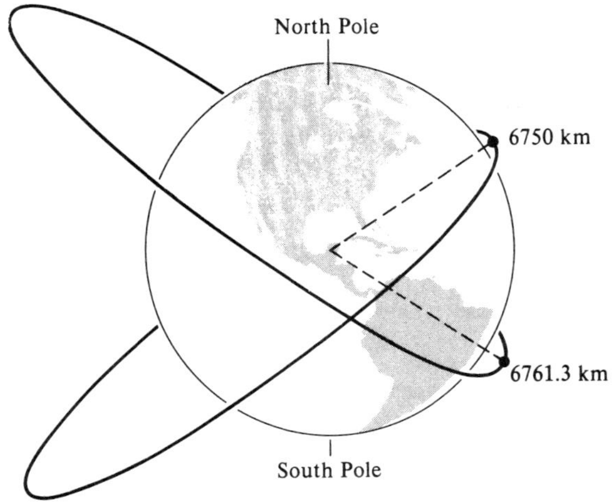

Fig. 1.9 The variation in the distance between an orbiting satellite and the earth's center when perigee is in the southern or northern hemisphere indicates that the earth is not uniformly shaped. (From "The Shape of the Earth" by Desmond King-Hele. Copyright © 1967 by *Scientific American, Inc.* All rights reserved.)

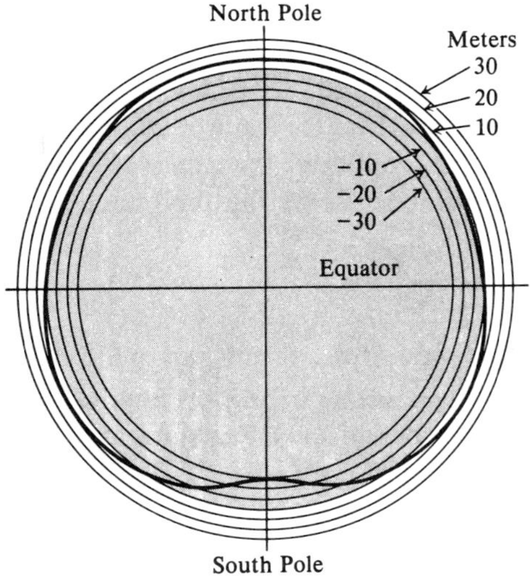

Fig. 1.10 An accurate plot of the deviation of the earth's shape from the theoretical geoid of revolution, as measured by satellite, shows the earth to be slightly pear-shaped. (From "The Shape of the Earth" by Desmond King-Hele. Copyright © 1967 by *Scientific American, Inc.* All rights reserved.)

elevated 15 m above the geoid at the north pole. At midlatitudes, the surface is depressed 8 m in the northern hemisphere and elevated 8 m in the southern hemisphere. These irregularities give the earth a slight "pear shape" (Fig. 1.10).

The deviations from the geoid that lead to the pear shape of the earth seem quite insignificant. Yet, in many cases these small differences and anomalies allow us to better understand the earth. Variations in the acceleration due to gravity, magnetic force, heat flow through the earth's crust, and travel speeds of seismic waves act in concert to furnish important clues to the scientist trying to increase his knowledge of the earth and its interior.

THE EARTH'S GRAVITY FIELD

The earth has a mean radius of 6371.2 km, a mass of 5.983×10^{24} kg, and therefore a mean density of 5.522 g/cm^3 (mean density is total mass divided by volume). The earth's gravitational attraction (acceleration due to gravity) on a unit mass is about 981 cm/sec^2 at sea level. This acceleration, like the earth's radius, varies with latitude. Table 1.1 lists typical values for the earth's radius and sea-level gravity at varying latitudes.

Table 1.1 Variation with latitude of the earth's radius and acceleration due to gravity

Latitude	*Radius, km*	*Acceleration due to gravity, cm/sec^2*
0°	6378.4	978.04
45°	6365.5	980.53
90°	6356.9	983.22
Average	6371.2	981.00

C.R.C. *Handbook of Chemistry and Physics, 34th* ed.

The variation in the earth's acceleration due to gravity has two principal sources. Newton's law of gravitation states that the force of attraction between two mass bodies is proportional to the product of their masses and inversely proportional to the square of the distance separating their centers of mass:

$$F_g \propto \frac{M_1 M_2}{R^2}, \tag{1.1}$$

where F_g is the gravitational force, M_1 and M_2 are the masses to the two bodies, and R is the distance between them. The constant of proportionality that allows us to write Eq. (1.1) as an equality is G, the universal gravitational constant. In the cgs (centimeter-gram-second) system of units, $G = 6.67 \times 10^{-8}$ cm^3 g^{-1} sec^{-2}. We can now rewrite Eq. (1.1) as follows:

$$F_g = G\frac{M_1 M_2}{R^2}. \tag{1.2}$$

In order to determine g_0, the Newtonian component of the acceleration due to gravity acting on an object on the earth's surface, we use Eq. (1.2), with M_1 the mass of the earth and M_2 the mass of the object, as well as Newton's second law of motion, according to which

$$F = ma;$$

that is, the force acting on a body is equal to the product of its mass times its acceleration, or, in terms of the Newtonian gravitational acceleration and M_2,

$$F_g = M_2 g_0. \tag{1.3}$$

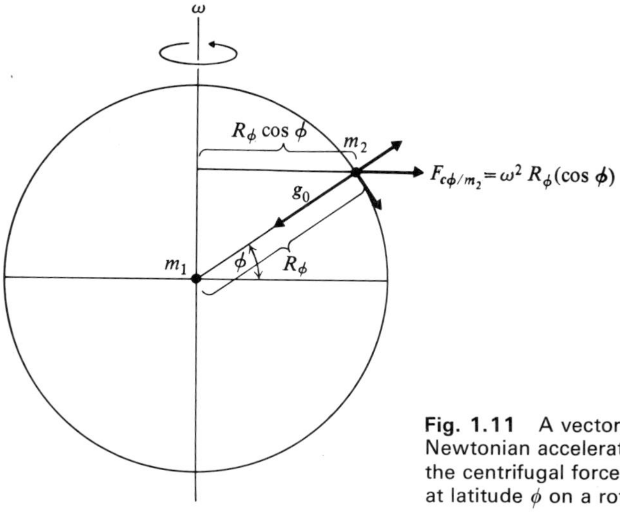

Fig. 1.11 A vector diagram showing the Newtonian acceleration due to gravity, g_0, and the centrifugal force that acts on a mass located at latitude ϕ on a rotating earth.

Since Eqs. (1.2) and (1.3) represent the same force, we can write

$$G\frac{M_1M_2}{R^2} = M_2g_0,$$

and, solving for g_0, we get

$$g_0 = \frac{GM_1}{R^2}$$

where g_0 is the Newtonian component of the gravitational acceleration, M_1 is the mass of the earth, and R is the distance from the earth's center to the center of the mass M_2.

Since the radius of the earth varies with latitude (Table 1.1), GM_1/R^2, or g_0, must also vary with latitude, increasing to a maximum at the poles, where R is smallest, and decreasing to a minimum at the equator, where R is greatest.

Another factor affecting the earth's acceleration due to gravity is the centrifugal force. This force is directed outward and perpendicular to the earth's axis of rotation and hence it has a component acting opposite to the gravitational acceleration g_0 (Fig. 1.11). The centrifugal force at any latitude acting on a unit mass M_2 is

$$F_{c_\phi} = M_2\omega^2R_\phi(\cos\phi). \tag{1.4}$$

The centrifugal acceleration

$$F_{c_\phi}/M_2 = \omega^2R_\phi(\cos\phi)$$

is the product of ω^2, the square of the rotation rate of the earth, R_ϕ, the radius of the earth at latitude ϕ, and the cosine of the latitude. This acceleration, there-

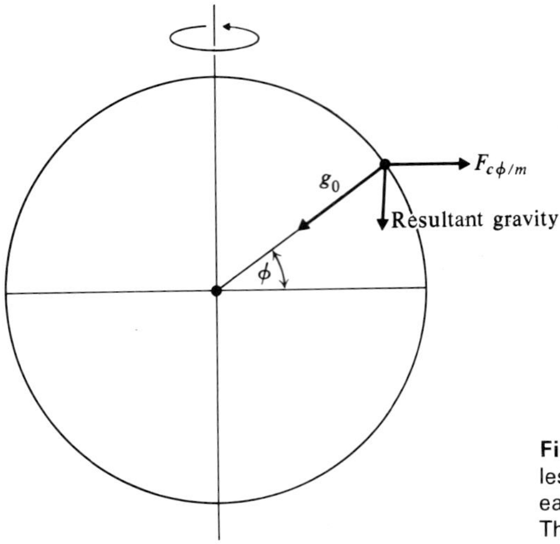

Fig. 1.12 The resultant gravity vector is less than g_0 and is displaced from the earth's center due to the centrifugal force. The effect is exaggerated for clarity.

fore, is greatest at the equator and zero at the poles, where $\phi = 90^\circ$. Hence a unit mass at rest at some latitude of the rotating earth is attracted by Newtonian gravitation toward the earth's center and at the same time is pulled away from the earth's axis by the centrifugal force of rotation. Consequently, the *direction* of the earth's gravity, the resultant of Newtonian attraction and centrifugal force, differs from the line directed toward the center of the earth. Centrifugal acceleration therefore reduces the gravitational attraction below the value obtained from Newton's law and deflects it from the earth's center. However, in Figs. 1.11 and 1.12 this effect is exaggerated, as we readily realize if we compare the centrifugal acceleration's maximum value of 3.41 cm/sec^2 at the equator with the gravitational acceleration of 980 cm/sec^2.

If our plastic earth were completely covered with water, so that its mobile surface could respond to the gravitational and centrifugal accelerations, its liquid surface would nearly coincide with the geoid mentioned earlier. The theoretical geoid surface of the earth is everywhere perpendicular to the effective or resultant gravity and hence, by definition, *level* (not to be confused with a plane, or geometrical surface, which is defined by three points in space). Figure 1.13 shows a level surface. Again, the figure is distorted, showing a disproportionately large effect of the centrifugal force.

One property of a level surface is that a mass can be moved about on it without gaining or losing potential energy. A glance at Fig. 1.13 shows that a mass moved from the polar areas to the equator is raised to a position farther from the earth's center. However, no work against gravitational forces is required because the earth's acceleration due to gravity decreases proportionately. We can calculate the effective acceleration due to gravity over the theoretical geoid surface and then use the values obtained and the shape of the

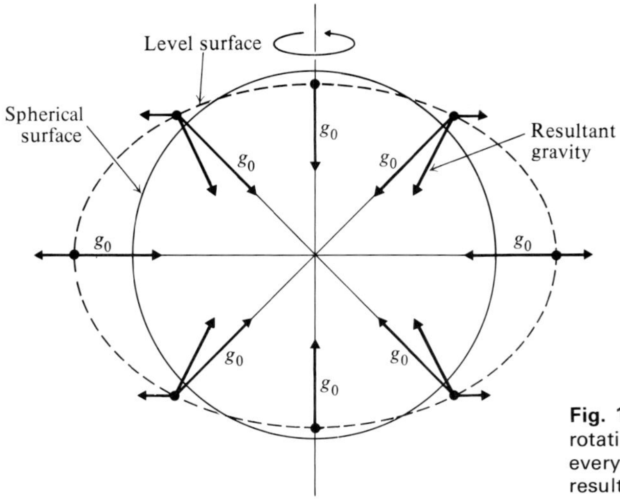

Fig. 1.13 The level surface of a rotating plastic geoid is everywhere perpendicular to the resultant gravity vector.

geoid as references with which we can compare actual gravity measurements taken in different locations and the shape of the real earth. As a matter of fact, it was in this way—by comparing the actual mean of the earth's surface with the theoretical geoid—that the pear shape of the earth was determined. Similarly, the differences between measured gravity and theoretical gravity over the geoid have provided us with insight into the physical make-up of the earth.

If the earth were made up of homogeneous materials or of concentric homogeneous layers conforming to the level surface shape, observed and reference gravity would be identical. Since the observed and the theoretical values differ, it follows that the earth must be constructed of inhomogeneous materials with a slightly irregular distribution. Measured gravity will be less than the reference gravity in mass-deficient regions (regions of light earth materials) and greater than the reference gravity in areas of mass surplus (regions of heavy earth materials).

In practice, it is not possible to measure gravitational acceleration over the geoid reference surface. The mountainous topography of the land precludes it. However, one can compensate for the changes in elevation by reducing the gravity readings taken at various elevations to a standard reference level. The reference level chosen is mean sea level, since this level best approximates the geoid surface. The measured values reduced to sea level are then compared with the theoretical values obtained for the geoid. The difference between the local measured gravity and the local theoretical gravity is called the *gravity anomaly* and is expressed in milligals. One milligal is 1×10^{-3} cm/sec^2. The distribution of gravity anomalies over the earth is indicative of the variation in the density of the materials underlying the earth surface.

Table 1.2 Acceleration due to gravity as related to elevation

Meters above sea level	*Gravitational acceleration, g, cm/sec² at 45° Lat*
0	980.6210
200	980.5593
400	980.4976
600	980.4358
800	980.3741
1000	980.2124

C.R.C. *Handbook of Chemistry and Physics*, 34*th* ed.

Table 1.3 Acceleration due to gravity at selected cities

City	*Elevation, m*	*Gravitational acceleration, g, cm/sec²*
San Francisco	114	979.965
New Orleans	2	979.324
St. Louis	154	980.001
New York	38	980.267
Minneapolis	256	980.597
Seattle	58	980.733
Denver	1638	979.609

C.R.C. *Handbook of Chemistry and Physics*, 34*th* ed.

Although the mean sea level surface is a good approximation to the geoid surface, it does deviate from it. The deviations are due to forces which act on the sea in addition to the gravitational and centrifugal accelerations, making it stand at varying elevations around the earth. Tables 1.2 and 1.3 show the effect of elevation and geographic location on observed gravity.

THE STRUCTURE OF THE EARTH

The average density of the earth is 5.522 g/cm³. Since the mean density of the earth's crustal material as established from deep well cores is about 2.82 g/cm³, it follows that huge quantities of material within the earth must have densities much larger than that of the surface material. Therefore, large changes in density must occur with depth. Since the gravitational acceleration varies by only approximately 0.5% over the sea level surface of the earth, it follows that the large density changes with depth must be uniformly distributed. This is our first clue to the internal arrangement of the earth; it indicates that the interior materials are distributed in concentric spherical shells or layers (Fig. 1.14).

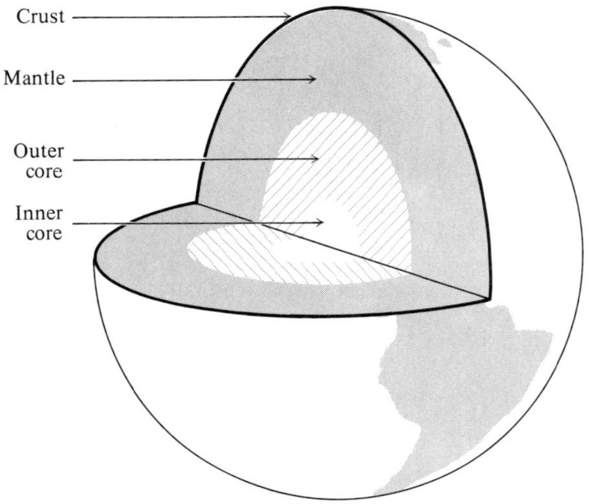

Fig. 1.14 The interior layers of the earth. (Adapted from M. G. Gross, *Oceanography*, Charles E. Merrill, Columbus, Ohio, 1967.)

Table 1.4 Layers of the earth and their densities

Layer	*Depth range, km*	*Density g/cm³* *Range*	*Approximate average*
Crust	0–33	2.5–3.3	2.8
Mantle	33–2900	3.3–5.7	4.5
Outer core	2900–5100	9.4–14.2	11.8
Inner core	5100–6370	16.8–17.2	17.0
Total earth	0–6370	2.5–17.2	5.5

From J. Spar, *Earth, Sea, and Air*, 2d ed., Addison-Wesley, Reading, Mass., 1965.

These layers have been designated as the *crust*, the *mantle*, the *outer core*, and the *inner core*. Table 1.4 shows the thicknesses and densities assigned to these layers.

The density of the material changes continuously in each layer; the table lists the range, as well as the average density for each layer.

How were the thickness and densities determined in the absence of actual measurements which to date have been limited to the material in the crust? Clue two to the earth's structure is obtained by studying the propagation of *seismic waves*, the shock waves from earthquakes which pass through the earth.

There are two types of seismic waves: *transverse*, or S-waves, and *compressional* (longitudinal) or P-waves. Transverse waves, like water waves, oscillate perpendicular to their direction of travel; compressional waves are similar to sound waves, i.e., they travel by compression and rarefaction oscillating in the

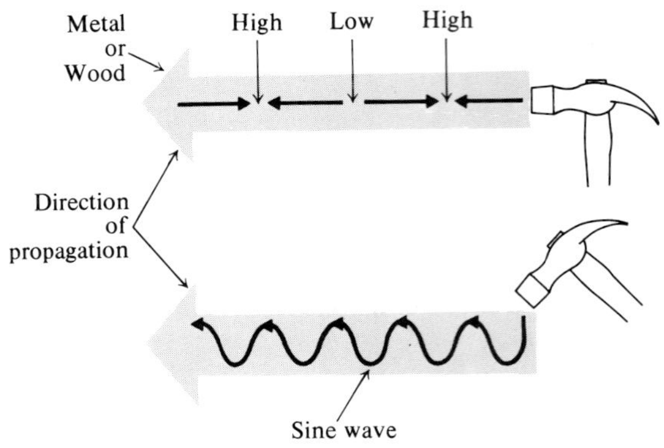

Fig. 1.15 Progressive waves of compressional type, P-waves above; and transverse, S-waves below. (Adapted from T. F. Gaskell, *World beneath the Oceans,* Aldus House, London, 1964.)

direction of travel (Fig. 1.15). Both of these waves will be bent (refracted) if they pass from one material into another having a different density and therefore speed of propagation.

From the refraction patterns of seismic waves the researcher can determine the thickness of each layer of the earth's interior, the speed at which the wave is propagated in each layer, and the density of each layer. Seismic refraction studies also indicate phase changes (from solid to liquid state) in the material through which the waves travel.

The fastest seismic waves are the short-period compressional P-waves. These travel at speeds ranging from 5 to 13 km/sec (average speed 9 km), increasing in speed as the density of the transmitting material increases. If P-waves pass from a solid into a fluid material, their speed abruptly decreases. The refraction of these waves as they pass through the earth encountering the core creates shadow zones on the opposite side of the earth, indicating that the outer portion of the core is in a liquid state (Fig. 1.16).

The transverse S-waves travel more slowly than the P-waves and are incapable of passing through a solid-liquid interface. Measurements of S-wave distributions show that they give rise to a large shadow zone antipodal to the disturbance center (Fig. 1.17). Data from both P- and S-waves therefore give a distribution of the density in the mantle and the size, density, and phase state of the earth's core.

In 1909, Andrija Mohorovicic, in his investigations of seismic wave propagations, discovered that the speed of propagation of seismic waves under land masses changed abruptly at a depth of about 33 km below the earth's surface, indicating a change (discontinuity) in the character of the earth's composition and density at this level. This discontinuity has been named after its discoverer, the *Mohorovicic discontinuity*, popularly shortened to *Moho* (Fig. 1.17). It is the boundary marking the upper limit of the mantle and the lower limit of the earth's outer crust. Thus, the discovery of the Moho is our final clue to the earth's structure. Our picture of the earth's concentric layers is now complete;

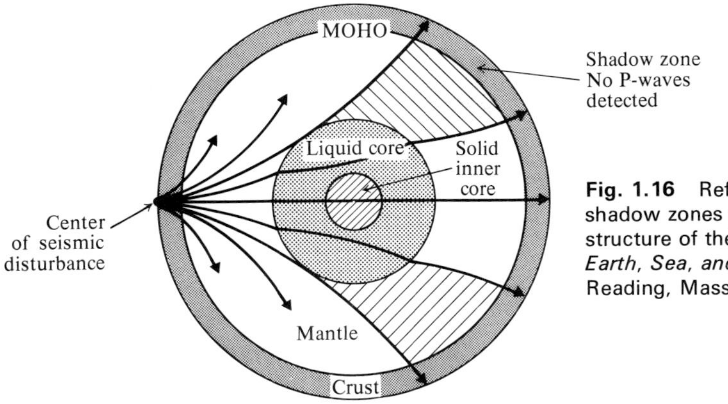

Fig. 1.16 Refraction of P-waves and shadow zones produced by the interior structure of the earth. (From J. Spar, *Earth, Sea, and Air*, Addison-Wesley, Reading, Mass., 1962.)

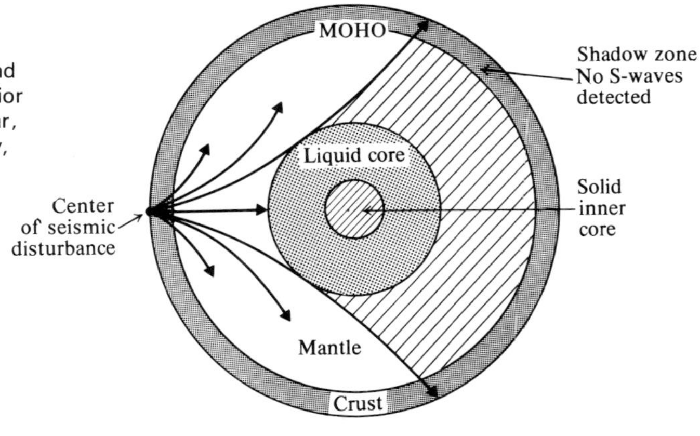

Fig. 1.17 Refraction of S-waves and shadow zones produced by the interior structure of the earth. (From J. Spar, *Earth, Sea, and Air*, Addison-Wesley, Reading, Mass., 1962.)

crust, mantle, liquid, outer core, and possibly a solid inner core—all determined from remotely sensed geophysical data.

Studies of gravity distribution and seismic wave propagation yield considerable information not only about the gross structure of the earth, but also on the fine structure of the earth's outer layers. For example, gravity measurements taken in the mountains yield values that are lower than expected even after correction to the sea level reference, indicating that the crustal material underlying mountainous areas must be of great depth and low density. Localized near-surface deposits of heavy minerals valuable as mining deposits can also be detected by the use of gravimetric surveys. On the other hand, measurements taken over oceanic areas indicate that the seas must be resting on crustal material nearly as dense as the land crust but thinner. Thus the Moho must lie

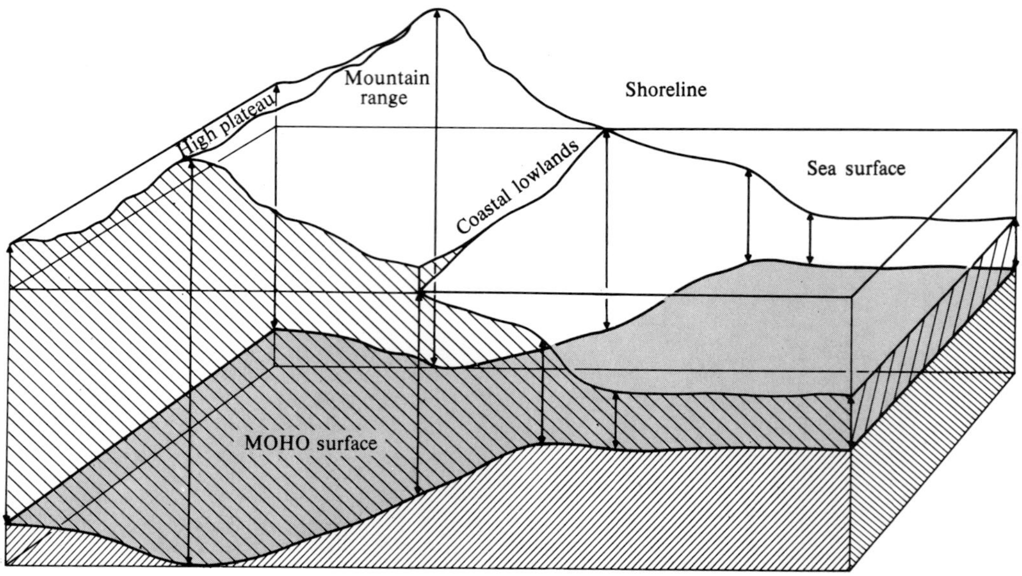

Fig. 1.18 The variation in depth of the Moho with change in the thickness of the earth's crust.

at a deeper depth under the land masses than under the oceans and, in particular, at a deeper level under the mountainous areas of the continents (Fig. 1.18). This is why it is advantageous to drill down to the Moho from the ocean floor instead of from land.

THE CONCEPT OF ISOSTASY

This distribution of crust thickness and mantle depression agrees with the theory of *isostasy* of the earth, according to which equal sectors of the earth contain almost equal masses of earth. That is, the surface layers (crust) of the earth float on the more plastic mantle, with each portion of the crust displacing the mantle according to its bulk and density. Denser crustal materials sink proportionately deeper into the mantle than crustal materials of low density. Similarly, regions of high elevations and therefore large vertical thickness also sink deeper into the mantle than do low regions. However, if the crust is formed locally of light material, a smaller percentage of its vertical thickness is embedded (Fig. 1.18). If true isostatic equilibrium is maintained, then locally the crustal materials should seek their proper level of flotation. However, the strength of the crust and the fact that a less dense material gives buoyancy to an adjacent denser material by mechanical attachment along their common boundary upsets the equilibrium slightly (Fig. 1.19).

We can show that this equilibrium exists by proving that the mantle under the land masses and oceans has approximately the same pressure distribution generated by the overlying column of earth material. Assuming that the materials are fluid, we can use the hydrostatic equation for pressure, $P = \rho g z$, where ρ is the density, g is the acceleration due to gravity, and z is the thickness of each layer. At a depth of 2900 km, the pressure divided by gravity is equal to the sum of the products of density and thickness for each overlying layer. Table 1.5 lists the values used in making the comparison between continental and oceanic areas.

Table 1.5 Pressure/gravity within the mantle

	Layer	Thickness, cm $\times 10^{-5}$	Average density, g/cm^3	$P/g \times 10^{-5}$, g/cm
Land	Crust above sea level	0.8	2.90	2.3
	Crust below sea level	35.0	2.90	101.5
	Mantle	2865.0	4.50	12,890.0
	All layers	~0–2900		~12,994
Ocean	Seawater	4.0	1.02	4.08
	Sediments and crust	6.0	2.7	16.20
	Mantle	2890.0	4.5	13,000.00
	All layers	~0–2900		~13,020

The shearing strength of the earth's crust, which can sustain some imbalance of isostatic equilibrium, is occasionally exceeded, and sudden shifts occur that are called *faulting*. These may occur in response to lateral or vertical forces working on the crust.

The concept of isostatic equilibrium is difficult to realize, especially when we look at the massive mountain systems that are apparently in balance with the lowlands and oceans. We are able to visualize it more readily when we consider the magnitude of the earth's features on the earth's scale. When we reduce the earth to a sphere 15.3 cm (about 6 in.) in diameter and scale down its surface features accordingly, they become roughnesses no greater than the small skin bumps on a grapefruit, and it is no longer difficult to accept the fact that the earth is capable of supporting irregularities of a similar scale.

THE MOBILITY OF THE EARTH'S CRUST AND ITS DETECTION

The geomagnetic properties of the earth that cause a magnetized needle to seek alignment with the north-south magnetic poles have been the subject of much debate. At one time the earth was thought to contain a permanently magnetized substance that aligned itself within the earth to create a magnetic field. This hypothesis, however, is not compatible with the concentric-layer structure of the earth, nor is it compatible with the current hypotheses of magnetic-field reversal and wandering of the magnetic poles. The modern concept of geomagnetism is based on the hypothesis that the earth's magnetic field is produced by electric currents generated in the outer core of molten iron. The principles underlying this concept are the same that underlie the operation of an electric generator, or dynamo. Thus the phenomenon of geomagnetism is also called the "dynamo effect." According to the dynamo theory, the energy source for the "dynamo" is convective movements in the earth's liquid outer core.

The orientation of magnetic materials in the crustal material distributed over the earth indicates that the earth's magnetic field has undergone changes through geologic time. The evidence is provided by geophysical and geochemical dating of crustal rocks and the study of paleomagnetism. Changes in the magnetic orientation in crustal material can be produced in two ways: either the magnetic force field of the earth changes and the land positions remain constant, or the land masses move relative to a fixed force field. There is evidence to support both mechanisms. (The latter is incorporated into the theory of continental drift.)

Recent research using paleomagnetic and dating techniques in the area of the Mid-Atlantic Ridge indicates that over the last 6,000,000 years the ridge has spread laterally at a rate of about 1.25 cm per year. However, this rate has not been constant over the entire ridge and time; expansion has been periodic rather than continuous. From similar measurements it has been estimated that the present lateral movement of the East Pacific Rise is four times as great. The lateral spreading is believed to be caused by the motion of convective cells in the mantle. These cells rotate about their axes, forcing the basaltic mantle material up and into the crust above the zone of ascending cell motion, where it then moves laterally, slowly dragging the crust along until it reaches a region of downward cell motion (Fig. 1.20). Although the presence of such cells provides us with a satisfactory mechanism for driving crustal spread, we now have to track down the factors that lead to accepting the presence of these cells.

The rate of increase in the earth's interior temperature with depth is known to be greater than would be expected if compressive forces were the only heat-generating factor. The difference is believed to be due to the presence of radioactive materials which release heat as they decay. It may therefore happen that a region of the mantle contains a zone of lighter or hotter materials; these would slowly start to rise. As they rise, their temperature would drop due to the decrease in the overlying pressure. However, this temperature drop would be small, and the rising mantle material would soon be much hotter than, and therefore less dense relative to, the material surrounding it at the higher level. This would cause the rising material to increase its rate of ascent, carrying its heat with it. The upward transport of mantle material into the crustal layer would cause a temperature increase with depth

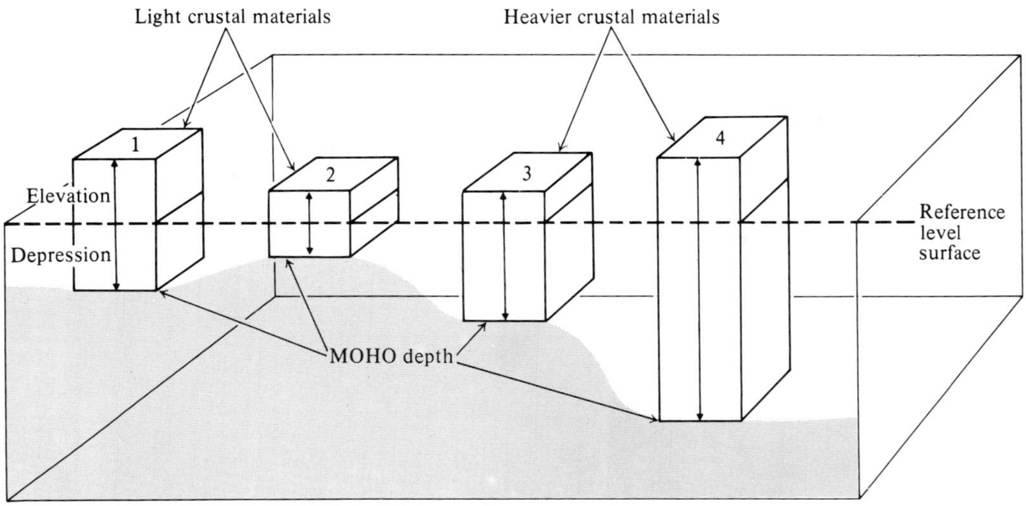

Fig. 1.19 Lighter crustal material can rise to a higher elevation with less depression of the Moho than heavier materials. Compare 1 to 4 and 2 to 3.

in the crust above the ascending zone that is greater than the increase found in an area of descending material. Hence an increase in temperature gradient in the earth's crust and therefore an increase in heat flux from the earth's interior through the earth's surface—if the thermal conductivity of the crust is constant—should occur centered above the ascending zone of a convection cell and coincide with the regions of the earth that are marked by lateral stretching.

The Mid-Atlantic Ridge system is frequently used as an example illustrating this relationship between spreading of the crust and large heat flux and substantiating the convection cell theory. However, the direct measurement of thermal gradients and thermal conductivity of the sediments required to determine heat flux is a difficult task. It is even more difficult to establish patterns of heat flux when measurements are few and the composition, sediment thickness, distribution of sediment types, and tilt of the sea floor vary.

THE HEAT FLUX FROM THE EARTH'S INTERIOR

The average heat flux through the sea floor is found to be about 1.3×10^{-6} cal cm^{-2} sec^{-1}. Values as high as 5.4×10^{-6} cal cm^{-2} sec^{-1} have been measured over localized hot spots while fluxes of 0.6 cal cm^{-2} sec^{-1} occur in less thermally active regions. Recent research on heat flow through the ocean floor has been used both to support and to defeat the hypothesis of a higher flux along the Mid-Atlantic Ridge as measurements of the flux in this region are highly variable. Measurements along the East Pacific Rise show that this region definitely has a higher heat flux. However, lateral stretching of the crust over

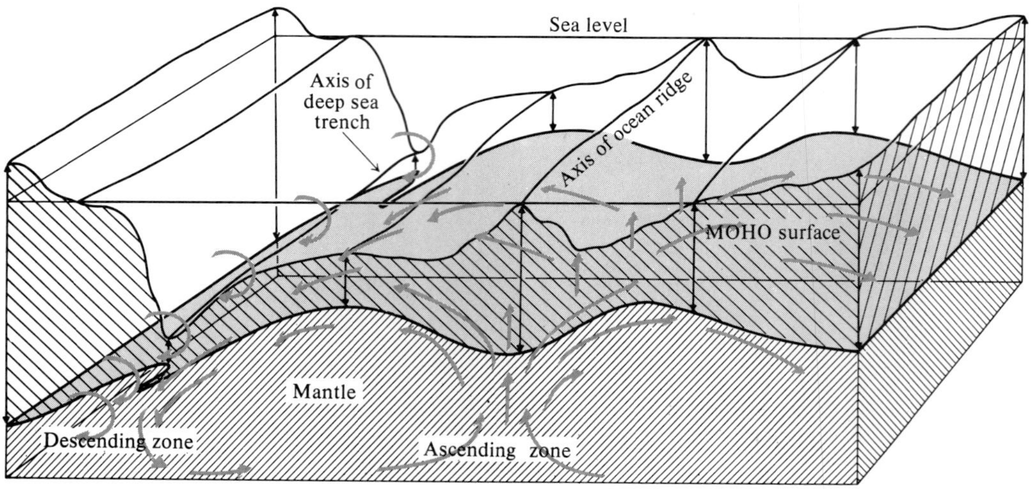

Fig. 1.20 Convection cells within the mantle contribute to crustal material above ascending zones and move the crustal materials laterally away from these zones toward descending zones.

this region is also nearly four times as great as that in the Mid-Atlantic Ridge system.

When many measurements of heat flux through the sea floor are available for averaging, then trends become evident. The heat fluxes for different oceanic regions listed in Table 1.6 have been statistically derived from measurements made at 2000 observation points. The large deviations of values found over ridges indicates that many low values can be obtained in these regions of supposedly high heat fluxes.

Table 1.6 Heat flux through the sea floor

Regional type	*Heat flux, cal cm^{-2} sec^{-1}*
Trenches	$0.99 \pm 0.61 \times 10^{-6}$
Basins	$1.28 \pm 0.53 \times 10^{-6}$
Oceanic ridges	$1.82 \pm 1.56 \times 10^{-6}$

Based on J. C. Maxwell, "Continental Drift and a Dynamic Earth," *American Scientist*, 56 (1), 1968.

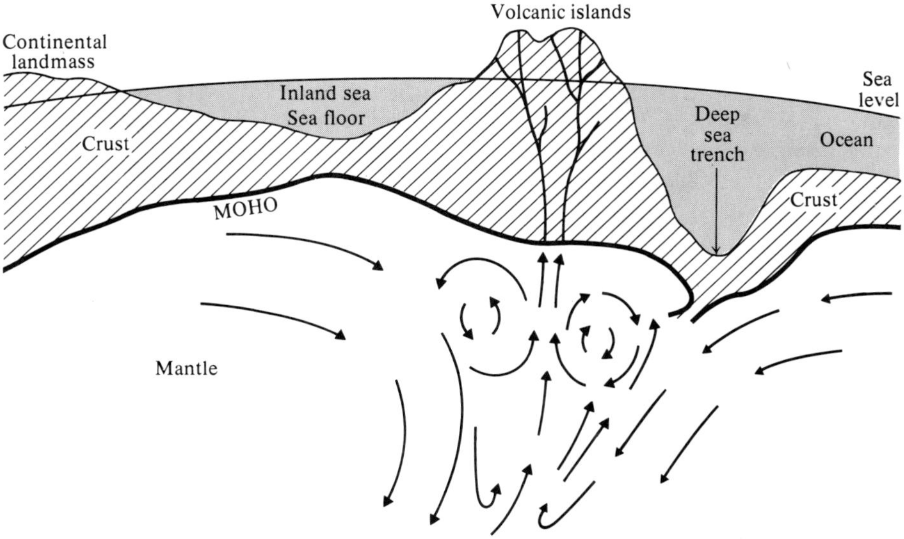

Fig. 1.21 Deep sea trenches and the volcanic island arcs that enclose inland seas are associated with the descending zones in mantle convection cells.

The deep trenches in the oceans owe their existence to the same mechanism that is at work in the buildup of ridges. However, while the ridges rise above areas of ascending cell motion, the trenches are formed by a depression in the crust and a removal of crustal material above areas of intense descending cell motion. Trenches in the sea are often associated with arcs of islands having volcanic activity. Figures 1.20 and 1.21 show the relationship between convection cell motion in the mantle and the formation of oceanic trenches.

Our knowledge of the structure of the earth is far from complete. However, each measurement made in the various fields of geophysics increases our understanding of the structure of the earth and its dynamics. Scientists combine the data from gravimetry, geomagnetics, thermodynamics, and seismology to develop theories (such as the convection cell theory) about the internal structure of the earth. When many independent measurements substantiate a theory, it can be accepted despite the lack of direct observational verification, whereas a theory postulating, say, a mobile earth interior from data gathered in only one area of geophysical research would be far less—if at all—acceptable.

STUDY QUESTIONS

1 Calculate the Newtonian component of the acceleration due to gravity acting on a unit mass located on the earth's surface at the poles, at 45° Lat, and the equator. Use the mean radius of the earth for the radius at 45° N or S and discuss the reason for the change in acceleration with latitude.

2 Calculate the centrifugal force per unit mass acting on a particle on the earth's surface at the poles, 45° Lat, and the equator. Calculate the magnitude of the resultant gravitational acceleration obtained by combining the centrifugal acceleration and the Newtonian acceleration due to gravity.

3 What other factors may cause the observed acceleration due to gravity at any point on the earth's surface to deviate from the values calculated in Problem 2?

4 Since a level surface is everywhere perpendicular to the direction of gravity, how must the earth be shaped to have its surface perpendicular to the resultant of the Newtonian and centrifugal forces of Problem 2?

5 Why is the heat flux through the sea floor found to be statistically higher under the oceanic ridge systems and lower under the deep-sea trenches?

6 How can a cylindrical projecting surface be oriented over the spherical earth to show *any desired* great circle as a straight line?

7 When it is 1245 hours standard time on the western edge of a time zone of uniform width, what is the actual local apparent sun time on the eastern edge of that zone?

RECOMMENDED READING

Beiser, A., and the editors of *Life*, *The Earth*, Time Inc., New York, 1962.

Benfield, A. E., "The Earth's Heat," *Scientific American*, **183**, 6 (December 1960).

———, "The Earth's Magnetism," *Scientific American*, **182**, 6 (June 1950).

Bowditch, N., *American Practical Navigator*, H. O. Publication No. 9, U.S. Naval Hydrographic Office, Washington, D.C., 1962.

Cromie, W. J., *Why the Mohole? Adventures of Inner Space*, Little, Brown, Boston, 1964.

Elsasser, W. M., "The Earth as a Dynamo," *Scientific American*, **198**, 5 (May 1958).

Gamow, G., "Gravity," *Scientific American*, **204**, 3 (March 1961).

Heiskanen, W. A., "The Earth's Gravity," *Scientific American*, **193**, 3 (September 1955).

King-Hele, D., "The Shape of the Earth," *Scientific American*, **217**, 4 (October 1967).

Raisz, E., *General Cartography*, 2nd ed., McGraw-Hill, New York, 1948.

Strahler, A. N., *The Earth Sciences*, Harper and Row, New York, 1963.

Wilson, J. T., "Continental Drift," *Scientific American*, **208**, 4 (April 1963).

CHAPTER 2

THE EARTH IN MOTION

INTRODUCTION

Although the earth's rotation upon its axis, its motion around the sun, the motion of the moon, and that of artificial satellites are well known to most of us, the role of these motions in determining the natural periods of time and their effects on the earth and its inhabitants are not as fully understood. Even less well known is the role of a reference in defining an event or the effects generated by using a reference frame on the earth as opposed to one fixed in absolute space.

In this chapter, we shall explore some of the effects produced by the earth's motion. We shall discuss the problem of defining a day and find that there are three types of days, each having a different length. We shall also see that arbitrary time intervals such as days, months, and years, established by man, differ from natural time periods, requiring the periodic adjustment of the calendar to the earth's pace.

Finally, we shall discuss the apparent forces arising when the motion of bodies is judged against the rotating earth. We will find that two independent motions, quite evident from a viewpoint in fixed space, combine when viewed from the rotating earth to cause bodies in motion across the earth's surface to undergo a deflection.

The motions of the earth produce changes in solar energy levels which in turn create the seasons of the year, generate atmospheric motions, and affect the circulation of the oceans. These motions also control the periodicity of the oceanic and atmospheric tides. Motions of the earth-sun-moon system are as fundamental to the understanding of what is happening on this earth as is the earth's structure.

THE MOTIONS OF THE EARTH AND THE REFERENCE FRAME

The earth, one of the nine planets orbiting about the sun in our solar system, has several modes of motion. The three most important are the translation of the earth about the sun, the rotation of the earth upon its axis, and a very slow wobbling motion of the rotational axis, called precession; nutation, a slight periodic irregularity superimposed on the precession, is a minor factor (Fig. 2.1a). How these motions appear to the observer depends on his vantage point. From a fixed point in distant space, he could readily observe the three modes of motion if given sufficient time; however, if his observations are made from the earth, the motion of the earth is not necessarily apparent. For all practical purposes, the observer on earth may not be aware that the earth is

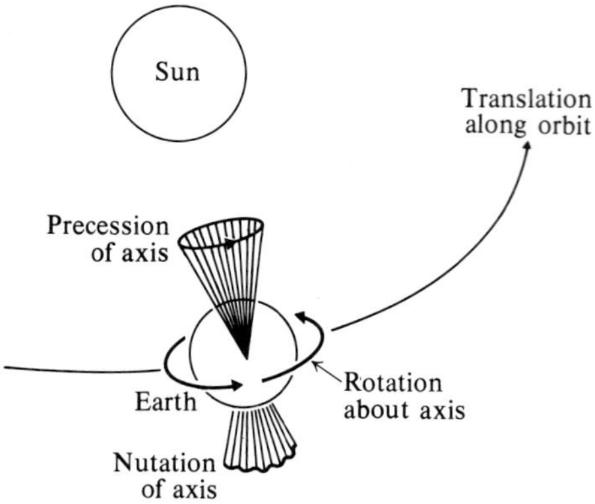

Fig. 2.1a Principal motions of the earth: (1) translation along its sun-centered orbit, (2) rotation on its axis, (3) precession of the axis, and (4) nutation.

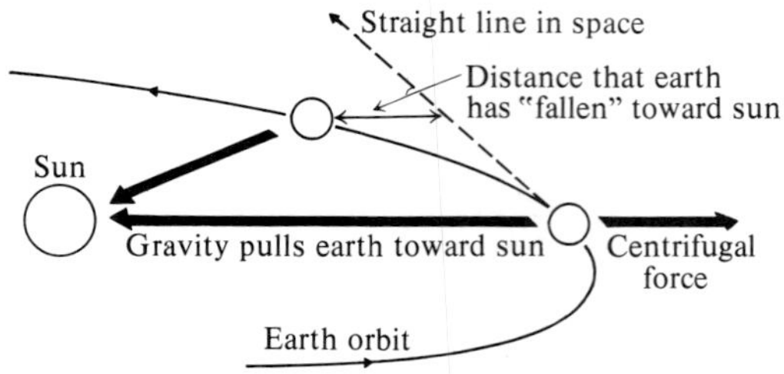

Fig. 2.1b The earth stays in its orbit due to a balance between gravitational and centrifugal forces.

rotating about its axis and traveling its path around the sun; thus for many centuries man assumed that the earth was fixed in space and that the sun and stars moved across the sky and behind the earth only to reappear. Man has since learned to differentiate between apparent and real motions.

In our study of the earth's motion, we shall consider the sun and stars to be fixed in space. The earth will move in an orbit about the sun. The plane described by the earth as it travels about the sun will be called the *plane of the ecliptic* and will always contain the centers of the earth and sun. The orbit described by the earth as it translates about the sun is elliptical, with the sun's center located at one of the foci of the ellipse. The earth is locked into this orbital path by a balance between the gravitational attraction acting as a centripetal force between the earth and sun, and the centrifugal force that acts on a revolving body such as the earth, trying to pull it out of its orbit and allow it to proceed on a straight-line flight into space (Fig. 2.1b).

THE EARTH'S ORBIT

The earth moves about the sun, completing one orbit in approximately 365.25 days, the *planetary year.* However, its progression, or translation, along this elliptical orbit is not uniform. Since the sun is located at one focus of the ellipse, the distance between it and the earth changes during one traverse of the orbit. When the earth is closest to the sun, it is said to be at *perihelion* (91.5×10^6 mi, January) and when it is farthest from the sun, it is at *aphelion* (94.5×10^6 mi, July). On an astronomical scale, the variation in distance between sun and earth at perihelion and aphelion is small, indicating that the

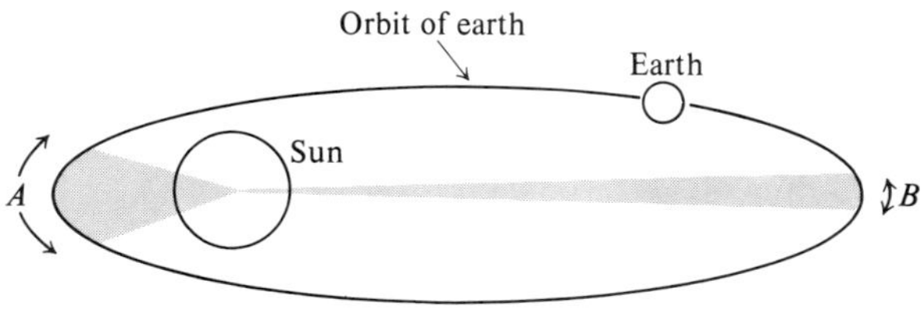

Fig. 2.2 The earth sweeps out equal areas in equal time periods. The angular speed of the earth in its orbit varies to accomplish this. At position A, ω_A = angle A/unit time. At position B, ω_B = angle B/unit time.

orbit is only slightly elliptical; this is indeed the case—its eccentricity is only about 0.017. However, this variation is sufficient to cause variation in the gravitational attraction between the two bodies and consequently changes in the rate at which the earth moves in its orbit. During January, when the earth is closest to the sun, it moves faster than during July, when it is farthest from the sun. This phenomenon is described in Kepler's* second law, which states that a line drawn between the earth's and sun's centers sweeps out equal areas in the ecliptic plane in equal times (Fig. 2.2). As can be seen in the figure, the angle swept by the earth in equal times must change appreciably at different positions about the orbit to maintain equal areas.

NATURAL PERIODS OF TIME IN THE EARTH-SUN-MOON SYSTEM

The earth describes one rotation upon its axis relative to the sun in the period of time known as the *apparent solar day*. It is defined as the time interval between two successive passages of the sun over the meridian of the observer (*solar transit*†). The length of the apparent solar day varies slightly during the year. The mean value is 24 hours, which determines our clock day or *mean solar day*. Relative to a star fixed in space, the earth completes one revolution about its axis in approximately 23 hours and 56 minutes. This time interval, called the *sidereal day*, is defined as the time required for a star to complete two successive transits of a meridian. The sidereal day is nearly constant in comparison with a solar day.

The rotation rate of the earth on its axis, or the time required for one rotation of the earth relative to a star, can be determined accurately only with independent and precise timing devices. The advent of atomic clocks

* Johannes Kepler (1571–1630). His three empirical laws of planetary motion were later confirmed by Newton's work.

† The passage of an astronomical body across the meridian of the observer is called a transit.

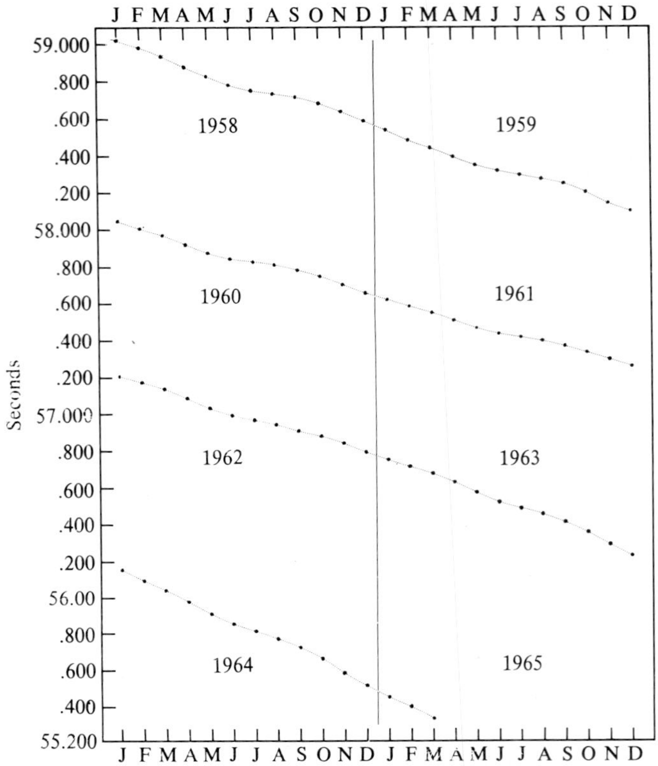

Fig. 2.3 The difference between monthly averages of Greenwich mean solar time and atomic clock time. A tabulated value of 58.890 is equivalent to −1.110 sec. This record shows that from about January 1958 to April 1963 the slowing of the earth's rotation was causing the earth to lose about 1.25 msec/day. After April 1963 the earth started to lose about 1.85 msec/day, ending at 1.98 msec/day in January 1965. This loss when summed up causes the earth to lose about one half-second per year in its rotation about its axis. (From *Journal of Geophysical Research* **72**, no. 20. Reproduced by permission.)

allowed accurate determinations of the sidereal day and showed that the earth's rotation rate is slowing slightly and that the earth's sidereal day is increasing by about 1.25 msec per day (Fig. 2.3). In April, 1963, this increase jumped to 1.85 msec per day for reasons still unknown, producing a loss of about half a second per year. This seems like little, but on the geologic or astronomical time scale of billion of years, it will amount to an increase of several hours in the length of a day. This variation in the sidereal day is a secular change,* which is caused in part by tidal friction, the slowing action affecting the earth's rotation as it turns within the tidal envelope. In addition

* A *secular* change is one which is so slow that for practical purposes it may be considered to be proportional to time over extended periods.

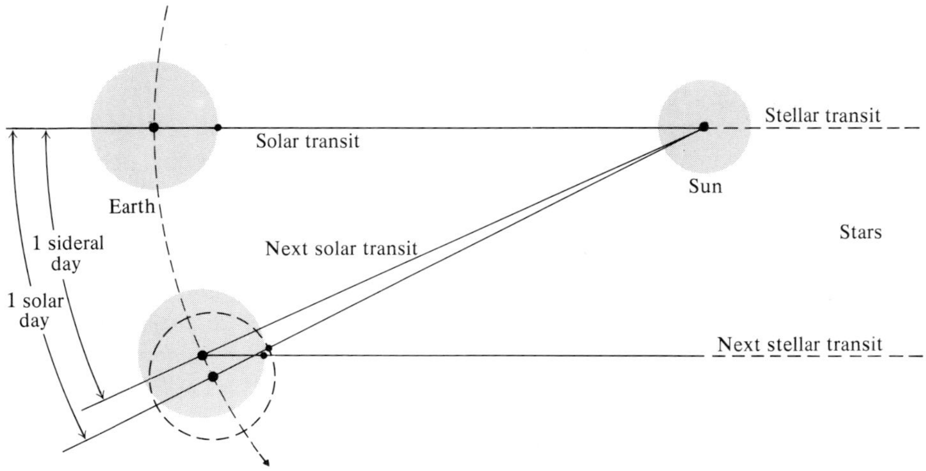

Fig. 2.4 Two successive stellar transits, demarking a sidereal day, occur before two successive solar transits that demark a solar day. The earth must move an additional distance along its orbit and, on the average, turn an additional 1° angle on its axis to complete a solar day after completing a sidereal day. The mean solar day is thus about 4 min longer than the sidereal day. (Redrawn from J. Spar, *Earth, Sea and Air*, Addison-Wesley, Reading, Mass., 1962.)

to the secular changes, there is a small periodic change in the length of the sidereal day. This is caused by small seasonal changes in the distribution of the mass of the earth and its atmosphere with respect to its rotational axis.

The two days mentioned thus far are the apparent solar day and the sidereal day. It has been stated that the mean value of the solar day is 24 hours, and that the sidereal day is about 23 hours and 56 minutes and is changing slowly. We shall now consider the variations occurring in the solar day. These are due to two factors: the eccentricity of the earth's orbit and the tilt of the earth's axis of rotation. We shall first discuss the effect that displacement along the earth's orbit has on the length of the day when the reference body is (a) the sun and (b) a star located at an infinite distance in space.

Consider the sun to lie directly above a meridian of longitude and at the same time in line with a star located infinitely far away. As the earth rotates and translates along its orbit, a new line drawn from the infinitely far star to the earth's center remains parallel to the initial line, while the new line from the sun's center to the earth's center does not (Fig. 2.4). The star will be found to lie exactly above the starting meridian after 23 hours and 56 minutes have elapsed; this time period is the sidereal day, the interval between two stellar transits. At the moment of completing the second stellar transit a line between the sun and earth centers intersects the earth's surface slightly east of the starting meridian; thus the solar transit or solar day is not yet completed. The earth must turn through an additional angle to complete the solar transit and thus the solar day.

On the average, this angle is about equal to $\frac{1}{365}$ of a circle (2π rad) and the time required for the earth to turn this angle is about $\frac{1}{365}$ of a day, or approximately 3 minutes and 56 seconds. The average solar day is therefore about 3 minutes and 56 seconds longer than the sidereal day. These are average values. The eccentricity of the earth's orbit has the effect that the earth advances faster and farther along its orbit near perihelion (winter) than aphelion (summer). Consequently the earth must sweep out a larger angle (turn farther) for the sun to repeat a meridian transit near perihelion. The mean solar day, 24 hours, forms the basis for our clock time, but the earth does not dance to this tune in perfect rhythm.

A similar phenomenon occurs between the earth and the moon. In one sidereal day the earth completes one revolution relative to the stars, while the moon completes one orbit relative to the stars about the earth in 27.3 days, the *sidereal month.* (This is also the time it takes the moon to complete one rotation about its axis relative to the stars.) Thus while the earth completes one sidereal day, the moon moves an angular distance along its orbit about the earth amounting to $(1/27.3) \times 360^\circ$. The earth must then turn an additional angle of about $360^\circ/27.3$ to complete one revolution relative to the moon, *the lunar day.* The time required for this additional turning is about 24 hours/27.3, or 0.88 hour. A lunar day is therefore about 53 minutes longer than a sidereal or a solar day. It is this time differential that is responsible for the fact that the lunar tide in the sea occurs approximately one hour later on each successive day. (We shall discuss this differential in greater detail in Chapter 17.)

The *sidereal month* and the *lunar* or *synodic month* are unequal for the same reasons. The earth sweeps out an angle in the ecliptic plane of approximately 1/12 of 360° in one sidereal month. If the sun, moon and earth are initially in line above an earth meridian, a period of time equivalent to the sidereal month plus 1/12 of the sidereal month, or about $2\frac{1}{2}$ days, will be required for the sun, moon, and earth to come back to the same position (from new moon to new moon). Hence the lunar or synodic, month is equal to about 29.5 days. Since the moon rotates on its own axis, completing one revolution relative to the stars in 27.3 days, the same time it takes for the moon to go once about its orbit (sidereal month), the same side of the moon always faces the earth. This latter relationship is not accidental, but is controlled by the asymmetry of the moon under the gravitational pull of the earth.

Considering these natural periods of time set by the astronomical bodies, with a year of 365 plus a fraction of days long, and with our clock system keyed to a rigid 24-hour day that is not adhered to by the earth, one would expect the calendar and the seasons of the year associated with it to eventually become seriously out of phase. Caesar, when establishing the Julian calendar in 45 B.C., provided for an extra day every four years, making the average length

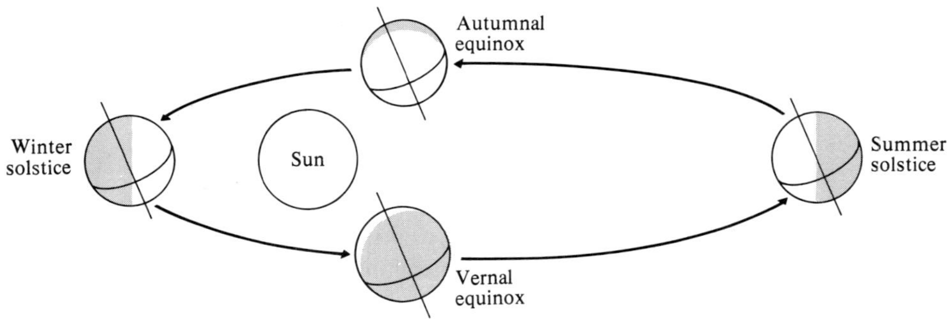

Fig. 2.5 The tilt of the earth's axis to the ecliptic plane makes the north pole of the earth point toward the sun during northern summer and away from the sun during northern winter. Thus the sun does not rise above the horizon at latitudes above the arctic circle during the northern winter solstice and does not set at these latitudes during the northern summer solstice.

of the Julian year 365 days and 6 hours. This seemed to be a logical solution, since the Julian year was based on the length of time the earth needs to pass from the vernal equinox through the summer solstice, autumnal equinox, winter solstice, and back to the vernal equinox positions on the orbit (Fig. 2.5). This period of time referenced to the sun is the *tropical year*, which is equal to 365 days, 5 hours, 48 minutes, and 45.7 seconds. The tropical year is shorter than the *sidereal year*, the period of time in which the earth completes one orbit around the sun in space, i.e., relative to the stars. This difference is due to the fact that the equinoxes slowly shift their position on the earth's orbit.

This shift, called the precession of the equinoxes, is caused by the precession of the earth's axis, which completes its circle of precession every 25,800 years. Its effect is a gradual shift in the equinoxes relative to a fixed calendar year such as that defined by the Julian calendar, causing the seasons as governed by the relative position of the earth to the sun to get out of phase with the calendar. Over the long period of time the Julian calendar was in use, the 11-min difference for each year between the Julian and the tropical years grew into an appreciable error; little by little the vernal equinox moved back into the calendar winter.

A wholesale adjustment was made by Pope Gregory in the 16th century to return the vernal equinox to the calendar spring. First, March 21 was set as the date of the vernal equinox. In addition, leap years were eliminated at every century mark except for those that are evenly divisible by 400. (Thus, 1700, 1800, and 1900 were not leap years, but 2000 will be one.) The Gregorian calendar, adopted in the United States in the 18th century, is still in use and the seasons are now nearly in harmony with the calendar, since the tropical year is now only about 26 seconds longer than the average calendar year.

THE TILT OF THE EARTH'S AXIS AND SEASONAL VARIATIONS IN SOLAR ILLUMINATION

The apparent path of the sun as seen from the earth is called the *ecliptic*, and the earth's orbital plane is called the *plane of the ecliptic*.

The system of latitude and longitude on earth can be projected onto the sky, the celestial sphere. Imagine the earth's axis of rotation extended in both directions into space. The points at which it intersects the celestial sphere are called the north and south celestial poles, and just as on earth, the celestial equator is that great circle on the sky all points of which are at the same distance from the two celestial poles.

The earth's axis is tilted at an angle of 23.5° with respect to the perpendicular to the plane of the ecliptic. This angle is called the *obliquity of the ecliptic*. Because of it, the ecliptic is tilted by the same angle to the celestial equator (Fig. 2.5). The ecliptic crosses the celestial equator at two points called the *equinoxes*. The *vernal equinox* is the point at which the sun crosses the equator moving from south to north on about March 21, the beginning of the spring season. On about Sept. 23, the sun moves from north to south, crossing the equator at the *autumnal equinox*, the beginning of fall. On about June 22, the sun is farthest north of the equator at the point called *summer solstice*, the beginning of summer; at noon, it appears directly overhead at 23.5° N Lat, the Tropic of Cancer, and everywhere north of the Arctic Circle, 66.5° N Lat, the sun remains above the horizon all day. On about Dec. 21, the sun has reached the southernmost point of its migration, *winter solstice*. It stands directly overhead at noon on 23.5° S Lat, the Tropic of Capricorn, and does not rise within the Arctic Circle, nor does it set within the Antarctic Circle (66.5° S). At the vernal and autumnal equinoxes, days and nights are equal (12 hours) everywhere, and the sun appears directly overhead at noon on the equator.

Observed from the earth, the sun appears to cross the equator twice, reaching its extreme zenith position at each of the tropics once a year. However, in reality the sun is not moving north and south nor is the tilt of the earth's axis changing appreciably in spite of the precession and nutation, but it is the combination of the tilt and the earth's orbit around the sun that creates the sun's apparent seasonal migration.

Although the tilt of the earth's axis remains essentially constant, its direction changes with time. Thus once every 25,800 years it swings around in a tight circle to trace a double conical figure, with the apexes at the center of the earth. At present, the earth's axis points close to the star Polaris (North Star), but it is slowly moving away. In 14,000 A.D. the new North Star will be Vega, but about 28,000 A.D. Polaris will again assume this role.

The path of precession is not entirely smooth. The gravitational pulls of sun and moon superimpose on it a slight bobbing effect called *nutation*. The period of nutation is 18.6 years, but its amplitude is small.

Those who live at moderately high latitudes are well aware of the seasonal fluctuation in hours of daylight and darkness, while those living in the tropics notice very little change. Although the mean solar day is by definition divided into 24 hours, the length of a "day" is loosely used as the time period extending from sunrise to sunset, while "night" extends from sunset to sunrise. Care must be used not to confuse the two meanings of the word *day*. The statement that "the days get longer in the summer" does not refer to a lengthening of the apparent solar day, which of course is longer during the northern winter, when the earth is near perihelion.

At low latitudes, the hours of daylight do not vary appreciably over the year and the hours of reduced light at dawn and twilight are brief, while at high latitudes the seasonal change in daylight hours is very large, from total darkness for 24 hours at one of the poles to total daylight for 24 hours at the other. The hours of dawn and twilight also lengthen as one moves toward higher latitudes during summer. Even the traveler along a north-south route within the continental United States notices that the evenings and early mornings are different. Usually, he does not try to fully reason out this difference until the shortness or length of twilight is casually mentioned.

THE EARTH-MOON SYSTEM AND ITS EFFECT ON THE EARTH'S ORBIT

The gravitational attraction between the earth and the moon prevents the moon from spinning off into space, just as gravitational forces hold the earth and the moon in their orbit about the sun. Gravitational attraction between two large masses, however, acts on both bodies; that is, the earth is attracted toward the moon with a force that is equal and opposite to the force that attracts the moon to the earth. If the earth were firmly anchored in space, then the gravitational force would hold the moon in its orbit relative to the anchored earth; but since the earth is not securely anchored in space, the gravitational force exerted by the moon on the earth acts to displace the earth. Although the forces exerted by the earth on the moon and the moon on the earth are of equal magnitude, the effect they have on each other is different for each.

The mass of the moon is only 0.012 the mass of the earth, and thus the acceleration of each body due to the same force is very different; that is, the effective gravitational pull exerted by the moon on the earth is only 0.012 that exerted by the earth on the moon. In spite of its smallness, the moon's gravitational pull would permanently displace the earth as it moves along its orbit around the sun unless it were counteracted by an acceleration of equal magnitude acting opposite to the moon's attractive force.

Consider a mechanical system consisting of a large and a small spherical mass connected by a rigid weightless bar. If this dumbbell-like system is rotated about its center of mass (i.e., the balance point of the bar), the centrifugal forces generated by the two spheres are equal and opposite at the balance point. The earth-moon system, linked together by gravitation, behaves like

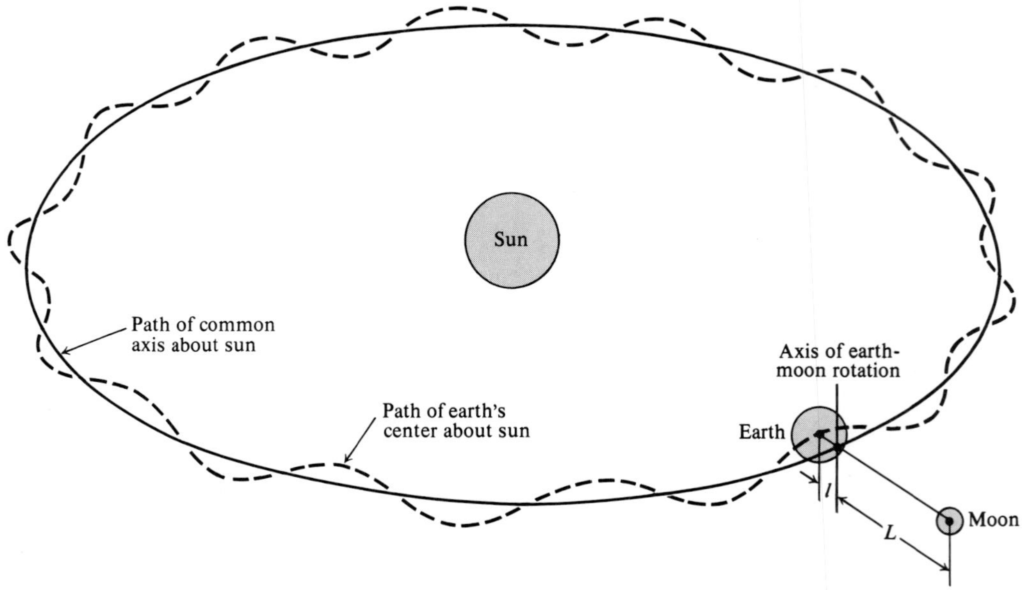

Fig. 2.6 The common axis of rotation of the earth-moon system travels a smooth orbit about the sun, while the earth's axis oscillates about this track due to the rotation of the earth-moon system.

such a dumbbell, and it is the center of mass of this asymmetrical dumbbell that actually traces the smooth elliptical path around the sun (Fig. 2.6).

Since the earth's mass is 80 times greater than the moon's, the center of mass of the earth-moon system is displaced toward the earth. If we consider the mass of the earth as 1, then the mass of the moon is 0.012. The center of mass is therefore displaced from the earth's center a distance equal to 0.012/1.012 times the distance separating the centers of the earth and the moon. Since the mean distance between the earth and moon is 384×10^3 km, the center of mass of the earth-moon system is 4550 km away from the earth's center but is still within the earth (radius of the earth is 6371 km). As the moon circles the earth, the center of the earth traces an S-curve along the path about the sun.

An adequate description of planetary motion requires refinement after refinement, taking into account the gravitational forces of all planets, etc. The moon moves in a predictable fashion relative to the earth's center, while the earth-moon system moves in a particular manner relative to the sun's center, and the sun's center moves relative to infinitely distant stars, which also may be moving. The determination of absolute motion and the effect of each planetary body on another is not a simple matter; indeed it may be an impossible task for lack of a truly stationary reference system from which motion may be judged.

APPARENT MOTION PRODUCED BY A ROTATING REFERENCE FRAME

On earth, bodies in motion are judged relative to time and the latitude and longitude system. The motion of an airplane flying from one city to another, an artificial satellite traversing the sky, even a person walking along a road are described relative to the earth's fixed coordinate system. If the earth is used as the reference frame, celestial bodies appear to move around the earth and migrate north and south while the earth is considered to remain stationary. However, these motions are relative rather than true absolute motions and come about from the choice of the rotating earth as the arbitrary reference frame. Any body fixed in space and independent of the earth will appear to move relative to an earth-fixed coordinate system. However, if the body is also moving in space, then the motion observed relative to the earth is a resultant of the body's space motion and the apparent, or relative, motion generated by the moving earth coordinate system. Also, any object moving over the surface of the earth is at the same time moving in fixed space, and its progress as determined against a coordinate grid attached to the earth has both real and apparent components. In the following discussion we will not worry about the apparent motion caused by the earth's translation about its orbit, but will concern ourselves exclusively with the rotational effect.

Bodies moving over the earth with high frictional contact do not appear to possess apparent motion caused by the earth grid's changing orientation relative to the moving body. The high friction between the moving body and the earth forces the body to move with the rotating grid system, and hence the apparent-motion component is insignificant. If a body moves over the earth's surface with very low frictional contact, the rotation of the earth will move the earth's fixed grid relative to the body, and the body will therefore appear to have an apparent motion. In other words, to an observer located on the rotating earth, it seems as though projectiles, winds, ocean currents, etc., were acted upon by a force pushing them to the right or left, depending on the hemisphere in which the observation takes place. In oceanography, meteorology, and other fields investigating the motion of bodies with low friction coupling to the earth, the concept of an apparent motion created by the rotation of the earth is extremely important since it governs the direction of travel of these moving bodies relative to the earth.

Let us consider an artificial satellite in polar orbit, i.e., one that circles above the earth passing directly over the two poles. This satellite is locked in the orbital plane that passes through the poles and the earth's center, and must move with the earth along its orbit around the sun. Relative to the earth, this translation of the satellite is identical to that of the earth and need not be considered. If the satellite circles the earth once every 90 minutes, then it moves over 40 degrees of latitude every 10 minutes, and during these 10 minutes, the earth rotates 2.5 degrees under the satellite. Thus a line dropped

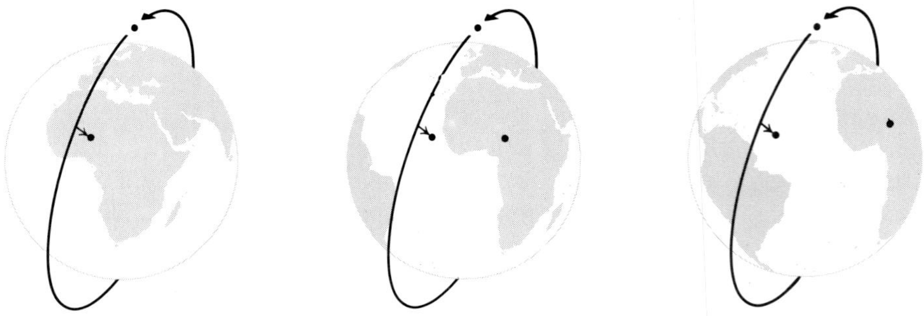

Fig. 2.7 Each time the satellite orbits the earth and crosses a given latitude it appears to be displaced westward due to the eastward turning of the earth within the satellite's orbit. (From Earth Science Curriculum Project, *Investigating the Earth*, Houghton Mifflin, Boston, 1967.)

from the satellite perpendicular to the earth's surface would describe not a straight but a curved path (Fig. 2.7). The deflection from the straight line is made up of two components of motion: *real* (motion created by the movement of the satellite in its orbiting plane) and *apparent* (motion created by the rotation of the earth upon its axis within the orbiting plane). Newton's law of inertia states that bodies in motion tend to move in a straight line unless acted upon by an outside force. The curvature in the satellite's observed track as it appears on the earth must therefore be produced by a force. Since it is the apparent motion that produces the deflection from a north-south track, the force involved is called an *apparent force* (strictly speaking, it is no force at all). It has been named *Coriolis force*, after Gaspard Gustave de Coriolis (1792–1843), who was the first to observe that an object bearing across a turning surface veers to the right or left, depending on the direction of rotation, and it is this Coriolis force which accounts for the displacement imposed by our use of a rotating coordinate system (an earth-fixed coordinate system) as a reference rather than a space-fixed, or inertial, coordinate system.

THE SOURCE OF THE CORIOLIS FORCE

Before discussing the effect that the Coriolis force has on the earth, we shall explain the concept of force as applied here. We saw in the example above that the satellite's curved track can be understood as resulting from the interaction of two motions: the satellite's orbital motion and the earth's rotation. Although no force concept is required to explain this phenomenon, it is intuitively more convenient to consider the deflection of the satellite's track from a straight line as the work of a force. This apparent (Coriolis) force is a centrifugal force. If we consider a particle of unit mass at rest at latitude ϕ, the centrifugal acceleration, or force per unit mass, can be written as $\omega^2 R$, where ω^2 is the square of the earth's angular velocity and R is the normal distance between the particle and the axis of rotation. Any stationary particle of unit mass (1 g in cgs units) on the earth therefore is subject to a centrifugal force that is defined by the earth's rotation rate ω (2π rad/sidereal day) and the distance between the particle and the earth's axis of rotation R = (earth's radius $\times$ cos ϕ).

If the particle of unit mass is now allowed to move east, south, north, or west over the spherical earth's surface, the centrifugal force acting on it changes. For example, since the earth spins toward the east, a particle moving eastward relative to the earth will move about the earth's axis of rotation at a rate that is faster than that of the stationary particle (and the earth beneath), and hence it is subject to a greater centrifugal acceleration. As a result, the moving particle is pulled outward from the earth's axis of rotation by a greater amount than a stationary particle is but, since the centrifugal acceleration is small compared with gravity, the moving body is constrained by gravity to stay on the earth. The excess force can only slip it along the earth's surface in a direction that increases R (i.e., toward the equator) in an attempt to match the centrifugal force of the moving body to that of a stationary particle. Thus the surplus centrifugal force of the moving particle acts at right angles to its motion, and an eastward-moving particle is deflected toward a lower latitude; that is, in the Northern Hemisphere, an eastward-moving particle is deflected in a southern direction and in the Southern Hemisphere, it is deflected north.

If the particle were moving westward across the earth's surface, its rate of rotation about the earth's axis would be slower than that of a stationary particle and hence its centrifugal acceleration would be smaller than that of the particle at rest (or the earth beneath). The westward-moving particle would therefore experience a deficit of centrifugal force, which would tend to move it poleward—closer to the earth's axis of rotation and toward a position at which its centrifugal force would match that of a stationary particle. The particle would again slip sideways but this time to a higher latitude; in the Northern Hemisphere, it would be deflected northward and in the Southern Hemisphere, southward. We see, then, that a particle moving east or west is acted upon by an acceleration that deflects the particle to the right of its direction of travel in the Northern Hemisphere and to the left in the Southern Hemisphere. We shall see later that this deflecting force plays a considerable role in the controlling path of winds and ocean currents which move with low friction. The difference in the centrifugal force possessed by a moving body and a stationary body on the earth comprises the Coriolis Force.

The deflecting effect of the Coriolis force is exactly similar for particles traveling in the north-south direction. A particle traveling northward in the Northern Hemisphere carries with it the centrifugal acceleration of the lower latitude at which it began its motions to a higher latitude, where the centrifugal force is weaker, because R, the distance from its axis of rotation, is smaller. Thus, at the higher latitude, the moving particle has a small surplus of centrifugal force. As a result, the particle rotates eastward about the earth's axis at a rate that is faster than that of a stationary particle at the higher latitude. Hence the northward-moving particle will be deflected to the east relative to the earth. A particle moving southward in the Northern Hemisphere will have a smaller centrifugal acceleration than a stationary particle at the lower latitude and will therefore rotate about the earth's axis at a slower rate than the stationary particle. Hence the southward-moving particle will be deflected to the west from its original course relative to the earth. In the Southern Hemisphere, it will be deflected to the east.

The above reasoning can be applied to bodies moving in any direction over the earth's surface; all we have to do is to resolve the direction of travel into north-south and east-west components. In general, to the observer located on the rotating earth, it seems as if bodies moving horizontally over the earth's surface are deflected to the right in the Northern Hemisphere and to the left in the Southern Hemisphere.

Before going on to a discussion of other topics, we shall briefly review what happens to particles moving in a vertical direction. A particle in vertical motion also suffers deflection. However, in dealing with the ocean environment, one nearly always neglects this aspect, because vertical velocities are usually very small. For moving masses of air, vertical velocities are greater, and hence deflections from the vertical do receive consideration in meteorological studies.

The deflection occurring during a body's vertical motion is limited to the east-west direction. If a body is dropped from the top of a tall tower and air turbulence does not affect its path, it will land to the east of the point through which the vertical from the release point should go. Descending bodies are deflected to the east. If a body is shot from the earth's surface to the top of the tower, it will pass to the west of it. The deflections can be explained as follows.

The distance from the earth's axis of rotation to the top of the tower is larger than that to the bottom of the tower. Hence, as the earth spins, the top of the tower must traverse a circle of larger circumference, and consequently must have a higher velocity (greater centrifugal force) than the bottom of the tower does. A body released at the top of the tower therefore has a higher velocity than does the ground at the base of the tower. During its fall, the body maintains its higher velocity and hence is deflected to the east. For a tower of a given height, the difference between the lengths of the radii at the top and bottom of the tower is greatest at the equator, while it is zero at the poles. The apparent force per unit mass which deflects bodies having a vertical velocity W is similar to the force acting on bodies in horizontal motion but is a cosine function of the latitude, $2\,\omega \cos \phi\, W$.

THE EFFECT OF THE CORIOLIS FORCE ON MOVING BODIES

When moving bodies are tracked against the earth's grid system, it is, in many instances, possible to observe the deflection of their paths due to the earth's rotation and therefore the presence of an apparent force. A projectile fired from the North Pole parallel to the earth's surface and along the prime (0°) meridian, traveling at a rate so as to strike the 45° N Lat circle in exactly one hour, will land at 45° N Lat and 15° W Long, because the earth has turned eastward through 15 degrees during the projectile's flight. The path of the trajectory as traced on the earth shows that the projectile was deflected to the right of its direction of motion by an apparent force, the Coriolis force. The projectile may be fired from the North Pole along any meridian and if one hour is required for it to reach the 45° N Lat circle, it will always be deflected 15 degrees of longitude to the right of the meridian it was aimed to follow. If the projectile is fired from the South Pole, it will again land 15 degrees west of the meridian along which it was aimed, but the deflection is now to the left of the direction of motion. In other words, the apparent force in the Southern Hemisphere acts in a direction opposite to that in the Northern Hemisphere. Since the apparent force changes its direction or sign from one hemisphere to the other, it must be zero at the equator, which separates the two hemispheres, and no deflection can occur at this latitude.

The change in sign of the apparent force with the change in hemisphere and its zero value at the equator indicate that the force is a function of latitude. At either pole, regardless of the direction over the earth in which the projectile is fired, the displacement from the original path in degrees of longitude is a function of flight time and the earth's rotation rate about the polar axis (360° or 2π rad per sidereal day).

Let us now consider a frictionless projectile fired in any direction along the earth's surface from a point A located at a latitude between the poles and the equator. If we knew the rotation rate of the earth about an axis through A and earth's center, we could proceed exactly as in the polar case; i.e., we would assume that our new axis was the polar axis with a new rotation rate, and we would draw a grid of meridians and latitudes over the earth with this new axis as the axis of rotation. We could then readily determine the deflected distance as a function of our new grid, new rotation rate, and projectile flight time. The touchdown point of the projectile obtained from this new grid could then be related to the real earth grid system. We would find, as in the polar case, that the deflection was to the right of the direction of motion in the Northern Hemisphere and to the left in the Southern Hemisphere. To see how much deflection there would be, all we would still have to do would be to determine the rotation rate about our new axis. We shall discuss in the next section how this can be done.

THE FOUCAULT PENDULUM AND THE PENDULUM DAY

In 1851, the French physicist Jean Bernard Léon Foucault (1819–1868) demonstrated the rotation of the earth by suspending a heavy iron ball at the end of a 200-ft wire and making it swing as a pendulum (Foucault pendulum). Once set in motion, a pendulum suspended from a near-frictionless pivot will continue to swing in a fixed plane, and, since the forces that act to maintain the pendulum in motion are independent of any rotation about a vertical dropped from the pendulum pivot, the pendulum can be used as an independent reference against which the rotation of the earth, or rather the "vertical component" of the earth's rotation, can be measured.

Foucault pendulums are often on display in museums or planetariums. To the observer, it seems that the pendulum changes orientation, swinging in a slightly different direction with reference to a pattern marked on the floor below it. In reality, it is the building, the floor, and the observer that rotate slowly about the pendulum's vertical and the fixed plane of swing. The observer, firmly attached to the floor by friction, is not aware of his motion, and hence it seems to him that it is the pendulum which moves relative to him and the building. If he could ride to and fro on the pendulum, he would see that the building slowly rotates.

To an observer on earth, a pendulum set in motion above the North Pole would appear to change its direction of swing by 15 degrees per hour. The time required for the earthbound observer to witness the swing plane traversing a full 360 degrees is one *pendulum day*, which at the pole is equal to one sidereal day. A pendulum suspended above any other point on the earth, excepting the equator, will also appear to change its direction of swing with time although the change in direction occurs more slowly. The pendulum day at any other location is equal to the sidereal day divided by the sine of the latitude of that location. The rotation rate about a vertical at any location on the earth is therefore 2π rad per pendulum day. Thus we have

$$\text{pendulum day at latitude } \phi = \frac{\text{sidereal day}}{\sin \phi},$$

$$\text{rotation rate at latitude } \phi = \frac{2\pi \text{ rad} \sin \phi}{\text{sidereal day}}.$$

That is, the rotation rate about any axis vertical to the earth's surface is equal to the rotation rate at the polar axis times the sine of the latitude of the new axis. At the equator, since the latitude angle and the sine of the latitude are zero, the pendulum day is infinitely long, and the rotation rate of the swing plane is zero.

In the projectile examples cited earlier we were interested only in determining the touchdown position after an hour's flight. However, it is possible to describe a projectile in its complete track over the earth by determining its positions at successive time intervals from its speed, direction of travel, and the earth's rotation about successive axes. We shall find that the displacement from the original path due to the deflecting force will vary if either the speed of the projectile is varied while the latitude is held constant or if the speed is held constant and the track is determined across a variety of latitudes. Thus the magnitude of the apparent force (Coriolis force) is a function of both the speed of the moving body and the latitude.

THE MAGNITUDE OF THE CORIOLIS FORCE

The Coriolis force (surplus or deficit of centrifugal force carried by the moving body) acting on a unit mass in motion is

$$\text{Coriolis force per unit mass} = 2\omega\,(\sin\phi)\,V,$$

where ω is the angular rotation rate of the earth about its polar axis (2π rad/sidereal day), $\sin\phi$ is the sine of the latitude of the unit mass, and V is its instantaneous horizontal velocity. It is evident from this equation that the Coriolis force increases with increasing latitude and increasing velocity, but the amount of deflection caused by it does not increase in direct proportion. If the body moves very fast (large V), the Coriolis acceleration is large and the body traverses a large distance in a short period of time; in this same time interval the earth rotates through only a small angular distance and hence produces only a small deflection from the body's intended direction of travel. If, however, the body moves slowly, it will, in a given time, cover only a short distance in the intended direction of travel and the angular distance through which the earth rotates in the same interval will be of an order of magnitude comparable to the distance traversed by the body in the original flight direction. Hence, in this case, the deflection will be great although the Coriolis acceleration is small.

The deflection distance at right angles to the original path caused by the Coriolis acceleration is determined by

$$D = \tfrac{1}{2} \cdot 2\omega\,(\sin\phi)\,V \cdot T^2,$$

and the distance the body moves in its original direction is

$$S = VT.$$

From these equations, it follows that if V is small, a longer time T is required to cover a fixed distance S. Since D varies with VT^2 and S varies with VT, then as T increases, D increases more rapidly than S. When V is large, a small time T is required to cover a fixed distance S, and D may be much smaller than S.

The Coriolis force acts, of course, on intercontinental missiles, long-range artillery shells, and satellites, and compensation for this deflection must be made in calculating flight paths. The Coriolis force also acts on cars or trains, but even at high latitudes it is not sufficient to overcome the friction between the earth and the vehicle and cause a lateral displacement from its intended course. For fluids, on the other hand, friction is small, and a mass of air or water moving at relatively low speed over the earth suffers deflection by the apparent Coriolis force whenever it moves over an appreciable distance.

GROSS ASPECTS OF THE CORIOLIS PROBLEM IN THE OCEANS AND ATMOSPHERE

In the Northern Hemisphere the moving surface water of the central oceans is deflected to the right, producing large clockwise gyres in the surface currents, while in the Southern Hemisphere deflection to the left produces counterclockwise gyres (Fig. 2.8). In the atmosphere, the outward flow of air from a cell of high pressure (a region of dense air) is deflected to the right in the Northern Hemisphere, producing clockwise motion about the high (*anticyclonic flow*), while in the Southern Hemisphere the deflection to the left produces counterclockwise motion about the high. Despite the change in direction in the two hemispheres, the circulation about a high-pressure cell is always called *anticyclonic.*

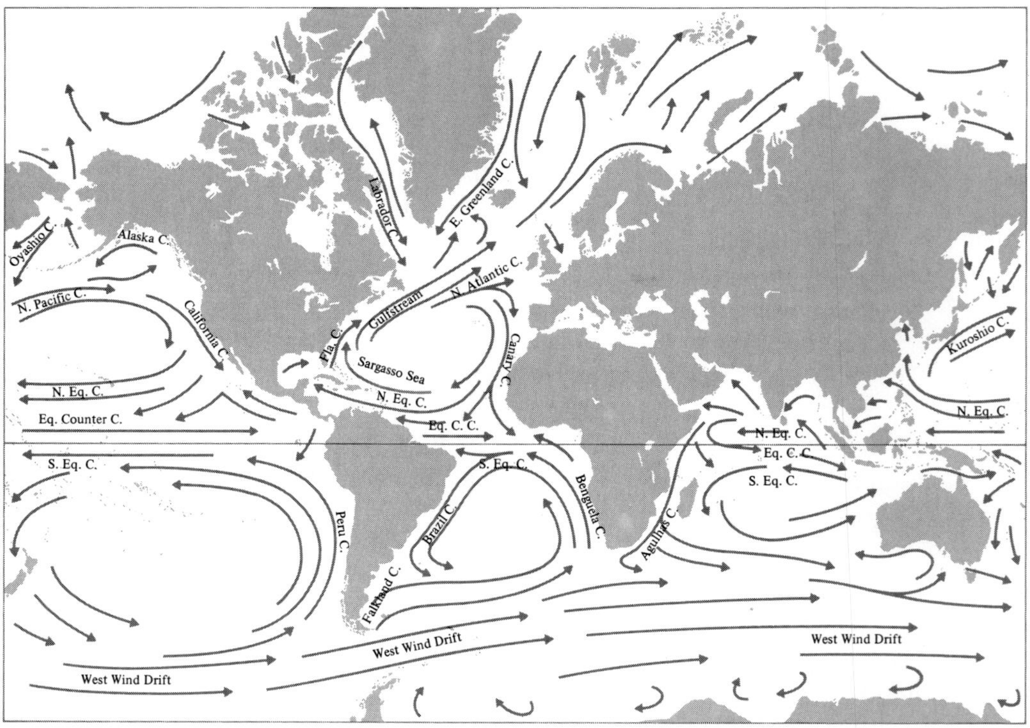

Fig. 2.8 Direction of flow of the principal oceanic surface currents.

The air moving into a low-pressure region (a region of low-density air) is also deflected, producing counterclockwise and clockwise flows in the Northern Hemisphere and in the Southern Hemisphere respectively. Circulation about a low-pressure cell, regardless of its direction, is called *cyclonic flow*. This deflection tends to orient the air flow parallel to the isobars (lines of equal pressure on a map) separating high and low pressure regions. At low latitudes, where the deflecting force is small compared to the effect of friction, or in regions of rough terrain, where the friction between the moving air and the earth is greatly increased, the air leaving the high-pressure cell or entering the low-pressure cell suffers only a small deflection. In this case, the flow tends to cross the isobars rather than to move parallel to them. Weather charts clearly show that as friction increases the importance of the Coriolis force decreases. The wind arrows in the region of friction-promoting topography, such as the Rocky Mountains, nearly always cross the isobars at a relatively sharp angle compared to the angle of these arrows found in regions of smooth topography and hence lower friction, such as the Plains States or the oceans.

The importance, to the study of the oceans, of the natural periods defined by the earth's motion and the apparent forces which deflect large masses of water and air cannot be overemphasized. We therefore suggest that you review this chapter very thoroughly to ensure that you fully understand the topics discussed in it. You will encounter them over and over again in the course of your study.

STUDY QUESTIONS

1 If the earth moved in its orbit around the sun but did not rotate about its axis, what would be the approximate length of one day as defined by successive periods of daylight and darkness?

2 If the precession of the earth's axis reversed its direction of motion, what would happen to the relationship between the sidereal year and the tropical year?

3 If the rotation of the earth upon its axis reversed its direction, what would happen to the relationship between the sidereal day and the mean solar day? Assume that translation along the orbit would not vary from the existing pattern.

4 Why are the twilight hours during the summer longer at higher latitudes than at low latitudes?

5 What are the mechanisms that prevent (a) the moon from permanently displacing the earth from its orbit about the sun, and (b) the earth from attracting the moon into collision?

6 Show how the period of the earth's rotation about any vertical (the pendulum day) varies with latitude.

7 Show how the rotation rate of the earth about any vertical varies with latitude.

8 Two bodies move over the rotating earth's surface at different speeds. Assume that there is no friction. Explain carefully and rigorously why the slower body will undergo a greater deflection from its direction of motion than the faster body does. Cite an example for each of the two cases.

RECOMMENDED READING

Beiser, A., and the editors of *Life, The Earth.* Time Inc., New York, 1962.

McDonald, J. E., "The Coriolis Effect," *Scientific American*, **186**, 5 (May 1962).

McLellan, H. J., *Elements of Physical Oceanography*, Pergamon Press, New York, 1965.

Spar, J., *Earth, Sea, and Air*, Addison-Wesley, Reading, Mass., 1962.

Strahler, A. N., *The Earth Sciences*, Harper and Row, New York, 1963.

Von Arx, W. S., *Introduction to Physical Oceanography.* Addison-Wesley, Reading, Mass., 1962.

CHAPTER 3

THE HYDROSPHERE, ITS DISTRIBUTION AND TRANSFER PROCESSES

INTRODUCTION

Water is found on earth in all three phases: solid (ice), liquid (water), and gas (water vapor). Although it is distributed among earth reservoirs of reasonably fixed capacity, it does not remain locked in these reservoirs; instead it continuously circulates in a closed cycle, moving from one reservoir to another.

In this chapter, we consider the creation and distribution of water on earth, the processes that govern its transfer among reservoirs, and the direction in which it is moving. The ramifications of water distribution and transfer are so far-reaching that it is impossible to do more than discuss the ones that are most relevant to the study of oceanography. However, although we are thus deliberately restricting the scope of our investigation, we shall never lose sight entirely of the magnitude of the issues involved in the presence or absence of water, the rates at which it is supplied and removed, and its effect on the earth and mankind.

Because this text is concerned with oceanography, it will place emphasis on the oceanic reservoir, since it is the largest reservoir of the hydrosphere and therefore the most important one on the global scale. The processes of evaporation and precipitation and the atmospheric conditions that enhance one process or the other will also be discussed in detail, for these processes control the direction of the hydrologic cycle* and are the only natural processes by which water can be delivered from the principal reservoir to man's land domain.

THE RESERVOIRS OF THE HYDROSPHERE

Water in its three phases (gas, liquid, and solid) within the earth-air system forms the hydrosphere. The waters of the hydrosphere are present as oceans, lakes, and rivers, groundwater seeping through the crustal rock strata, water vapor in the atmosphere, and as snow and ice in seasonal storage or in the more permanent form of glaciers. The water is not static, but circulates continually, changing its location and passing from one phase to another as it travels from reservoir to reservoir through the hydrologic cycle. The transfer of water in the hydrologic cycle varies over the earth and controls in part the features of the

* The transportation or movement of water through the reservoirs on the earth is known as the *hydrologic cycle.*

land and ocean surfaces. However, at any one time there is a relatively uniform volume of water tied up in each of the subareas of distribution, or reservoirs (Table 3.1).

Table 3.1 Distribution of the earth's water by volume, km^3

Water vapor and condensate in the atmosphere	15.3×10^3
Rivers and lakes	510.0×10^3
Groundwater	$5,100.0 \times 10^3$
Glacial and other land ice	$22,950.0 \times 10^3$
Oceanic water	$1,369,350.0 \times 10^3$

That most of the earth's water is found in the oceans is not surprising, but the volumes of glacial ice (not to be confused with sea ice, which is considered part of the oceanic water volume) and groundwater compared with those of rivers and lakes may, at first glance, seem unexpectedly large. The volume of water in each of these reservoirs is frequently expressed in terms of *sphere depth*, the depth that a volume of material would have if it were spread uniformly over a smooth sphere of the same surface area as the earth, 510×10^6 km^2. Sphere depth equals water volume divided by earth's area. The volumes of water in Table 3.1 converted to sphere depths are listed in Table 3.2.

Table 3.2 Sphere depths of the earth's water, m

Water vapor and condensate in the atmosphere	0.03
Rivers and lakes	1.0
Groundwater	10.0
Glacial and other land ice	45.0
Oceanic water	2685.0

From C. W. Wolfe *et. al.*, *Earth and Space Science*, D. C. Heath, Boston, 1966.

The earth's water has not always been distributed in the proportions given in Table 3.2, as we can readily imagine from the following example. Let us assume a drop in the volume of seawater of about 7.5% (equivalent to a drop of about 200 m in the present sea level), and let us further assume that the water removed from the oceans would become entrapped in land ice. We would find that this relatively modest change in the seawater volume would result in a more than 400% increase in the present volume of land ice. In other words, a small change in the oceans' volume would produce large changes in other reservoirs. Our example is not unrealistic when we consider the changes that have taken place in geologic time. It has been estimated that the changes due only to the recent ice ages have brought about variations in the sea level in excess of 100 m, and the changes produced by all factors affecting the hydrologic cycle are well in excess of 200 m.

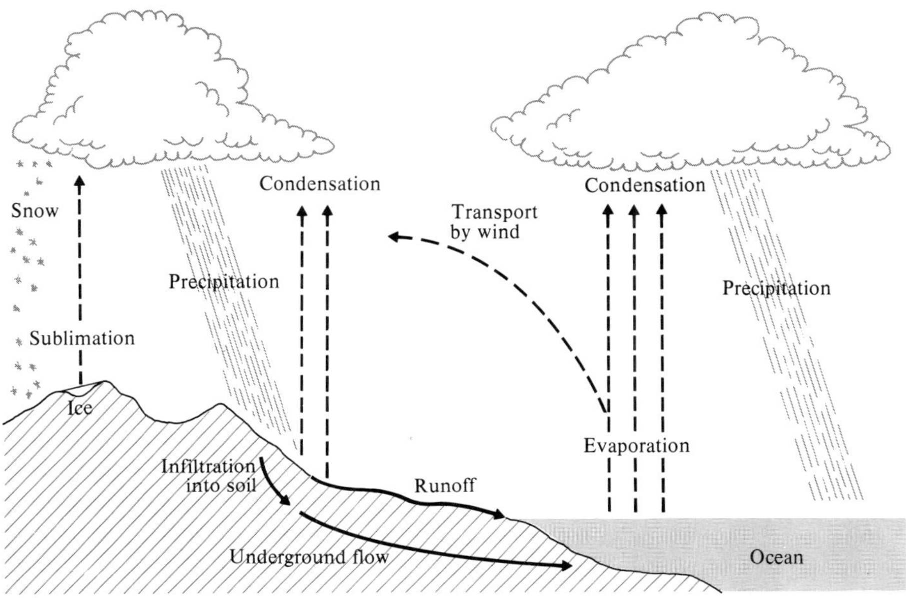

Fig. 3.1 The circulation of water in the hydrologic cycle. (Redrawn from J. Spar, *Earth, Sea, and Air,* Addison-Wesley, Reading, Mass., 1962.)

THE HYDROLOGIC CYCLE

The *hydrologic cycle*, the cycle through which water is transferred from one reservoir of the hydrosphere to another while maintaining the reservoir distribution, i.e., the total volume of water, is a complex process (Fig. 3.1). Although the water recirculates constantly through the reservoirs, the cycle is not uniform in space over the earth and varies with time, having both short- and long-term variations. *Evaporation* of the surface waters supplies the air with its moisture, while *precipitation* removes moisture from the air and returns it to the earth. Moving air carries the moisture around the earth, and allows moisture removed from one location to be deposited in another. Evaporation and precipitation are concurrent processes, but their rates vary; a net trend is produced when one process exceeds the other.

Averages of precipitation and evaporation determined over all oceanic areas show that evaporation exceeds precipitation by 7 cm/year. Since the seas do not suffer any permanent depletion, it follows that the excess evaporation must fall on the land and return to the sea by way of rivers and as groundwater. Evaporation and precipitation also occur over land, but on the average precipitation must exceed evaporation to compensate for the loss of water from the sea and to allow it to be returned by the rivers and streams. Assuming that the storage volumes in the different reservoirs remain relatively constant when averaged over long periods of time, we find that the volume lost by one reservoir must be gained from another. Although it is possible to determine the average volume of water transferred through this cycle each year, it must be remembered that at any one instant this volume is part of one or the other reservoir.

As stated above, the annual excess evaporation over the earth's sea surface is 7 cm. This corresponds to a volume of

$$(7 \text{ cm/year}) \times 71\% \times \text{earth's area in cm}^2 = 25.3 \times 10^3 \text{ km}^3\text{/year of water.}$$

This volume of water must be transferred in the hydrologic cycle and deposited on land as excess precipitation, amounting to

$$(25.3 \times 10^3 \text{ km}^3\text{/year}) \div (29\% \times \text{earth's area in cm}^2) = 17 \text{ cm/year.}$$

Many factors enter into the process of moisture exchange between earth and atmosphere in the hydrologic cycle, the most important being the temperature of the air, the difference in temperature between the air and water, the wind speed, and the moisture content of the air. Warm air can contain a great many water molecules, because the higher the temperature, the faster molecules move. When molecules are moving fast enough, many of them can be jammed into a given space. If the air is cooled, the molecules slow down and as they strike surfaces or collide with other molecules, they tend to stick—the vapor begins to condense. Thus cold air holds less water vapor than does warm air.

Air that is saturated with water vapor will give up its moisture when it is cooled but may take up additional water when it is warmed. If unsaturated warm air passes over a cold water surface, evaporation occurs slowly, provided that the warm air is not cooled to the saturation point and condensation (fog) does not set in. What happens in this case is that the cooler, denser air next to the water's surface acts as a barrier between the more freely moving warmer air above the cool layer, preventing the warm air from picking up an appreciable amount of moisture except by molecular and small-scale turbulent transfer processes. An air column in which cold air underlies warm air is said to be *stable* and resists natural convective overturn (vertical mixing) that would promote transfer of water to the atmosphere. Furthermore, in this situation the temperature of the air increases with height. An increase of temperature with altitude is called *inversion.* This phenomenon creates smog over industrial cities. The low turbulence and mixing present in a stable atmosphere retard the transfer of ground-produced aerosols to the upper atmosphere, where they can be dispersed.

If cool, unsaturated air moves over a warm water surface, evaporation is greatly enhanced. The water warms the air at the base of the air column; as a result the air becomes less dense than the overlying air and can hold more moisture. In this case, the temperature falls with altitude, and the air column is unstable; the resulting free convection of the air promotes the large-scale transfer of water. The large-scale turbulent mixing which takes place during convective turnover of the air column brings new cool air in contact with the sea surface, which again picks up heat and moisture, giving rise to an increase in evaporation. We see then that the evaporation rate varies over geographic areas, depending on the temperature gradient between the air, the land, and the sea surfaces.

THE WATER VAPOR OF THE ATMOSPHERE

If the water-vapor content of the air at the surface of the earth is considered rather than the processes of water-vapor transfer, we can make some generalizations concerning the latitudinal distribution of the lower atmosphere's water vapor. As we saw above, the capacity of the air near the earth's surface to hold moisture is dependent on its temperature. At high latitudes the air is cool, dense, and, as mentioned earlier, has a low moisture content, while in low latitudes the air is warm and capable of holding more moisture. The concentration of water vapor in the atmosphere varies from almost zero to 4%. There are several ways in which the water content of the air can be expressed.

1 *Vapor pressure* is that portion of the total atmospheric pressure that can be attributed to the gaseous water vapor; i.e., it is the partial pressure of the water vapor expressed in millimeters of mercury. For any given temperature and pressure, there is a limit to the quantity of water vapor the air will hold. At this limit, the air is said to be *saturated.* The vapor pressure at this limit, at which vapor is in equilibrium with liquid water, is called the *saturation vapor pressure.* Saturation vapor pressure is dependent on temperature: the lower the temperature, the lower the saturation vapor pressure.

2 *Absolute humidity* is the density of the water vapor, i.e., the number of grams of water in a given volume of air.

3 *Relative humidity* is the ratio of the actual vapor pressure to the saturation vapor pressure. Relative humidity is expressed in percent. If one lowers the temperature of a body of air having a given water vapor content, the saturation vapor pressure will drop, leading to a rise in relative humidity. Conversely, increasing the temperature of this air raises its saturation vapor pressure and lowers the relative humidity.

4 *Dew point* is the temperature to which air must be cooled (at constant pressure) to produce saturation conditions. Condensation begins once the temperature has dropped to just below the dew point.

The standard (sea-level) atmospheric pressure due to all the gases in the atmosphere is 760 mm of mercury or 1015 millibars (1 bar $= 10^6$ dyn/cm^2). The vapor pressure (partial pressure of water vapor) accounts for only a very small portion, about 0.5 to 4%, of this total standard pressure. The saturation vapor pressure of air for different temperatures is shown in Table 3.3.

Table 3.3 Variation of saturation vapor pressure of air with temperature

Temperature, °C	*0*	*5*	*10*	*15*	*20*	*25*	*30*
Saturation vapor pressure, mm of Hg	4.58	6.54	9.21	12.78	17.51	23.69	31.71

O. G. Sutton, *Micrometeorology*, McGraw-Hill, New York, 1953.

Even though the water vapor content of the atmosphere is small, water vapor is the most important single factor in determining the earth's weather, the transfer of heat from one region to another, and the ability of the earth to sustain life.

THE EFFECT OF LATITUDE ON DIRECTION OF WATER TRANSFER

On the earth there are fairly well-delineated latitudinal zones of predominantly ascending or descending air. In a zone of ascending air, the air containing moisture is cooled as it rises; the relative humidity increases until saturation occurs. At this point, moisture starts to condense about dust particles in the atmosphere, forming clouds, and ultimately precipitation results. These zones of ascending air are regions of cloudiness, high precipitation, and predominantly low atmospheric pressure. One such zone circles the earth approximately at the equator above the tropical rain forests; another is centered between the latitudes 50° to 60° N and S above the temperate forest areas. There are also zones of descending air. In these zones, cool dry air from the upper atmosphere descends and is warmed in the process, suffering a decrease in relative humidity. However, as this air sinks and flows along the earth's surface toward the zones of rising air, it gathers moisture from the earth's surface and its relative humidity increases. These zones of descending air are regions of low precipitation and high evaporation with little cloudiness and predominantly high atmospheric pressure. They are found at latitudes 30° N and S and are centered above most of the larger deserts and savannas. Less well known are the high-pressure systems of the polar regions. These areas also receive little precipitation; at very high latitudes the temperature of the descending air does not appreciably increase at the earth's surface, and evaporation remains low in comparison that found at 30° N or S.

Precipitation-evaporation and the water vapor content of the air change with latitude not only because the mean temperature of the air changes with latitude, but also because the large-scale vertical circulation of the atmosphere varies with latitude. Besides the latitudinal control of evaporation and precipitation, there is also a control on moisture exchange that is governed by the dissimilar properties of ocean and continental land masses. In the next section we will discuss some of the additional factors contributing to the difference in moisture exchange over these environment.

INFLUENCE OF LAND AND WATER ON TRANSFER OF WATER

The difference between evaporation and precipitation over land and sea environments is governed by additional factors, two of which will be discussed in detail.

1 The land is elevated above the sea, and therefore the air moving across the earth's surface must rise from the sea surface to pass over the land masses of the continents and descend again to the sea surface on the opposite edge of the continent. As the air rises, it cools and loses moisture over the land, regaining it by evaporation from the ocean as it descends over the sea on the opposite side. This control that land topography exerts on atmospheric processes is called the *orographic* effect. On a smaller scale, it is evident in the precipitation patterns prevailing in mountainous regions. Precipitation is heavier on the windward side of mountains, whereas a rain shadow is often produced on the lee side. In the State of Washington, for example, the predominately southwest winds lose their moisture on the western and southern sides of the Olympic Mountains. This area receives as much as 14 ft of rain a year, whereas the lee side of the mountains near Sequim and Port Townsend receives about 18 in. of precipitation a year. A distance of about 60 mi separates the two extremes.

2 The land surface warms and cools much more readily than does the surface of large bodies of water, and hence the diurnal and seasonal range of surface temperature is much larger over land than over the oceans. Therefore, air moving from the sea toward the land will be warmed or cooled (made to rise or made to sink) depending on whether the land is hotter or colder than the sea. The vertical motions of the atmosphere and its direction of horizontal flow—land toward sea or sea toward land—as well as the increase or decrease in its moisture content will, in part, be determined by the temperature difference between land and water at the boundary between the two environments.

Although the points made above were greatly oversimplified, they do help to explain why the water in the atmospheric reservoir is not uniformly distributed over the earth. (The nonuniform distribution of land ice hardly needs mentioning.) Water falling as precipitation in one area may come from another area in which evaporation exceeds precipitation, or it could be a return of moisture to the same area from which it evaporated. In general, however, the moving air acts as a carrier of water in the hydrologic cycle. No other mechanism is available to supply water to the land-based reservoirs, and in its absence all the waters of the earth would eventually be confined to the seas and the hydrologic cycle would be broken.

The water cycling through the land reservoirs (glaciers, groundwater lakes, and rivers) must be resupplied from the oceans by the atmospheric route. Latitude plays an important role—as does elevation—in determining the reservoir through which and the rate at which the cycling water will pass.

At high latitudes or elevations, water may become tied up in land ice or permanent snow for prolonged periods before it appears again as melt water. The sources of lakes and streams are primarily found in regions of high precipitation; they are often fed by groundwater that carries off excess moisture below the surface in the same manner that surface streams carry off surface water toward the sea.

WATER RESERVOIRS IN ARID REGIONS

Lakes found in arid regions are supplied by underground or surface streams or both. If the outflow of water by streams from a lake is less than the supply, the water will be saline and the water balance will, in part, be maintained by evaporation. Whenever evaporation is the principal mechanism for removing river-supplied water and if it proceeds at such a rate that water removal is equal to or greater than water influx, the lake will become very salty, as for example, the Great Salt Lake in Utah, whose water is about 27% salt by weight and almost eight times saltier than average seawater.

Large rivers are occasionally found in dry hot areas, winding their way to the sea from a source in the wetlands (e.g., the Nile River). These rivers lose considerable amounts of water to evaporation and, if located in permeable soils, help to supply the groundwater system. Rivers in arid regions have been dammed to create storage reservoirs for irrigation purposes. Damming, however, increases the surface area of the river exposed to incident solar radiation and evaporation and usually results in increasing the river temperature and its salt content.

Since the scanty precipitation in arid regions does not provide appreciable quantities of water to percolate through the soil into the groundwater system, the water table is usually deep and the groundwater flow is slight.

The situations described above are greatly simplified, and there are, of course, exceptional cases that do not fit into this generalized scheme. However, our generalized and simplified approach helps us to visualize how water traverses the hydrologic cycle under varying climatic conditions.

THE OCEANIC RESERVOIR AND ITS DISTRIBUTION WITH DEPTH

The oceans of the world make up the largest, and in many ways the most important, of the water reservoirs. The volume of 1.37×10^9 km^3 of water within the oceans is distributed over the earth from the North Pole to about 60° S, passing through the latitudinal belts of predominant precipitation or evaporation and subject to latitudinal trends in the earth's surface temperature. However, this distribution is not uniform in area or depth. In 1921, the German geographer Erwin Kossinna published tables showing the distribution of oceanic waters with depth for the oceans and adjacent seas (Table 3.4).

Table 3.4 Percent ocean area between various depth zones

Depth zone, m	Including adjacent seas				Excluding adjacent seas			
	Atlantic	*Pacific*	*Indian*	*All oceans*	*Atlantic*	*Pacific*	*Indian*	*All oceans*
0–1000	20.4	8.8	7.3	11.9	9.6	3.9	5.9	5.9
1000–2000	5.3	3.9	3.4	4.2	3.6	3.4	3.1	3.4
2000–3000	8.8	5.2	7.4	6.8	7.6	5.0	7.4	6.2
3000–4000	18.5	18.5	24.0	19.6	19.4	19.1	24.4	20.4
4000–5000	25.8	35.2	38.1	33.0	32.4	37.7	38.9	36.6
5000–6000	20.6	26.6	19.4	23.3	26.6	28.8	19.9	26.2
6000–7000	0.6	1.6	0.4	1.1	0.8	1.8	0.4	1.2
> 7000		0.2		0.1		0.3		0.1

Although our knowledge of the bathymetry of the oceans has greatly increased since 1921, only very small changes have had to be made in Kossinna's data.

In 1966, H. W. Menard and S. M. Smith published an article in the *Journal of Geophysical Research* on the hypsometry of oceanic basins. Using new information and measuring techniques, they revaluated the percentage of ocean area in various depth zones. A comparison of their findings with Kossinna's values is given in Table 3.5. The agreement is remarkable and confirms the care with which Kossinna performed his work.

Table 3.5 Percent ocean area between various depth zones, comparing Menard and Smith's data with Kossinna's

Depth zone, m	*Menard and Smith*	*Kossinna*
0–1000	11.91	11.9
1000–2000	4.38	4.2
2000–3000	8.50	6.8
3000–4000	20.94	19.6
4000–5000	31.69	33.0
5000–6000	21.20	23.3
6000–7000	1.23	1.1
7000–8000	0.10	
8000–9000	0.03	
9000–10,000	0.01	0.1
10,000–11,000	0.00	

H. W. Menard and S. M. Smith, "Hypsometry of Ocean Basin Provinces," *Journal of Geophysical Research* **71**, no. 18 (Sept. 15, 1966).

The data in Table 3.4 may be used to determine the distribution of water within each ocean. For example, the table shows that, including adjacent seas, over 20% of the area of the Atlantic Ocean lies between the depth range of from 0 to 1000 m, compared with 8.8% for the Pacific Ocean and 7.3% for the Indian

Ocean. In fact, the Atlantic contains more extensive shallow areas than the Pacific and Indian Oceans combined. The shallow regions of the Atlantic, the adjacent seas such as the North Sea, the Baltic Sea, and Baffin Bay, and the broad continental shelves are some of the shallow areas. There are considerably fewer adjacent seas and broad continental shelves in the Pacific and Indian Oceans. Even if the adjacent seas are not included, the Atlantic contains more extensive shallow areas than the Pacific. Only the Pacific Ocean has large areas whose depths exceed 7000 m.

The volume of ocean water for each of the 1000-m depth intervals can be calculated from the percentages listed in Table 3.4 or 3.5. At depth 0, we have the sea surface which, including all oceans and adjacent seas, represents 100% of the oceans' area (the area of all the oceans is 361,000,000 km^2). According to the table, 11.9% of the total ocean area lies in the 0–1000-m depth interval. Therefore the area of the oceans at depths greater than the 1000-m isobath equals $100 - 11.9 = 88.1\%$ of the total ocean area or, if we take the average over the 0–1000-m interval, we obtain $88.1 + (11.9/2) = 94.1\%$ as the mean ocean area for the layer above the 1000-m isobath. The volume of water in this upper 1000-m layer is approximately

$$0.941 \times (361 \times 10^6)\ \text{km}^2 \times 1\ \text{km} = 338.2 \times 10^6\ \text{km}^3.$$

We can calculate the volume of water in the next depth layer (1000–2000 m) by determining an average area between the 1000-m isobath and the 2000-m isobath, $88.1 - (4.2/2)$, and then proceeding as above:

$$0.86 \times (361 \times 10^6)\ \text{km}^2 \times 1\ \text{km} = 310.46 \times 10^6\ \text{km}^3.$$

One can, of course, continue this process to determine the volume of seawater for each successive depth interval.

From Tables 3.4 and 3.5, one might expect at first glance that there should be a direct relationship between the tabulated area percent and water volume. However, the large area percentages for the middepth intervals indicate only that large areas of the ocean floor lie between these successive depth contours.

The statistical distribution of the topographic relief of the entire world can also be represented by a so-called *hypsographic curve*, which shows the distribution of land elevation and ocean depth as functions of the earth's total land and ocean areas (Fig. 3.2). We see from the figure that the mean land elevation is 840 m and that the land covers 29% of the earth's surface. The volume of land above sea level is therefore

$$0.84\ \text{km} \times 29\% \times 5.1 \times 10^8\ \text{km}^2,$$

where 5.1×10^8 km^2 is the total surface area of the earth. If all the irregularities on the earth's surface were smoothed out, we would obtain a level surface lying 2440 m below the reference sea level. This depth of 2440 m relative to the reference sea level is called the *mean earth sphere depth*. The mean depth of the sea ($\sim$3800 m) is the average depth of all land below the reference sea level

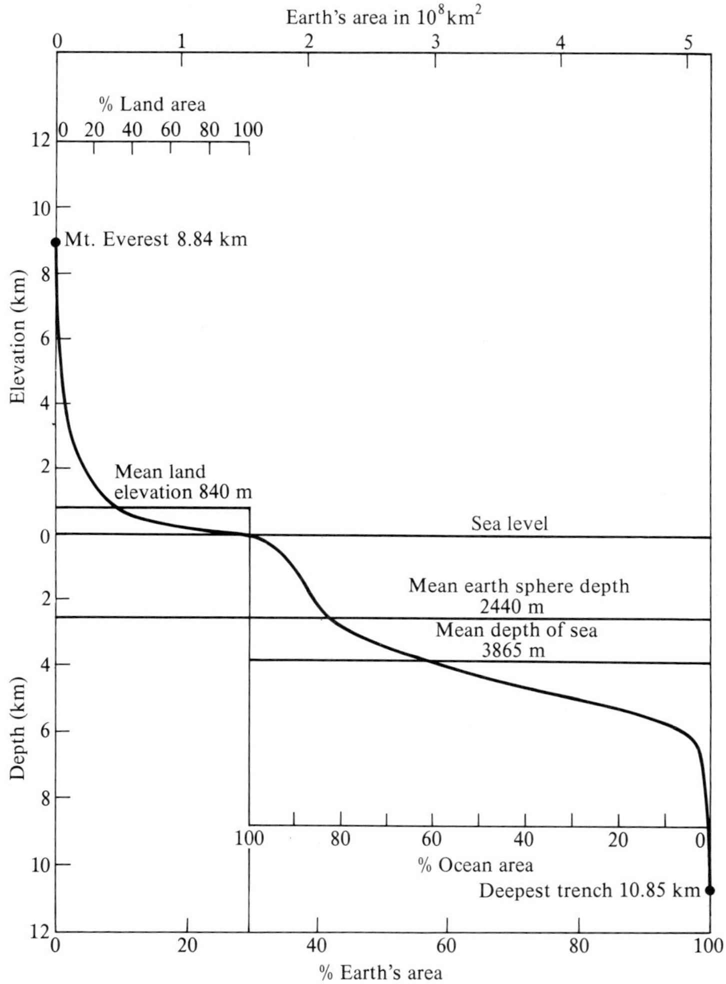

Fig. 3.2 The hypsographic curve.

over the ocean basin area. It is interesting to note that if all land above sea level were dumped into the ocean and the earth were smoothed off to its mean earth sphere depth, the water level would rise 244 m above the original reference sea level. The value of 244 m is the sphere depth of solid land currently above sea level. The distance from the new elevated sea level to the mean earth sphere depth would be 2684 m, which is the sphere depth of the oceans mentioned at the beginning of this chapter.

The distribution of ocean area over the depth intervals shown in Table 3.4 can also be determined from the hypsographic curve. From Fig. 3.2 we see that approximately 24% of the ocean area lies at a depth exceeding 5000 m, and approximately 57% at a depth exceeding 4000 m. The area of the oceans lying between these two depths is then 33% as shown in Table 3.4. Any area blocked

Table 3.6 Area, volume, and mean depth of oceans and seas (after Kossinna, 1921)

Body		*Area, 10^6 km²*	*Volume, 10^6km³*	*Mean depth m*
Atlantic Ocean	(excluding	82.441	323.613	3926
Pacific Ocean	adjacent	165.246	707.555	4282
Indian Ocean	areas)	73.443	291.030	3963
All oceans (excluding adjacent seas)		321.130	1322.198	4117
Large mediterranean seas		29.518	40.664	1378
Small mediterranean seas		2.331	0.402	172
All mediterranean seas		31.849	41.066	1289
Marginal seas and straits		8.079	7.059	874
All adjacent seas		39.928	48.125	1205
Atlantic Ocean	(including	106.463	354.679	3332
Pacific Ocean	adjacent	179.679	723.699	4028
Indian Ocean	seas)	74.917	291.945	3897
All oceans (including adjacent seas)		361.059	1370.323	3795

Based on H. U. Sverdrup, M. W. Johnson, and R. H. Fleming. *The Oceans*, Prentice-Hall, Englewood Cliffs, N.J., 1942.

out on the hypsographic curve diagram represents a volume. The area bounded by sea level, mean land elevation, and the distance from 0 to 29% land area represents the volume of land above sea level, since the area of this rectangle is the product of the land area and the mean elevation. It is a good exercise to determine why the lines representing mean earth sphere depth and mean sea depth are placed as they are on the hypsographic diagram, and how they relate to the volumes of earth materials.

The distribution of water is not the same for each of the major oceans and its adjacent seas (see Table 3.6 for details). Note that the Indian and Atlantic Oceans have very similar volumes and mean depths, while the Pacific Ocean is roughly twice as large in area and volume and slightly deeper than either of the others. The importance of adjacent seas and their effect on the Atlantic quickly become evident.

THE VARIATION OF THE OCEANIC RESERVOIR ON THE EARTH'S SURFACE

The unequal distribution of water in the oceans and the oceans' geographic location affect the properties of the oceanic reservoir, the climatic structure of the earth, and even man's population distribution. Table 3.7 shows the distribution of land and water in the Northern and Southern Hemispheres for 5-degree increments of latitude. The values listed are in percent of the available earth's

Table 3.7 Distribution of sea and land for zones of 5 degrees of latitude (in percent, according to Kossinna, 1921)

Latitude zone	*Northern Hemisphere Water*	*Northern Hemisphere Land*	*Southern Hemisphere Water*	*Southern Hemisphere Land*
90–85°	100.0	0.0	0.0	100.0
85–80°	85.2	12.8	0.0	100.0
80–75°	77.1	22.9	10.7	89.3
75–70°	65.5	34.5	38.6	61.4
70–65°	28.7	71.3	79.5	20.5
65–60°	31.2	69.8	99.7	0.3
60–55°	45.0	55.0	99.9	0.1
55–50°	40.7	59.4	98.5	1.5
50–45°	43.8	56.2	97.5	2.5
45–40°	51.2	48.8	96.4	3.6
40–35°	56.8	43.2	93.4	6.6
35–30°	57.7	42.3	84.2	15.8
30–25°	59.6	40.4	78.4	21.6
25–20°	65.2	34.8	75.4	24.6
20–15°	70.8	29.2	76.4	23.6
15–10°	76.5	23.5	79.6	20.4
10– 5°	75.7	24.3	76.9	23.1
5– 0°	78.6	21.4	75.9	24.1
90– 0°	60.66	39.34	80.92	19.08

From H. U. Sverdrup, M. W. Johnson, and R. H. Fleming, *The Oceans,* © 1942. Reprinted by permission of Prentice-Hall, Englewood Cliffs, N.J.

area in each latitude zone. A glance at the table shows that at midlatitudes the Northern Hemisphere (the temperate zone) has considerably more land area than the Southern Hemisphere. The averages over the 90° to 0° Lat range given at the bottom of the table also show the difference between the two hemispheres. It may be said that the Northern Hemisphere is the "terrasphere," while the Southern Hemisphere is the "hydrosphere."

Environmental factors such as surface temperatures, extent of land area, and the quantity of water in its various physical states throughout the hydrologic cycle control in part the availability of food and hence the population distribution. The temperate zones of the earth at midlatitude offer the best climatic conditions for man and his needs. However, only in the Northern Hemisphere are there vast expanses of land in these zones. Man also exists in regions that offer few natural comforts. These regions, however, are sparsely populated except for those that man has made more habitable by artificially producing a suitable environment—for example, by irrigation. In addition, a variety of natural factors, such as the presence or absence of elevation, land structure, and water supply, have created localized oases that are quite out of character with the rest of the zone in which they are located.

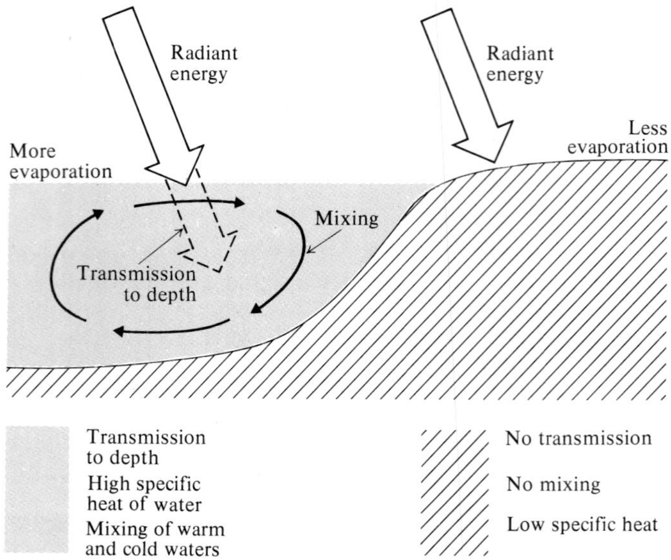

Fig. 3.3 The difference in properties of land and water result in a contrast in their ability to respond to heating and cooling. (Adapted from A. N. Strahler, *Physical Geography*, Wiley, New York, 1960.)

In this context, we wish to point out that the ratio of land area to ocean area and its variation over the earth, as well as topography, produce important deviations within the general climatic zones. For example, land heats and cools more readily than water because of its lower specific heat and its inability to store heat to depth (Fig. 3.3). Therefore the presence of land rather than water may drastically affect distributions and processes that are wholly or in part temperature controlled. The influence of land in producing climatic anomalies would be immediately apparent if one considered a completely water-covered earth model as a gauge to determine the latitudinal distributions and compared it with the real earth. Latitudinal climatic zones are not regularly distributed around the earth, but their boundaries shift continually as the zone passes from water to land, from low to high elevation and back down again. Topography and the land/water ratio are also the reasons for the many exceptions found in the wet and dry zonal belts of the world. If the earth were totally water covered, or if all the material on the earth's surface had the same specific heat, heat-storage capacity, and reflective index, and changes in elevation were minor, the climatic zones would be nicely defined and considerably more uniform. There would be no difficulty in tracing them around the world through both the ocean and land environments along a constant latitude.

THE IMPRINT OF THE HYDROLOGIC CYCLE ON THE OCEAN SURFACE

Despite the deviations in the climatic belts caused by the land masses breaking up the expanses of oceanic water, definite latitudinal trends in evaporation, precipitation, and surface temperature can be found and expressed quantitatively for the central oceanic areas. The characteristics of the ocean waters are controlled by surface processes, for it is only at the sea surface that heat or freshwater is added or removed in quantity. Thus the salinity of the midocean surface waters and their temperature are determined by the distribution of evaporation and precipitation processes as well as solar radiation. In certain zones, extremes in these processes produce salty and cold water of high density which sinks to the depths of the sea to mix gradually with other deep water and rise elsewhere. This circulation is caused and sustained by the unequal distribution of energy and the addition or removal of water at the earth's surface; in other words, the mechanism underlying the vertical circulation in the ocean is also a surface process.

The surface salinity of the world's oceans is controlled by the addition of water by precipitation and the removal of water by evaporation. Evaporation or precipitation alone does not determine the midocean surface salinity, but the excess of one process over the other does. In fact, there is a nearly perfect

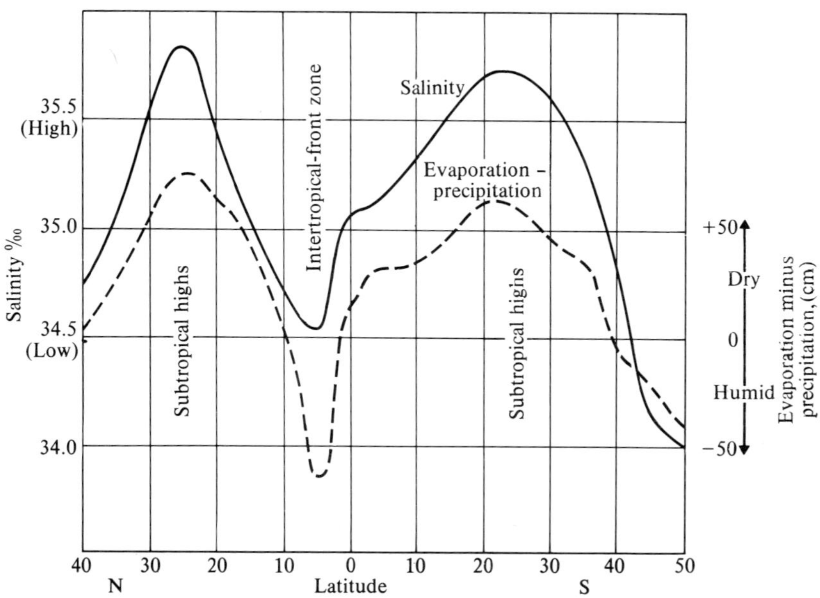

Fig. 3.4 The change in average mid-ocean surface salinities in response to latitudinal changes in evaporation and precipitation. (Based on data of H. U. Sverdrup. From A. N. Strahler, *Physical Geography*, Wiley, New York, 1960.)

TABLE 3.8 Average values of surface salinity, evaporation, and precipitation for all oceans as a function of latitude

Latitude	*S (‰)*	*P (cm/yr)*	*E (cm/yr)*	*E − P (cm/yr)*
40°N	34.54	93	94	−1
35	35.05	79	106	27
30	35.56	65	120	55
25	35.79	55	129	74
20	35.44	65	133	68
15	35.09	82	130	48
10	34.72	127	129	2
5	34.54	177	110	−67
0	35.08	102	119	17
5°S	35.20	91	124	33
10	35.34	96	130	34
15	35.54	85	134	49
20	35.69	70	134	64
25	35.69	62	124	62
30	35.62	64	111	47
35	35.32	64	99	35
40	34.79	84	81	−3
45	34.14	85	64	−21
50	33.99	84	43	−41

Based on H. U. Sverdrup, M. W. Johnson, and R. H. Fleming, *The Oceans*, Prentice-Hall, Englewood Cliffs, N. J., 1942.

linear relationship between evaporation minus precipitation and surface salinity. When evaporation minus precipitation is a large positive number, the surface salinity is high; when it is a large negative number, the surface salinity is low.

A tabulation of average oceanic surface salinity and average evaporation and precipitation by latitude is given in Table 3.8. This tabulation and its graphical representation (Fig. 3.4) show that when oceanic surface salinities, evaporation, and precipitation are averaged over narrow latitudinal belts in the equatorial regions, the latitude at which surface salinity is lowest and the negative value of evaporation minus precipitation is largest does not coincide with the earth's equator but lies 5 degrees to the north. At this latitude, the average surface salinity is 34.54 parts per thousand (34.54 g of salt per 1 kg of seawater) and evaporation minus precipitation is −67 cm per year, the displacement from the earth's equator being due to the imbalance of land and water distributions in the two hemispheres and the resulting imbalance in the earth's surface temperature. This zone of low-salinity surface water marks what may be called the *oceanographic* or *meteorological equator* of the earth. High-salinity belts appear in the regions of high evaporation and low precipitation at latitudes 25° N and 20° S, and a second zone of low salinity occurs in each hemisphere at latitudes 40° N and 50° S.

The data in Table 3.8 were averages obtained for all oceans. However, the distribution of surface temperatures and salinities with latitude is different in each ocean, since each geographic area has its own set of climatic conditions that locally affect the temperature and salinity. Table 3.9 shows a comparison of the average surface temperatures and salinities within 10° zones of latitude for the Atlantic and Pacific Oceans. The trends in the north-south distribution of salinity are similar to those shown in Table 3.8. However, when we compute separate averages of salinity for the Atlantic and the Pacific, we find that the Pacific Ocean's surface waters are less saline and at low latitude its temperatures are slightly higher than those of the Atlantic. The values obtained for each of the oceans also differ from the world oceanic averages of Table 3.8.

Table 3.9 Average surface temperature and salinity in the oceans between parallels of latitude

	Atlantic Ocean		*Pacific Ocean*	
Latitude	*Temperature, °C*	*Salinity*	*Temperature, °C*	*Salinity*
70–60°N	5.60			
60–50°	8.66		5.74	
50–40°	13.16		9.99	
40–30°	20.40	36.35	18.62	34.17
30–20°	24.16	36.71	23.38	35.88
20–10°	25.81	36.00	26.42	34.61
10–0°	26.66	35.42	27.20	34.48
0–10°S	25.18	35.96	26.01	35.11
10–20°	23.16	36.59	25.11	35.55
20–30°	21.20	36.15	21.53	35.57
30–40°	16.90	35.24	16.98	35.00
40–50°	8.68	34.26	11.16	34.36
50–60°	1.76		5.00	
60–70°	– 1.30		– 1.30	

Based on H. U. Sverdrup, M. W. Johnson, and R. H. Fleming, *The Oceans*, Prentice-Hall, Englewood Cliffs, N.J., 1942.

Tabulations such as those given in Tables 3.8 and 3.9 do not reflect the distribution completely, since they do not reveal variations over the longitudal extent of the sea. Figures 3.5 and 3.6 show the gross features of the oceanic surface temperatures and salinities. The north-south trends which we discussed are clearly evident. The east-west trends in the distributions are much less distinct and appear almost negligible in comparison. Most of the east-west changes in surface values are caused by the displacement of surface water by north-south surface currents and the differences between maritime and continental climates on the windward and leeward sides of the land masses at temperate latitudes.

In our future discussions of oceanic water properties and circulation and their interrelationships, the geographic distribution of temperature and

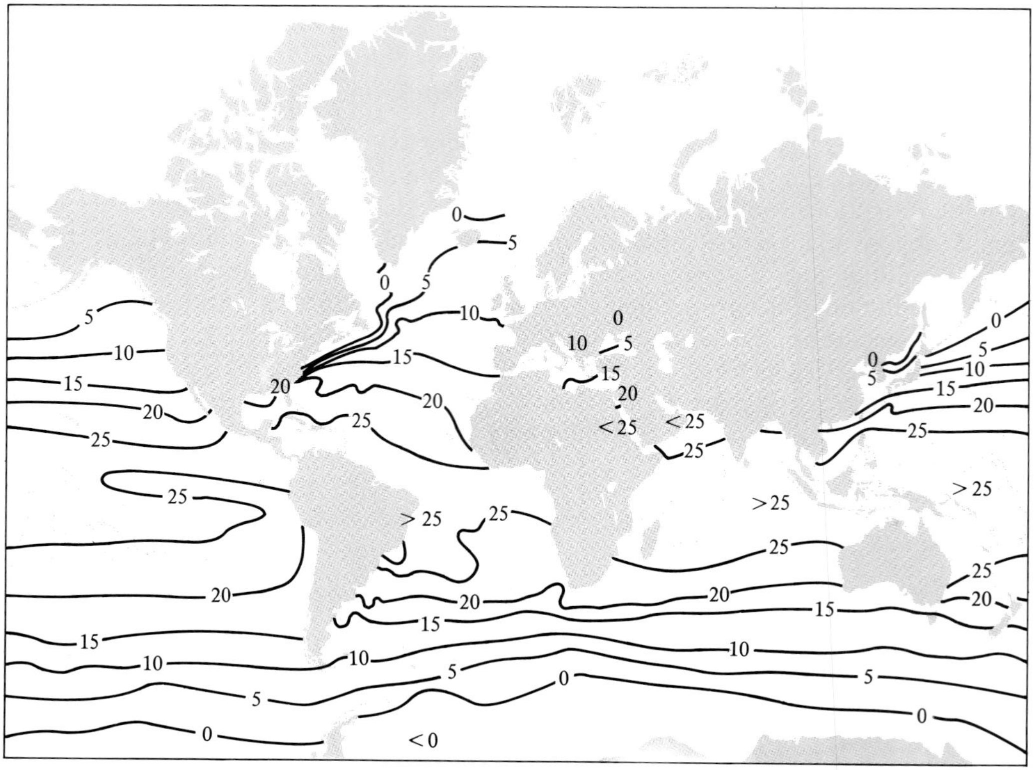

Fig. 3.5 Sea surface temperatures in degrees Celsius, northern winter.

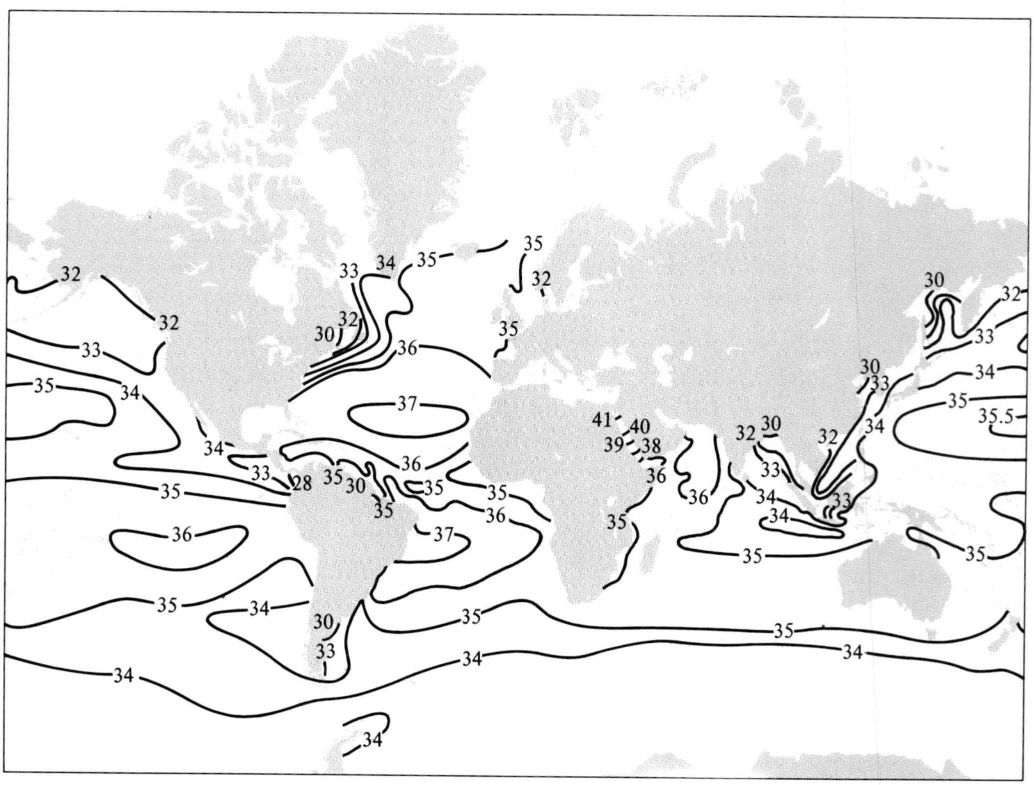

Fig. 3.6 Sea surface salinities in parts per thousand, $^0/_{00}$, northern summer.

salinity, and the use of these parameters as indicators identifying the type of water and its region of surface formation under specific conditions of evaporation, precipitation, and air temperature, will become increasingly clear. Data collected over depth in the oceanic environment that show low salinity associated with high temperature, low salinity with low temperature, high salinity with low temperature, or high salinity with high temperature give considerable information as to the climatic conditions existing when the sampled water was last in contact with the surface and the possible latitudes at which the contact occurred.

STUDY QUESTIONS

1 Calculate the rise in sea level that would occur if all glacial and land ice were to melt and be added to the world oceans without increasing the oceanic area. Assume the density of ice is 0.92 g/cm^3 and that isostatic adjustments of the earth's crust do not occur.

2 The excess water removed from the sea annually by evaporation is 7 cm/year, while the excess precipitation on land is 17 cm/year. What basic law is represented by this seeming imbalance?

3 What would happen to the hydrological cycle and indeed the land masses of the world if water had no gaseous phase at normal earth surface temperatures?

4 Why do small fluctuations in the storage capacity of the large water reservoirs cause important changes on the earth?

5 Determine the relationship between mean land sphere depth and mean ocean sphere depth and explain fully why the latter is greater than the former.

6 What does an area on a hypsographic curve represent?

7 Calculate the volume of water in each kilometer layer of the sea.

8 Why must the mean depth of the Pacific Ocean be greater than that of the Atlantic Ocean? In answering this question, consider the large-scale features of the bottom topography.

9 Plot a graph (E–P) in cm/year versus surface salinity of the sea and discuss the importance of your result. Consider how climatic factors influence surface salinity.

RECOMMENDED READING

Revelle, R., "Water," *Scientific American*, **209**, 3 (September 1963).

Sverdrup, H. U., M. W. Johnson, and R. H. Fleming, *The Oceans: Their Physics, Chemistry, and General Biology*, Prentice-Hall, Inc., Englewood Cliffs, N.J., 1942.

Wolfe, C. Wroe *et al.*, *Earth and Space Science*, D. C. Heath and Co., Boston, 1966.

CHAPTER 4

THE OCEAN'S BASINS AND BORDERLANDS

INTRODUCTION

The oceans of the world occupy the deepest interconnecting depressions in the earth's crust. Since its formation on our earth, water, driven by gravitational force, has sought to flow downhill to occupy these depressions and thus form oceans. The sea floor has features similar to the mountain chains, volcanic peaks, table mounts, canyons, hills, and valleys found on the land surface. However, these features of the ocean floor are hidden from view and have a slightly different nomenclature. The presentation of the topographic features of the sea floor is in a sense a lesson in the anatomy of the unseen portion of the solid earth's surface that lies beneath the sea.

If the water were removed from the ocean basins, where would the land end and the true sea-floor begin? This question is raised and discussed briefly to point out that features other than sea level may be used to distinguish between oceanic and continental environments; for example, geophysical properties of the crust may define the boundary between these two regions.

Sea level has not been constant over geologic time; the boundary between ocean area and land area has changed considerably, and areas of the earth's surface have changed from one environment to another due to uplift or submergence. Modern geophysical measurements indicate that continents drift, the ocean floor stretches in one place and contracts in another, and land rises in one place and sinks in another. With all this going on, are the ocean basins permanent features of the earth? The answer is both "yes" and "no" depending on your time scale. In this book we are concerned primarily with the modern ocean and relatively recent time. Within this time span the ocean basins can be considered permanent, and changes that do occur are insignificant.

THE MAJOR BATHYMETRIC PROVINCES OF THE OCEANS

As we can see from the hypsographic curve in Fig. 3.2, the oceans cover 70.5% of the earth's surface. The curve also shows the area of the different sea-floor zones of the earth's oceans. A shallow zone at depths from 0 to 200 m occupies about 7.6% of the total ocean area. In this zone are the *continental shelves*, the shallow apron of land surrounding the continental land masses. The next-deeper zone, ~200 to 3000 m, takes up about 15.3% of the oceanic area. Within this zone are the *continental slopes* of the oceanic floor, the regions in which the continental land blocks drop off rapidly to the deeper sea floor. The greatest portion (75.9%) of the sea floor lies at depths ranging from 3000 to 6000 m, the zone of the large oceanic depressions called *basins*. Only 1.2% of the sea floor lies at depths ranging from 6000 to 11,000 m; this is the area of the deeper oceanic basins and the deep-sea *trenches*, the steep-sided depressions that dip down sharply into the earth's crust. Since the extent of these zones has not been accurately determined, all values are approximations. Features such as the oceanic ridge systems, which rise from the sea floor but do not pierce the sea surface, occupy the same position on the hypsographic curve as the slopes. Despite this overlap, the areas on the curve devoted to specific depth ranges provide a reasonable picture of the extensiveness of each bathymetric division.

In this chapter, we describe the bathymetric regions of the sea floor with the intent of providing the reader with as clear a picture of its appearance as he has of the continental prairies, plains, hills, plateaus, and old and new mountains.

THE DIVISION BETWEEN OCEANIC AND CONTINENTAL ENVIRONMENTS

Since we are going to discuss the land beneath the sea, we must decide on the criterion that determines the boundary between the oceanic and continental regions of the earth. The presence of seawater and its mean level against the shore do not necessarily mark the separation between continental and oceanic environments. Overlap of these two environments exists at the fringes of the sea, depending on the criterion chosen to distinguish one environment from the other. From a geophysical point of view, the zone where the Mohorovicic discontinuity changes depth represents a good criterion for distinguishing between oceanic and continental structures of the earth's crust. Since this zone is located between the shore and the deep-ocean basins, corresponding roughly to the region of the continental slope, we find that the boundary between the continental land masses and the oceans can be located seaward of the mean sea level line. This zone offers perhaps the best concept of a geophysical boundary between these two quasi-permanent environments.

Seaward of this boundary zone the oceanic basins are assumed to be quasi-permanent (that is, they have existed over long geologic time spans), as are the continental land blocks shoreward of the zone. That part of the continental land block located shoreward of the zone but overlain by seawater encompasses the regions of marine environment that are less permanent from the viewpoint of geologic time spans. The less permanent character of these regions is due to the fact that fluctuations of sea level in the geologic past have alternately covered or exposed large parts of them, allowing their features to be altered by both aerial and subaerial processes. In addition, although these shelf regions are under water, they are highly affected by continental conditions, and hence are not truly oceanic in nature.

Parameters other than the geophysical structure of the earth reinforce the choice of a seaward boundary between purely oceanic and purely continental environments. The fringes of the sea are affected by the continental land mass in several ways; land drainage of freshwater, for example, changes the chemical composition of seawater. Suspended materials of terrigenous origin carried seaward and deposited in this border region produce a sediment type more indicative of a continental than oceanic environment. Even the biota in this region of overlapping influence are different from those of the open ocean. Thus, we shall consider that there are three regions or environments on the earth: true oceanic, true continental, and a region of overlap.

THE OCEAN BASINS

Within the true oceanic environment lie the oceanic *basins*, the large, oval-shaped central depressions that cover the greatest area of the deep-sea floor. Each ocean may have several basins separated from one another by a *ridge* or *rise* system, an elongated narrow elevation of the sea floor. The nomenclature applied to the features of the sea floor is descriptive only, since precise definitions of types are not available. However, the nonexistence of clearly defined terms is not surprising if we think for a moment of the variety of elevations on land, ranging from small bumps to majestic Alpine peaks, that are loosely lumped together under the name "mountain." To identify any one surface feature, a geographic name is added to the generic term; for example, Mid-Atlantic Ridge.

If the oceans were contoured in 1000-m increments, the oceanic basins would appear as the dominant topographic features of the sea floor (Fig. 4.1 — pp. 72–73). The major basins are listed in Table 4.1.

The basins are the deeper recesses of the sea floor, excluding the oceanic trenches. In certain regions climatic conditions at the sea's surface produce cold, dense, and relatively salty water of high density that sinks to the sea floor and gradually moves along to fill the basins. The most accessible basin will fill first until the dense water reaches the lowest point of the ridge or rise surrounding the basin. At this point, it will begin to spill over into the adjacent basin. This low point in the walls forming the basin, called the *sill depth*, determines the direction in which the water will travel. It is also the deepest point at which waters from two adjacent basins can mix freely. Below the sill depth the water in two adjacent basins remains effectively isolated unless

Table 4.1 Major ocean basins of the world. The numbers are keyed to areas marked in Fig. 4.1.

	Western side	*Central*	*Eastern side*
Atlantic Ocean	1 Labrador Basin 2 Newfoundland Basin 3 North American Basin 4 Western Caribbean Basin 5 Eastern Caribbean Basin 6 Guiana Basin 7 Brazil Basin 8 Argentina Basin 9 South Antilles Basin		10 North Polar Basin 11 West Europe Basin 12 Iberia Basin 13 Canaries Basin 14 Cape Verde Basin 15 Sierra Leone Basin 16 Guinea Basin 17 Angola Basin 18 Cape Basin 19 Agulhas Basin 20 Atlantic-Indian Antarctic Basin
Indian Ocean	21 Arabian Basin 22 Somali Basin 23 Mascarenes Basin 24 Madagascar Basin 25 Atlantic-Indian Antarctic Basin		26 India-Australia Basin 27 South Australia Basin 28 Eastern Indian Antarctic Basin 29 South China Basin 30 Sulu Basin 31 Celebes Basin 32 Banda Basin
Pacific Ocean	33 Philippines Basin 34 Caroline Basin 35 Solomon Basin 36 Coral Basin 37 New Hebrides Basin 38 Fiji Basin 39 East Australia Basin	40 North Pacific Basins 41 Mariana Basin 42 Central Pacific Basins 43 South Pacific Basin	44 Guatemala Basin 45 Peru Basin 46 Pacific Antarctic Basin

an influx of denser water added to one basin displaces the water in the basin over the sill to the next basin, thus starting a chain reaction of flushing through successive basins.

An example of the sill depth controlling the direction of travel of dense bottom water and therefore controlling in part the circulation and distribution of deep-ocean water is found in the southern Atlantic Ocean. Dense water formed during the winter season in the Weddell Sea region moves downslope into the southern Atlantic. This water moves northward through the successive basins along the western side of the Atlantic Ocean and eastward toward the Indian Ocean. The height of this dense water above the sea floor is not sufficient to rise above the barrier ridges surrounding the Angola Basin and thus is excluded even though the water extends northward all the way to the tropical regions on the western side of the ocean.

Fig. 4.1 Major ocean basins of the world. Numbers are keyed to basins listed in Table 4.1.

10
10
40
29
33
41
30
42
31
22
21
34
34
32
35
36
17
26
26
37
38
23
24
38
18
27
39
19
28
43
20,25

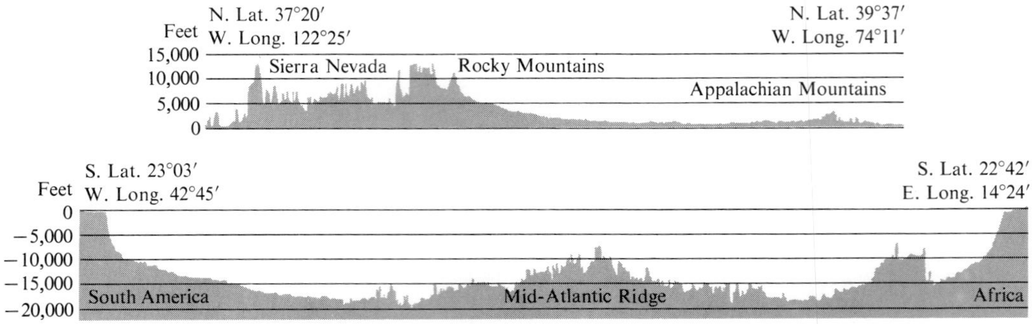

Fig. 4.2 Profiles across the United States and the Atlantic Ocean drawn to the same scale and showing the similarity in size and ruggedness of land and undersea mountains. (From F. P. Shephard, *Submarine Geology*, Harper, New York, 1948.)

OCEANIC RIDGES

Because its topography is in general smooth, without sudden sharp changes in depth, the basin floor is also called the *abyssal plain*. The depth soundings required to describe the basin with some degree of accuracy need not be as closely spaced as those needed to chart a ridge, the topography of which changes suddenly and markedly. Indeed, the ridges separating the deep ocean basins are more rugged topographically than either a rise or swell, which is usually well rounded, resembling the worn-down smooth chains of hills. The ruggedness of some ridge systems, such as the Mid-Atlantic Ridge, rivals that of major mountain systems (Fig. 4.2). Only the placement of the submerged ridge or elevated mountain chain relative to mean sea level distinguishes ridge from mountain systems in appearance.

The Mid-Atlantic Ridge system, with spurs reaching off into the Indian Ocean, has received the most study of any major ridge. At present it is the best-described submarine ridge of the oceans. Its central cleavage is known as a *rift valley*. The latter has been traced from Iceland southward through the Atlantic and around into the Indian Ocean (Fig. 4.3). It is thought to be the marine counterpart of the rift valleys on land, and is created by the lateral motion of the ridge material outward from the ridge axis. The rift-valley concept is therefore compatible with the convection-cell theory, lateral motion in the earth's crust, and anomalies in heat flux and magnetic structure. The extremely jagged appearance of the Mid-Atlantic Ridge in cross-section suggests that there are many rifts parallel to its axis and that lateral expansion of the crust has been occurring in this region for a considerable period of time.

Basins, ridges, swells, rises, and trenches are not uniformly distributed throughout the oceans. Rugged ridge systems appear primarily in the Atlantic Ocean running into the Indian Ocean, while deeper trenches are characteristic of the Pacific Ocean. Not only is the Pacific Ocean larger than the Atlantic,

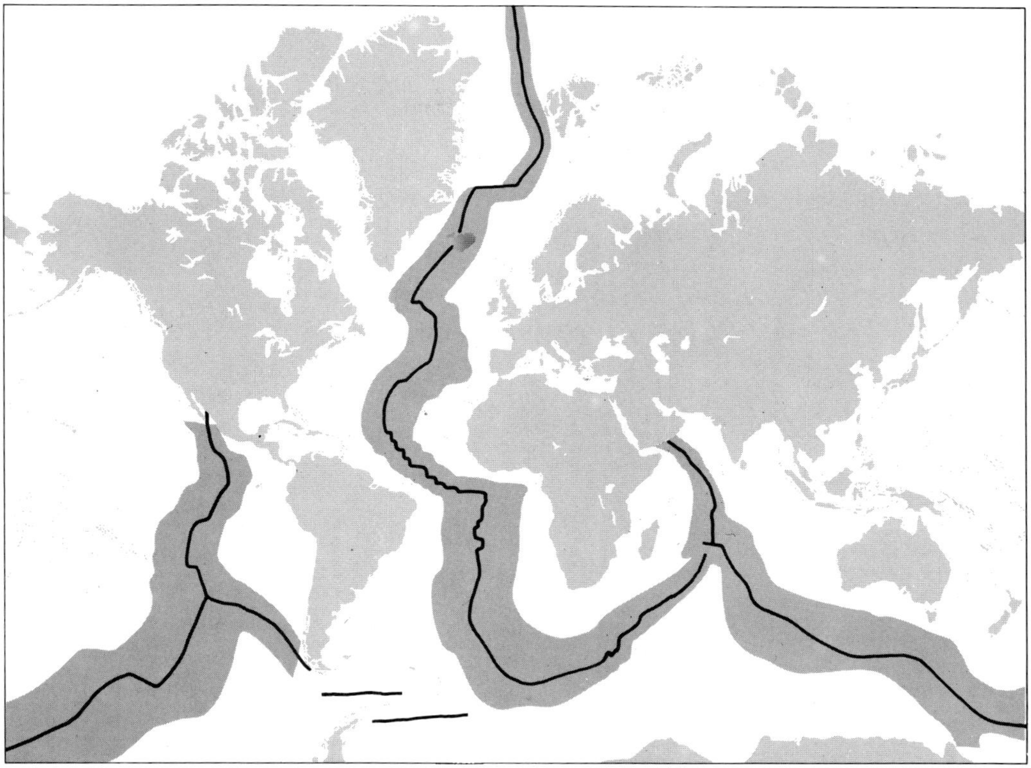

Midoceanic ridge — Mean position of rift valley

Fig. 4.3 Distribution of the major ocean ridges and their central rift valley. (From R. W. Fairbridge, ed., *The Encyclopedia of Oceanography*, © 1966 by Reinhold Pub. Corp. Used by permission of Van Nostrand Reinhold Co.)

but its less rugged rises do not occupy as proportionately large an area as do those in the Atlantic. This fact and the presence of numerous trenches deeper than 8000 m cause the Pacific to have more area of greater depth than any other ocean.

DEEP-SEA TRENCHES

The steep-sided, V-shaped *trenches* or *troughs* are usually associated with island arc chains. The trenches are believed to result from a downwrapping of the earth's crust which is thought to be related to the downward transport of crustal and mantle material in a convection cell. They usually lie on the outside of the curve of the island chain separating the islands' roots from the basin floor. Island arcs and deep-sea trenches are regions of seismic disturbance and are usually volcanically active. The greatest depths in the sea are found within the trenches. Depths in excess of 8000 m are not uncommon, and occasional deep holes in the trenches, called *deeps*, penetrate to depths in excess of 10,000 m.

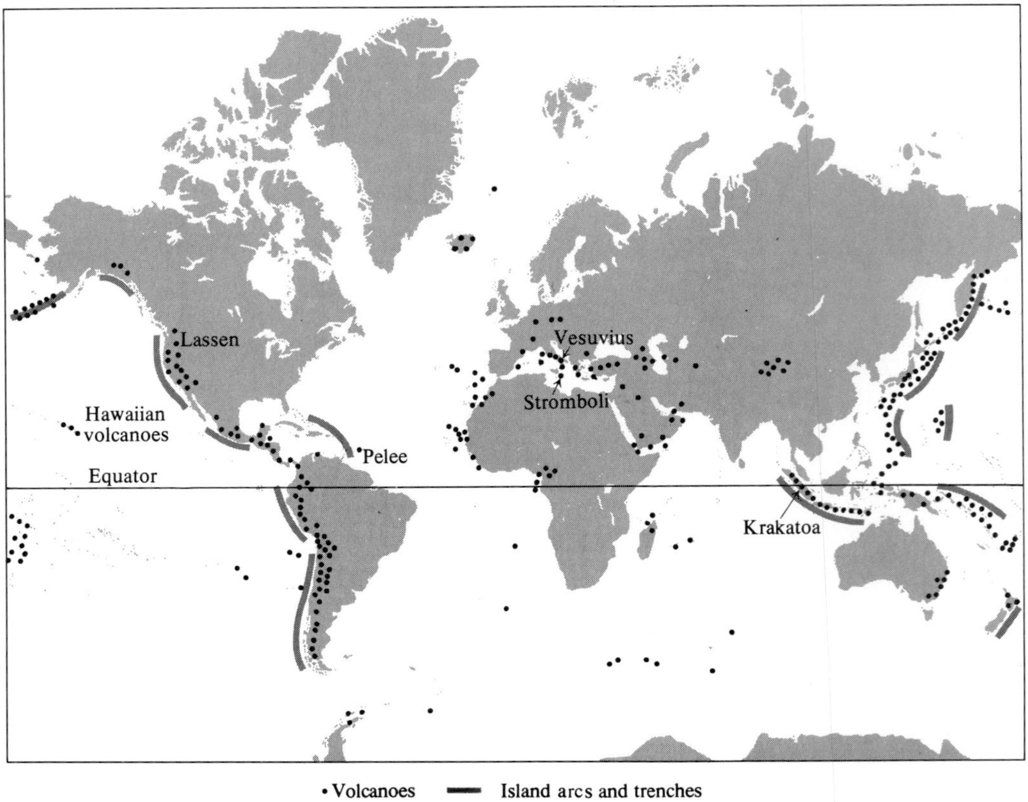

Fig. 4.4 Distribution of major volcanoes, island arcs, and deep sea trenches.

The Pacific Ocean is ringed by volcanic zones, island arcs, and trenches; its borders, in fact, are often referred to as "the rim of fire" (Fig. 4.4). The other two major oceans, the Atlantic and the Indian, have only one or two comparable features. The latter has only one deep trench, the Sunda or Java Trench. In the Atlantic, the Puerto Rico and Cayman Troughs or Trenches in the Caribbean region, as well as the South Sandwich Trench of the South Atlantic, are associated with island arcs. The Romanche Trench of the Sierra Leone Basin is not associated with an island arc.

The geologic structure of the earth under the trenches, island arcs, and inland (*epicontinental*) seas indicates that shoreward of the island arc, the geophysical pattern is continental, while seaward of the island arc, it is typically oceanic. Therefore the trenches are in the oceanic province.

SEAMOUNTS

The larger features of the basins (ridges and trenches) are the gross topographic forms of the deep-sea floor; smaller-scale features are superimposed on these major forms. Small rounded hills found as singular features of the basin floor

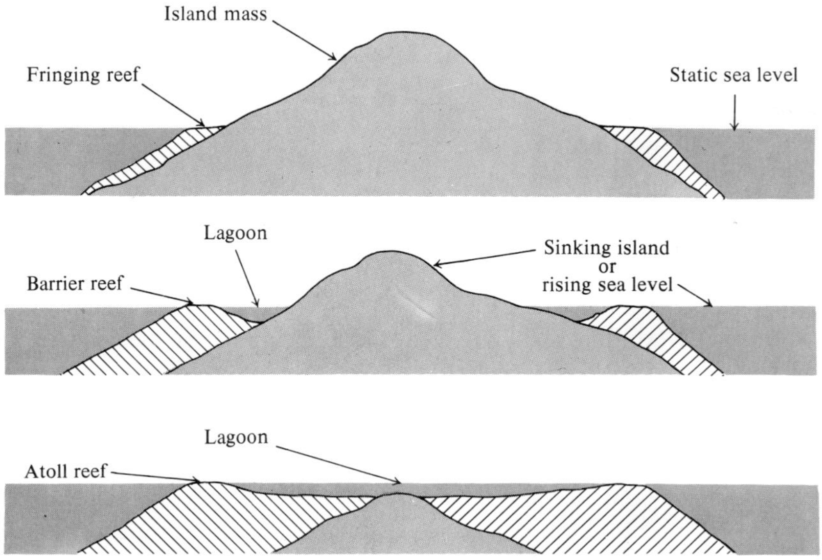

Fig. 4.5 The steps in the formation of coral atolls, according to Sir Charles Darwin's theory.

are known as *knolls*; if they form the equivalent of foothills in the large ridge systems, they are called *abyssal hills*. *Seamounts* are large singular hills or peaks that rise to a considerable height above the sea floor. Nearly all formed volcanically; they are further divided on the basis of whether they have a sharp cone or a flat top. The latter are known as *guyots* or *tablemounts*. Seamounts may rise directly from the flat ocean-basin floor or may be singular peaks jutting up from a ridge or rise. In some cases they break the sea surface and form midoceanic islands. Because erosional processes beneath the sea are not as violent as those above, seamounts tend to have smaller bases in proportion to their heights than do the singular volcanic mountains found on land.

Seamounts cause a localized weight loading of the earth's crust; after forming, they may depress the crust and subside. The flat-topped guyots found scattered over the oceans are thought to be volcanic peaks that were planed off by wave activity during the period that sea level and peak coincided. Changes in sea level and subsidence of the cone have lowered these seamounts to a point at which their flat tops are now 1200 to 1600 m below sea level.

In the region of the world where coral grows in abundance and annual sea-surface temperatures are above about 18°C, a seamount rising to the surface furnishes a shallow-water substrate on which the plants and animals that form coral reefs can flourish. Coral reefs then form as a *fringing reef* surrounding the emergent peak. If the seamount subsides, or if sea level rises at a rate at which coral can grow and build upward, the fringing reef becomes a *barrier reef* that is separated from the peak by a shallow lagoon. Further subsidence or rise in sea level and erosion of the peak will cause the reef, which has no remaining central peak above sea level, to form an *atoll* system (Fig. 4.5). Mechanical erosion of the coral growth creates a supply of coraline material to the center of the reef ring and fills the void left by the disappearing peak, forming a shallow lagoon. This hypothesis of atoll formation, first advanced by Charles Darwin and generally accepted today, suggests that there has been considerable subsidence or rise in sea level in the past. Drill holes and cores from coral atolls have determined that the coral substrate overlying the basaltic peaks can be as deep as 700 m.

THE MAGNITUDE OF OCEANIC TOPOGRAPHY

If exposed to view, the massive ridges of the oceans, the deep basins, and the deeper trenches would be a magnificent sight, certainly one as spectacular as the earth structures we now see above sea level. Man's perspective, however, is limited; in actuality, depths as great as 11,000 m and mountains as high as 8750 m are relatively insignificant in comparison to the earth. If the earth were scaled down to a spheroid 1 m in diameter, these crustal features would become depressions and elevations of 0.85 and 0.68 mm respectively, and would be hardly discernible. Therefore, it is not surprising that the earth has little difficulty supporting its textured surface and occasionally even modifying it.

Because they are small compared to the horizonal dimensions of the earth, it is hard to portray the crustal features to scale. Thus the vertical relief is almost always exaggerated relative to horizontal distance. Since a relief of the Atlantic sea floor drawn to true scale would fit along the edge of this page, we must draw the profile with an exaggerated vertical scale in order to depict the distribution of topographic features (Fig. 4.6). This distortion of the vertical scale, 500 times that of the horizontal, though necessary, produces a false impression of the real topography; thus one should always try to keep in mind the degree of exaggeration so as not to be misled by cross-sections.

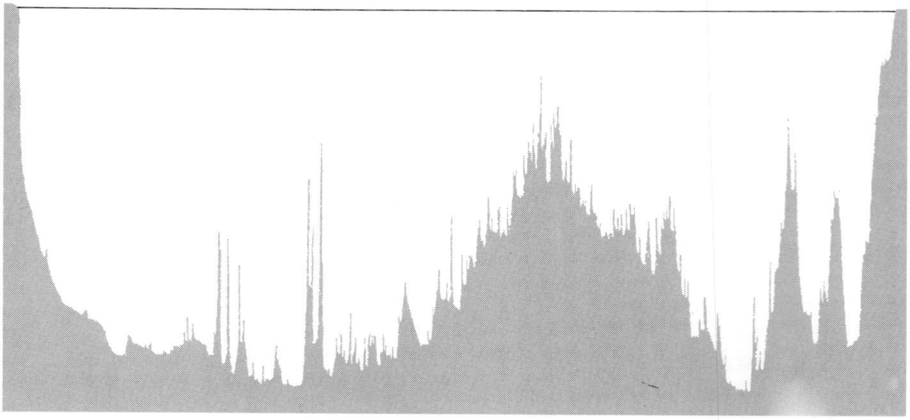

Fig. 4.6 A profile of the topography of the sea floor across the Atlantic Ocean, using a vertical scale 500 times the horizontal scale. This exaggeration in the vertical distorts the topography, but permits distributions of water properties to be portrayed.

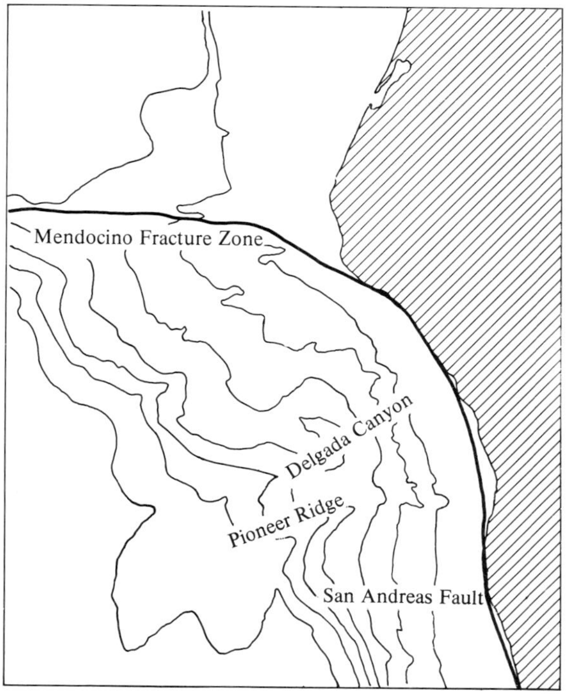

Fig. 4.7 Coastlines and submarine features are often associated with fault systems. (Redrawn from F. P. Shephard, *Submarine Geology*, 2d ed., Harper and Row, New York, 1963.)

CONTINENTAL SLOPES AND DEEP-SEA CANYONS

Moving from the true oceanic environment to the region in which the continental structure of the crust becomes oceanic, we come to the continental slopes, which extend from the outer edge of the relatively flat continental shelves fringing the land masses to the basin floor. The region where the inclination of the bottom changes from the flat shelves to the steep slope is called the *shelf break*. It can be found at depths less than or greater than 200 m, the depth that arbitrarily defines the outer edge of the shelf. The continental slope is similar in appearance to the flanks of rugged mountains: it is commonly deeply scarred by valleys, often has exposed rock outcroppings, and is usually not heavily carpeted by sediment because of its steepness. The average slope angle is approximately 3°, but slopes of up to 50° may be found along coasts lined with steep volcanic mountains. The continental slopes give the impression of being the sharp edge of the continental land blocks. Some investigators believe that they are regions of faulting that mark the line along which the earth's crust fell and then rose in great blocks to produce the land masses and ocean basins. The similarity between the land features formed by faulting and the continental margin along the California coast is often cited to illustrate this possibilty (Fig. 4.7).

The continental slope, like the deep-sea floor, has many superimposed features. Among the most impressive are *submarine canyons*, deep scars cut

Table 4.2 Major submarine canyons of the world

Pacific Ocean	*Atlantic Ocean*	*Indian Ocean*
Tokyo Canyon	Oceanographer Canyon	Indus Canyon
Bering Canyon	Judson Canyon	Ganges Canyon
Columbia or Astoria Canyon	Wilmington Canyon	Mocambo Canyon
Juan de Fuca Canyon	Norfolk Canyon	
Monterey Canyon	Congo Canyon	
Arguello Canyon	Sao Francisco Canyon	
Scripps Canyon		
Coronados Canyon		

into the slope which often terminate at their lower end on the basin floor in a pattern resembling that of a branching valley. They often have large deposits of debris which have been carried down the canyon and left on the sea floor in a fan at the lower terminus. The heads of the canyons are not confined to the shelf break, but can extend as a deep cut up onto the continental shelf and even inland, being associated with the land structure above water level. Some canyons are associated with large rivers; some have axes pointing directly down the continental slope, while others change direction and follow the faulted structure. Canyons occur throughout the world and do not appear to have a direct relationship to the geologic history of coastal areas; indeed, regions of volcanism, sedimentation, and coastal submergence or emergence all have canyons. Although canyons may or may not be associated with rivers, there is a strong correlation between the presence of large rivers and the larger canyons. A partial listing of the major canyons is found in Table 4.2.

Submarine canyons, like seamounts, have counterparts on land. Figure 4.8 is a cross-section of the Monterey submarine canyon drawn to the same scale as a cross-section of the Grand Canyon of the Colorado River. There is little difference between the magnitude of the two features; only their elevations relative to sea level distinguishes them.

Scientific literature contains considerable discussion of the origin of submarine cayons. Some scientists believe that rivers stream-cut the upper portions of canyons under subaerial conditions when sea level was considerably lower than it is now. However, some canyons are not associated with any existing structure that indicates the present or previous existence of a stream course. Other scientists think that canyons were scoured by erosional processes occurring below sea level. This theory holds that turbid suspensions of terrigenous origin flowed downslope, gradually abrading a channel beneath the sea surface. Along some coastal regions sand transported from source areas to maintain beaches is periodically directed away from the coast and may cascade down canyons to be deposited on the deep-sea floor (Fig. 4.9). Most canyons are sharply V-shaped in profile, with loose material often filling the bottom of the channel to produce a flat floor. Unconsolidated material from shallower depths can be channeled along a canyon to add to the deep-sea deposits,

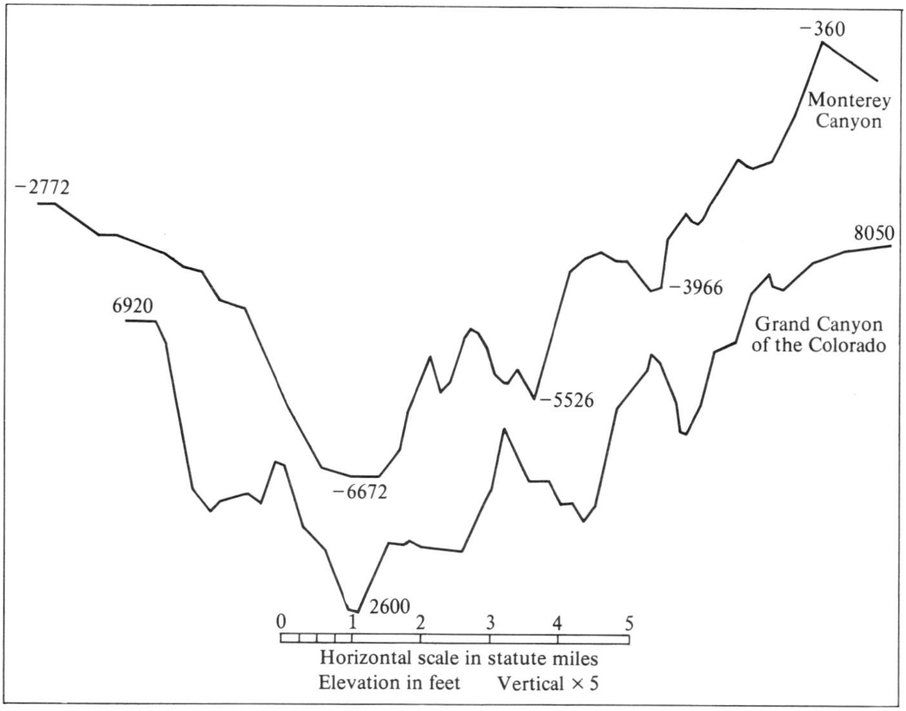

Fig. 4.8 A comparison of vertical profiles across the Grand Canyon of the Colorado and the Monterrey submarine canyon, drawn to the same scale. These two features are similar in magnitude but separated by their vertical position on the earth. (Redrawn from F. P. Shephard, *Submarine Geology*, 2d ed., Harper and Row, New York, 1963.)

Fig. 4.9 Sand fall in San Lucas submarine canyon. Official U.S. Navy photograph by R. F. Dill.

thus helping to perpetuate the canyon whether or not it is associated with a river.

THE CONTINENTAL SHELF

The last major region of the sea floor is the continental shelf, a flat apron lying at depths of less than 200 m. Consisting of sedimentary material more common to continents than the the ocean, the flat shelf areas represent land that has been alternately covered and exposed because of changing land elevation and sea level. The world average surface slope of the shelf is about 7 min, an angle that the unaided eye cannot discern as deviating from a level surface. The slope of the shelf at any given point, however, may vary quite markedly from this average, just as the width of the shelf, which averages 56 km, can range from 0 to 1500 km.

The recent international agreement that awarded ownership of the sea floor of the continental shelves to countries possessing the adjacent open coastal regions provided the United States with a land area greater than that of the Louisiana Purchase. This land, valuable for its mineral and oil resources, lies mostly around Alaska in the shallow arctic and subarctic seas. Figure 4.10 (see pp. 84–85) shows the distribution of continental shelf areas throughout the world.

The shelves are the fringing flats surrounding the continents at nearly sea level; thus they are underlain by the light crustal rock of high silica and aluminum content known as *sial.* Terrigenous materials carried seaward are deposited on top of the shelf base as unconsolidated material which can be sculpted and moved from place to place by the dynamic processes in the shallow overlying water. Currents scour channels in the shelf materials, removing material from one place and depositing it in another to produce submerged bars and banks. Storm waves can rapidly alter the nearshore regions of the shelf, completely changing the boundary between water and shore within days. Further seaward on the shelf changes do not occur as rapidly or as often, since the increased water depth acts to isolate the bottom from the active processes at work at the sea surface.

The shelf, like the slope and basin floor, is superimposed with distinctive features. We have already stated that some heads of submarine canyons penetrate into the shelf province. Other processes of the geologic past have also left their mark. At mid- to high latitudes large glacial troughs with their typical U-shaped profiles can be found where arms of glaciers have moved from the land out over the shelf province. In some instances, moraines deposited by these glaciers at the trough termini are discernible, as are stream and river channels formed at a lower sea-level stand.

The continental shelves reflect the history of the adjacent coastal regions. We can determine variations in erosional patterns, and the supply of sediments, for example, from the terrigenous deposits of the shelf. The eastern and southern coasts of the continental United States are piedmont alluvial plains, large continuous deltas that are formed by the outwash of materials from a source farther inland. Areas such as the northern coasts of Siberia, Alaska, and Canada and the regions off the large rivers of the Southern Hemisphere (the Plata, Amazon, and Congo) are also piedmont plains. These regions of extensive shelves are direct continuations of the present bordering land areas. Very thick deposits of sediment laid down by rivers over the earth's crust can cause it to sink deeper into the mantle and produce the drowned river valley systems characteristic of such regions. The extensive deltas formed off rivers can also load the crust to produce geosynclines, which in the future may become weak points in the earth's crust and fracture to form mountains.

The continental shelves along shores consisting of newly emerged mountain systems, such as those along the west coast of South America, are considerably reduced and in some cases almost nonexistent; thus the mountains appear to rise directly from the deep sea floor. These youthful coasts are not well supplied with terrigenous deposits washed seaward from the land, and the depositional load on what little shelf may be present is slight. The shelves of coasts that have undergone intense glaciation by the continental ice masses, as well as those with older mountains which are not separated from the sea by alluvial outwash plains, also have reduced sediment deposits. The removal of the ice sheet covering the former during interglacial stages has allowed the land to rise, thus reducing the shelf area. In the latter case, the older mountain systems may have depressed the shelves to a deeper depth or may even have completely obliterated the still older shelves they were formed from.

A discussion of all aspects of the relationships between continental shelves and their adjacent land masses, including their geologic history, is beyond the scope of this book. Although we have described some general associations, there are nearly as many exceptions to these relationships as adherents. It will suffice here to emphasize that the shelves have been formed and modified by processes of both land and marine origin and that their configuration is nonpermanent. The rise and fall of sea level and the buckling of the earth's crust have alternately added and subtracted substantial areas of these borderlands to and from the land masses.

OCEAN BEACHES

The region of the sea's contact with land that is most familiar to us is the shore or coast, the zone in which present sea level meets the land at the air-land-sea interface. Here again there is a problem of descriptive terminology, since words such as *shore*, *coast*, *beach*, and *seaside* are often used synonymously but need not have the same meaning. In scientific studies of the contact zone between sea and land, these terms have more precise connotations. For the purpose of our discussion, however, *coast*, *seaside*, and *shore* all refer to the contact zone between the sea and land, including the area between the low-

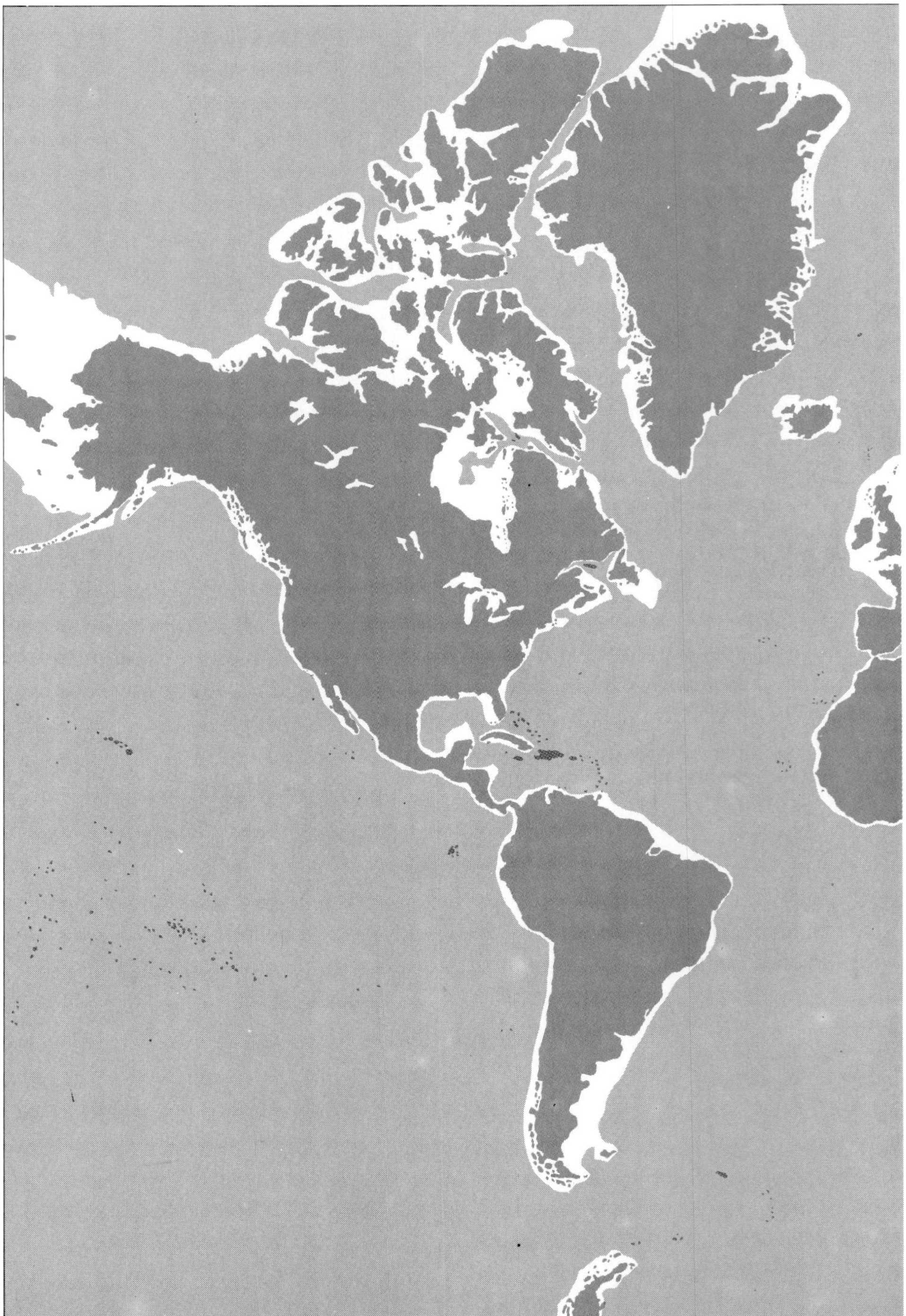

Fig. 4.10 Distribution of the continental shelves of the world. These shelves are the regions between the 200 m depth contour and the land masses.

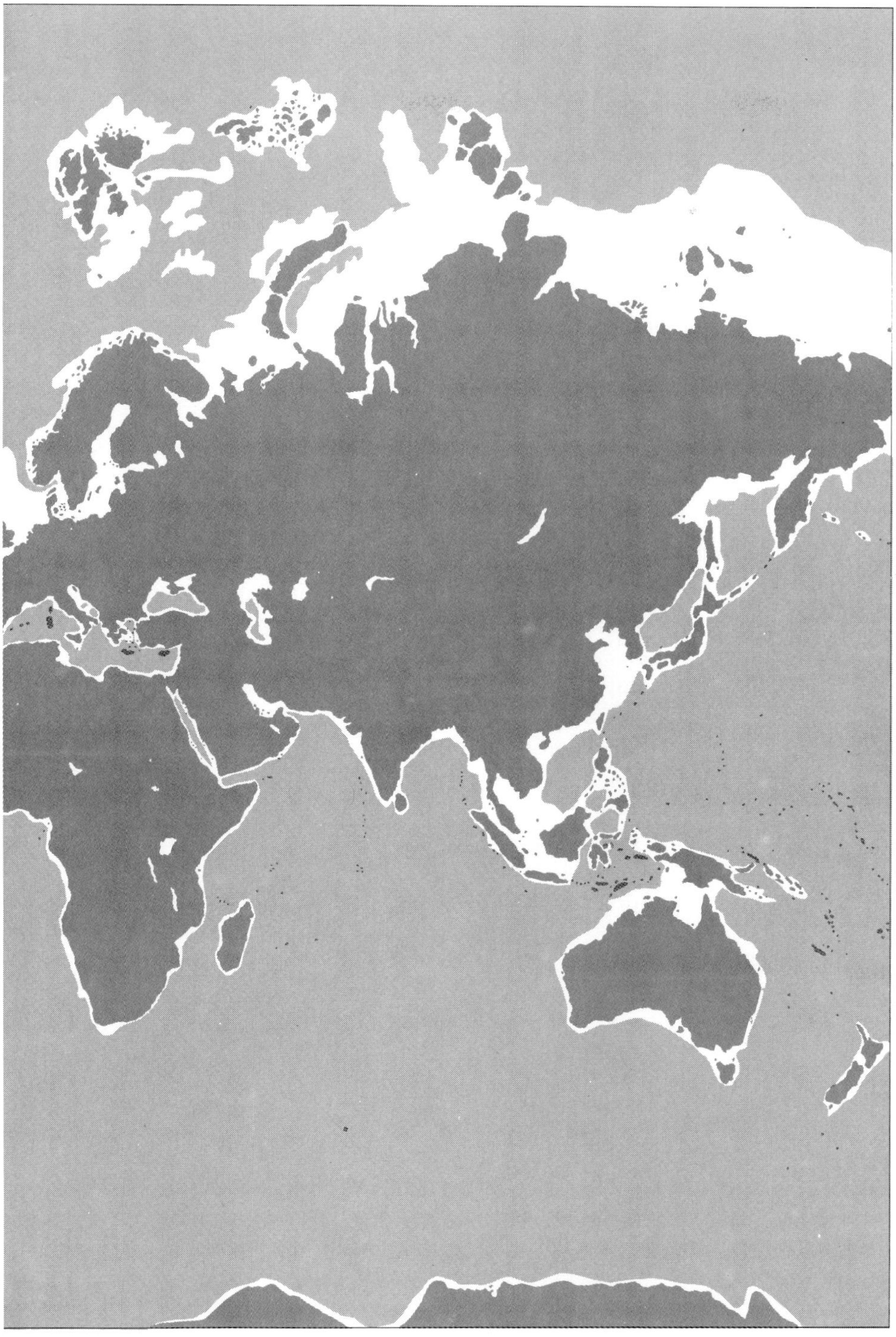

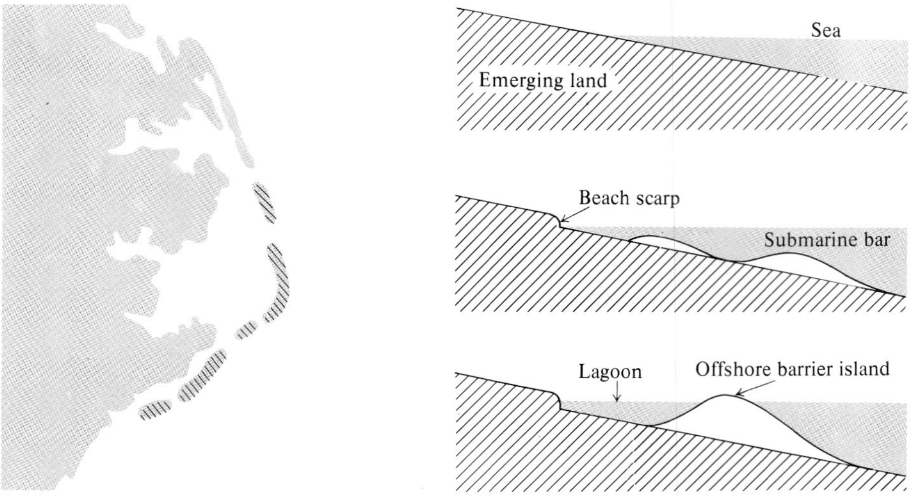

Fig. 4.11 The formation of offshore bars and barrier islands along low-lying coasts having shallow water and an abundance of sand. (After C. H. Cotter, *Physical Geography of the Oceans,* American Elsevier, New York, 1966.)

water stand and the distance shoreward directly influenced by the sea in both the near past and present. The terms apply to coastal regions openly exposed to oceanic conditions and regions within the protected arms of the sea.

Within the contact zone of sea and land are subareas with definite features which we can label more easily. Seaward of the sea's low-water mark is the *offshore* province; landward, to the point at which wave and seawater action no longer alter the land (i.e., a semipermanent cliff face), is the *shore* or *beach* province. Within the offshore province *sandbars* and *troughs* may be formed by wave and current action. The former, which are usually submerged, sometimes become sufficiently large to form *barrier islands* seaward of the coast (Fig. 4.11). Within the beach or shore province are the *low-*, *mean-*, and *high-water* marks of the sea. *Scarps*, abrupt changes in the beach slope formed by the erosional action of waves, are also present.

If there are two scarps on a beach, the more shoreward or higher one is usually formed by waves during storm periods. Plunging and breaking waves, often occurring with an extreme tide, eat away the beach and throw material into suspension to be carried either seaward or along the coast by currents. When the storm waves subside, smaller waves at normal tide stands cut an additional beach scarp closer to the high-water mark. This lower scarp is often changed, but the upper scarp will remain until another major storm alters its configuration. Usually a small ridge, the *berm crest*, and a flat area, the *berm*, form upslope from the scarp. The sloping *beach face* lies between the lower scarp and the low-water mark. This section of the beach is nearly featureless because of the constant reworking of its material by moving water. At the lower end of the beach face is a *low-tide terrace*, a flat formed

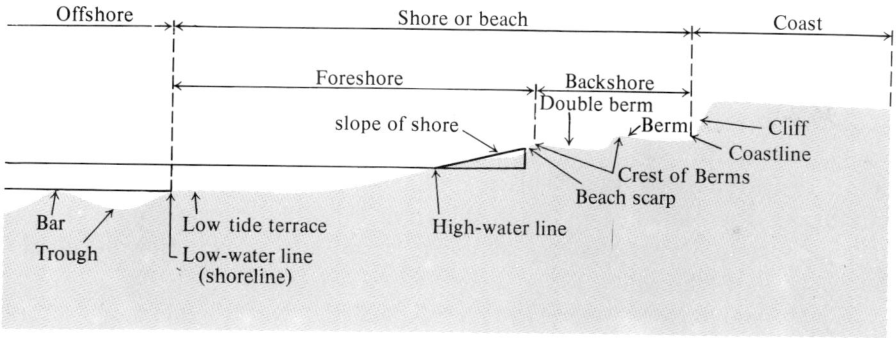

Fig. 4.12 The principal features and subdivisions that can be found on open-coast beaches. (From F. P. Shephard, *Submarine Geology*, 2d ed., Harper and Row, New York, 1963.)

by the deposition of material. Sediment being moved around by small waves at all levels of the tide is dropped in the quieter water below the low-tide line to form this flat.

Figure 4.12 depicts the anatomy of a beach region. Not all the features shown are likely to be found on a single beach. Their formation is affected by such factors as the degree of exposure to the open ocean, the composition and homogeneity of the beach materials, the steepness of the beach's slope, the range of tide, and changing wave patterns.

Although the continental shelves change with time, their upper limits, the shores, are even less static. Some beaches undergo considerable seasonal variation, losing their sand covering under wintertime wave conditions and regaining their sand under summertime conditions (Fig. 4.13). Others seem

Fig. 4.13 Seasonal changes of LaJolla Beach: (left) winter, (right) summer. (Courtesy of F. P. Shephard.)

to grow continuously, accumulating sand at the expense of other beaches, which waste away. The processes that form, maintain, and destroy beaches are subtle, poorly understood and easily disturbed. We know that beaches are dynamic and that they are dependent for their existence on a source of beach material, the supply and removal of which are in balance. If one of the factors controlling this balance is modified, the beach will change until a new equilibrium is attained. The beach shown in Fig. 4.13 is an example of how seasonal variation in the energy supplied by the waves can alter the supply and removal of material.

Waves and water transport of beach material are not the only factors controlling beaches. Some beaches lose sand because wind drives it inland, while others gain sand from interior desert regions. Under special circumstances, wind transport can affect supply and demand as much as water transport.

Man is also continually interfering with processes that control beaches. The damming of rivers which normally supply sediment for beaches creates large settling basins behind the dams where suspended materials are trapped and can no longer be delivered to the sea. Although the dam alters the supply of material, however, it does not greatly affect material removal; thus the balance is altered.

Along open coasts man has constructed breakwaters, groins, jetties, and other harbor facilities. These structures modify the wave-energy distribution along the coast and interfere with the *longshore currents* (currents flowing parallel to the shore) that transport materials to a beach while removing material from it to serve a beach downstream of the current. A groin built perpendicular to a beach decreases the current flow on the upstream side; thus sand that may be required to maintain a beach further downshore is deposited in the quieter, less turbulent water (Fig. 4.14).

Breakwaters or jetties that protect beaches and facilities from wave action but do not necessarily interfere with longshore currents can also alter beach configuration. Wave energy and water turbulence caused by wave motion and currents act to stir up beach materials and hold them in suspension while coastwise currents transport them to another location. Since there is less turbulence in an area protected from wave energy by a manmade structure, particles settle out in quiet water rather than remain in suspension. The protected areas thus become sediment traps and gradually fill in, becoming useless as harbors. The expense involved in removing and resuspending the settled material so that it can again supply the next area downcurrent is great.

Since beaches are controlled by dynamic processes and must maintain a delicate balance between constantly changing factors, it is nearly impossible to predict the effects of a coastal structure on beach configuration, even with extensive study prior to installation. Scale model studies offer the best guides for evaluating the impact of manmade structures on coastal dynamics, but even they can be misleading.

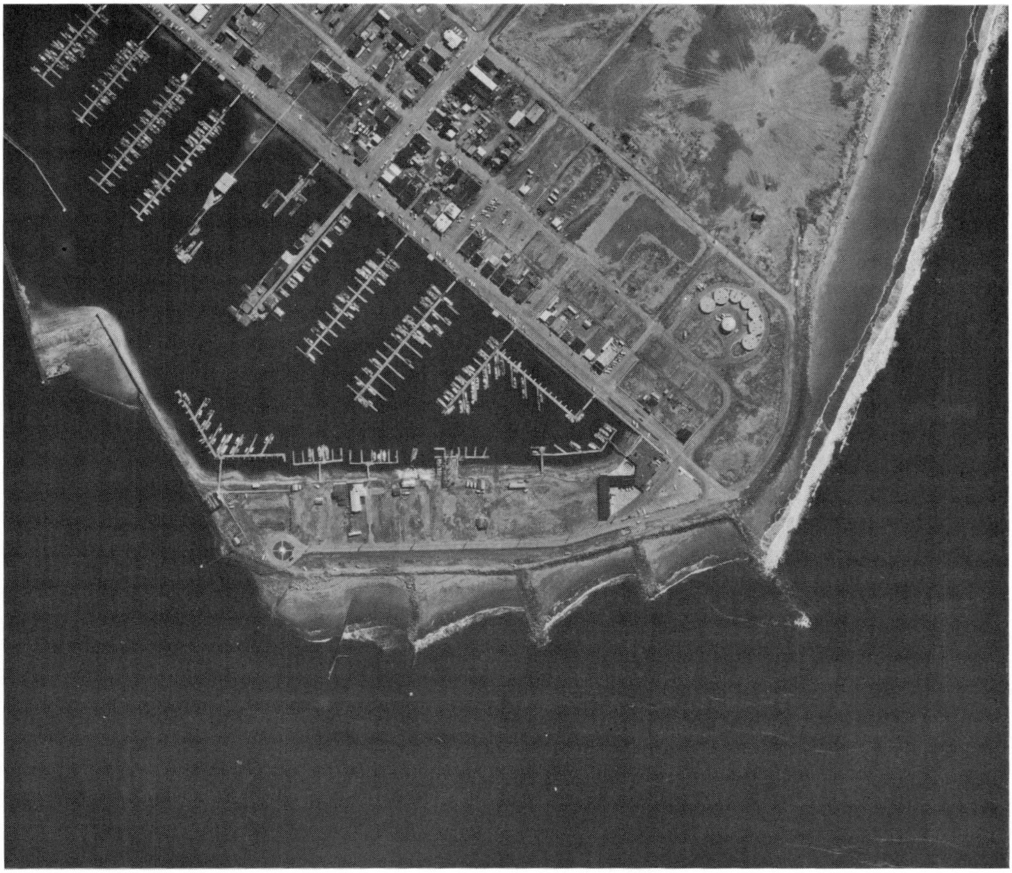

Fig. 4.14 A beach being stabilized by the use of groins which trap sediment. (U.S. Army Corps of Engineers photo.)

The expenditure of wave energy against the shore not only modifies the beaches but affects the entire coastline, slowly eroding its features and tending to give it a uniform appearance. The refraction of waves as they move shoreward over a changing depth of water concentrates energy on the headlands and disperses energy in bays, eventually wearing down the former and filling in the latter. The uneven distribution of crustal materials exposed to wave action along the coast, however, tends to counteract the progression toward uniformity. Since exposed hard rock is more resistant to wave erosion than are unconsolidated materials, bays are cut into the soft materials and rocky headlands jut out into the sea. Thus there is a constant battle between the effects of wave energy and coastal material in shaping the coastline. Along older coasts, where the exposed coastal material is uniform, we can see long straight sections of beach.

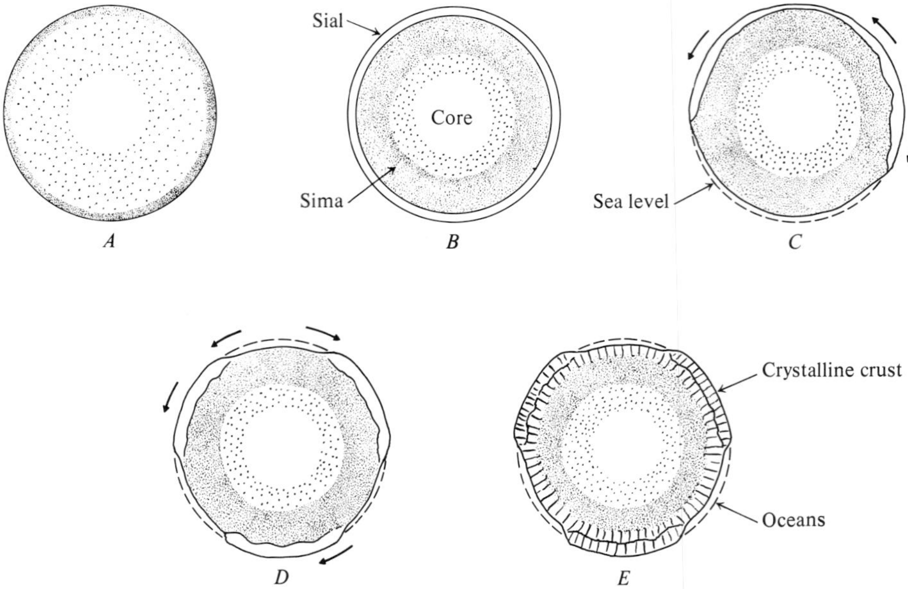

Fig. 4.15 Schematic representation of the origin of continents and oceanic basins in the early pre-Cambrian period by a process of buckling and drifting of an originally siallic layer enveloping the whole earth. (From Ph. H. Kuenen, *Marine Geology*, Wiley, New York, 1950.)

THE PERMANENCE OF OCEAN BASINS

If the shallow regions of the sea floor are likely to be modified over time, can we then assume that deeper regions of the sea floor are more permanent? Certainly the structural configuration of the earth's crust, with the deepening of the Mohorovicic discontinuity under the continents and the large thicknesses of sediments overlaying the basaltic layer under the oceans, tends to support this conclusion. Little is known, however, about the rates at which the more plastic internal layers of the earth readjust in response to changes in the earth's crust. We may use existing rates of sedimentation or denuding the land to establish the time period required to lay down the present thickness of marine sediments. But although this method establishes an approximate age for the oceanic basins, both the assumption that present deposition rates are representative of past rates and the relationship of the age and permanency of the basins to the earth's history are open to question.

In 1912, Alfred Wegener, a German researcher, proposed a theory of land displacement that has become known as the continental drift theory. He suggested that originally an outer layer of the light acidic silica-aluminum crustal rocks (sial) covered the entire earth. Folding then caused this crustal cover to split and the sial became localized on one side of the earth, forming a supercontinental land mass called Pangea. Wegener estimated that this

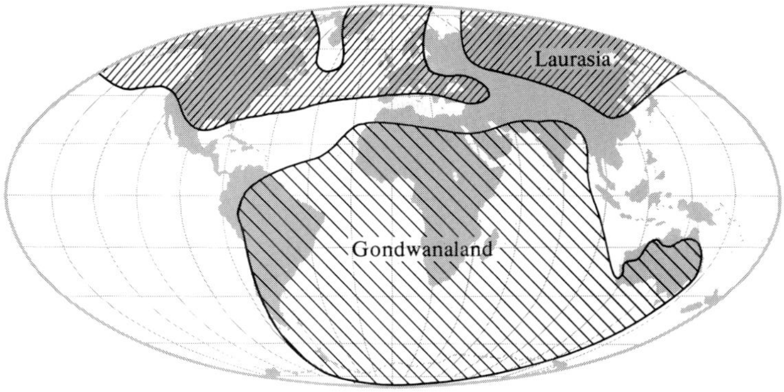

Fig. 4.16 The land masses of the world in late Paleozoic time, showing the large southern continent (Gondwana Land) and the northern continents which make up Laurasia. (After E. W. Sinnot and K. S. Wilson, *Botany—Principles and Problems*, 6th ed., McGraw-Hill, New York, 1963.)

event occurred in the Precambrian period. The thick layer of sial was then thinned by tensile forces acting along fault zones and Pangea partially separated, thinning out the sial to form the Atlantic basins and leaving the Pacific basins nearly devoid of sial (Fig. 4.15).

If the original distribution of these land masses was asymmetrical with respect to a round rotating earth, the drifting apart of the supercontinent was an effort to attain equilibrium, causing the oceans and land masses to become somewhat antipodal. Loss of heat from the earth's interior then gradually caused the crust to solidify to greater depths and slowed the migration of the land masses.

There are many variations of the continental drift theory, most of them refinements of the original hypothesis. Although the concept fell from favor after its initial acceptance, it is now back in vogue because direct observations of magnetic orientation anomalies of materials in the earth's crust, related geological structures on widely separated continents, and similarities of fossil bioforms on now-separated land masses all point to the earlier existence of a joined supercontinent. The modern convection-cell theory also supports the drifting continent concept since it explains the transportation of the masses of sial involved. The geophysical evidence supporting continental drift is presented in Chapter 1.

One variation of the Pangea theory deals with the supercontinents of "Gondwana" and "Laurasia" (Fig. 4.16), which contained most of the southern and northern land masses respectively. The connections between these masses were nonpermanent, being groups of island and isthmuses rather than continental in nature. Although the large masses of land that were close to each other have since drifted apart, separating at their weaker joints, their shapes have been retained to a certain degree. Evidence that supercontinents existed has been established geologically, geophysically, and paleontologically.

Mountains are believed to have been formed when the moving land masses collided with each other and buckled the earth's crustal skin (e.g., the Alps and the Himalaya Mountains).

No one can say what actually happened on the earth in man's prehistory. That the oceans and their basins have always been where they are now is as tenuous an assertion as that the basins have shifted halfway around the globe. But although scientists' hypotheses of these phenomena are often based on fragmentary evidence, sufficient data are available to make several theories plausible. One should keep an open mind to all possibilities and hope that future findings will supply the missing pieces of the puzzle.

Although it is possible that the ocean basins are nonpermanent through geologic time, within man's lifetime they are certainly permanent relative to the borderlands of the oceans, which change rapidly. Thus, on a shortened time scale, we shall consider the basins "permanent."

STUDY QUESTIONS

1 Produce a generalized list of the topographic provinces and features of the sea-floor that one would find as he progressed seaward from a continental land mass. Terminate your list at the central portion of a mid-oceanic ridge system.

2 If all fine-scale topography of the sea floor were ignored, which major topographic features would remain?

3 What is the counterpart of an oceanic sill on the continental land masses?

4 What is the importance of sill depths in controlling the distribution of the deeper water in the sea?

5 What features of and data obtained from the seafloor tend to support the concept of continental drift?

6 What surface topographic features and subsurface geophysical structures of the sea floor can you use to demark the boundary between the continental land blocks and the oceanic region?

7 Discuss a coastal region that you are familiar with. Identify the discrete anatomy of the beach and nearshore structure of this region in terms of the generalized anatomy of beaches presented here.

8 Describe the factors that maintain the equilibrium state of the beach cited in Problem 7.

RECOMMENDED READING

Bascom, W., *Waves and Beaches: The Dynamics of the Ocean Surface*, Anchor Books, New York, 1964.

Dietz, R.S., "The Pacific Floor," *Scientific American*, **192**, 3 (April 1952).

Fisher, R. L., and R. Revelle, "The Trenches of the Pacific," *Scientific American* Offprint No. 814, New York, November 1955.

Heezen, B. C., "The Origin of Submarine Canyons," *Scientific American*, **195**, 2 (August 1956).

Heezen, B. C., "The Rift in the Ocean Floor," *Scientific American*, **203**, 4 (October 1960).

Heirtzler, J. R., "Sea Floor Spreading," *Scientific American* Offprint No 875, New York, December 1968.

Hill, M. N., ed., *The Sea: Ideas and Observations on Progress in the Study of the Seas*, vol. 3, Interscience Publishers, New York, 1965.

Kuenen, P. H., *Marine Geology*, John Wiley & Sons, New York, 1950.

Menard, H. W., "Fractures in the Pacific Floor," *Scientific American*, **193**, 1 (July 1955).

Raff, A., "The Magnetism of the Ocean Floor," *Scientific American*, **205**, 4 (October 1961).

Shephard, F. P., "Submarine Canyons," *Scientific American*, **180**, 5 (April 1949).

Shephard, F. P., *Submarine Geology*, 2nd ed., Harper and Row, New York, 1963.

Stetson, H. C., "The Continental Shelf," *Scientific American*, **192**, 3 (March 1955).

Sverdrup, H. U., M. W. Johnson, and R. H. Fleming, *The Oceans*, Prentice-Hall, Inc., Englewood Cliffs, N.J., 1942.

Wegener, A., *The Origin of Continents and Oceans*, Dover Publications, New York, 1966. (This is a translation of Wegener's original work, published in 1929.)

CHAPTER 5

SEDIMENTS OF THE DEEP SEA

INTRODUCTION

The basins of the world's oceans are the final resting places for all solid particles that can be moved about on the earth's surface. In this respect, they are traps which record the earth's history bit by bit as each particle is delivered to their keeping.

Particles of solid material, each having its own composition, origin, and mode of delivery, form the sediments. Since these characteristics determine the composition of the sediments, they are used to describe them. Particle size, for example, may describe the sediment type, such as sand, mud, and clay. The source of the particle or its method of formation may also be used to identify it, e.g., lithogenic (rock source), volcanic, biogenic (biologically produced), and hydrogenic (formed by chemical precipitation). Sediments may also be classified by the biotic form that produces their major biogenic component, e.g., diatom ooze, radiolarian ooze; or by color and organic content. In short, there are many methods of identifying sediments, each with its adherents and detractors.

Regardless of how sediments are classified, they exist in recognizable distribution patterns in all the global oceans. In this chapter, we will discuss the distribution of the major oceanic sediment types, as well as their accumulation rates and the processes that determine depositional rates and patterns.

THE SOURCES OF SEDIMENTS

The deeper parts of the ocean basins are the lowest levels on earth to which eroded crustal materials can be transported by gravitational forces. They are the resting grounds for most solid particles washed seaward from the land, although some land materials are reduced to a soluble state by mechanical and chemical processes and thus contribute to the chemical content of seawater rather than to its sediments.

On land, it is easy to observe the effects of erosion and the transportation and deposition of eroded material. Outwash alluvial fans issue from the eroded hills of unconsolidated earth materials and large accumulations of talus (broken rock fragments) can be found at the bases of the harder rock bluffs. Muddy rivers endlessly carry their silt load toward the sea, while streams that appear clear and devoid of suspended material are still capable of slowly rolling sand and larger rock fragments along riverbeds. The transportation of material to the sea can be seasonal, as in the cases of heavy periodic rains that wash topsoil into swollen streams and the fracturing of rocks by winter ice forming in small cracks. One of the more subtle seasonal processes is the contribution of rock flour to rivers from alpine glaciers. During late spring and summer the glacial meltwater washes out a flour of ground-up rock so fine that it often stays in suspension in the water for a considerable time. The famous emerald-green appearance of Lake Louise and its outflow streams in Alberta, Canada, is due to fine rock dust held in suspension. The lake maintains this color through the summer and into the fall; when ice forms on the lake it also has a greenish cast. During the winter the water flowing from the glaciers ceases, ending the supply of rock flour. Since the ice layer over the lake prevents the wind from mixing the water, the fine rock particles gradually settle to the lake bottom. In the spring, the water under the melting greenish-white ice is clear and blue; the increasing warmth renews the supply of meltwater and rock flour, completing the cycle.

Examples of both organic and inorganic materials working from high ground toward the sea are all around us. Inorganic material may be igneous or sedimentary, in which case it was deposited at some earlier time in the bottom of a lake or in part of an ancient inland sea. Both types are rock and therefore contribute to the *lithogenic* materials carried toward the sea.

Erosional processes occurring on the exposed land masses are referred to as *subaerial* because they occur at the land-air interface. In the sea erosional processes occur beneath the water except for those at the tide line. Not all the mechanisms for wearing away rock from the elevated land are present in the ocean, and the effects of those that are are greatly reduced.

The sediments of the sea are not entirely derived from the continental land masses or from the shallow shelves that surround them; skeletal remains of oceanic plants and animals also make important contributions. Broken coraline materials form large calcereous deposits surrounding the steep slopes of volcanic pedestals in the warmer latitudes. The siliceous and calcareous skeletons of small, floating plant and animal plankters form deposits which in some cases are only slightly contaminated by land-derived materials. Even hard calcereous components of larger marine animals, such as sharks' teeth and the ear bones of whales, can contribute to the sediments.

The story of the source, transport, type, and distribution of sediments in the sea is complex and still not fully understood. It will be outlined in the remainder of this chapter. We can classify sea sediments according to their five basic sources as follows:

1 *Land-derived materials (terrigenous).* Both organic and inorganic materials from the elevated land masses (material derived from inorganic rock is lithogenic).

2 *Marine plant and animal materials (biogenic).* May include inorganic material, but only that created by marine organisms.

3 *Submarine volcanic materials (volcanic).* Volcanic material given directly to the sea floor.

4 *Materials crystallized from seawater (hydrogenic).* Manganese nodules and other forms of precipitated materials.

5 *Particles of extraterrestrial origin (cosmic).* Cosmic dust, meteoritic particles.

TRANSPORTATION OF SUSPENDED MATERIALS

The process of sedimentation entails the movement of materials from these five sources and their eventual deposition on the sea floor. We shall first consider transportation. Sea sediments are found in a large range of particle sizes. If we trace them from their sources to their farthest limits, a recognizable pattern of particle-size distribution emerges. This is particularly true of sediments of terrigenous origin, which display a large variation of particle size. The largest sediment particles are found close to their sources, while the smallest are the most remote.

On land, turbulent flowing streams and rivers have sufficient speed to move sizable rock fragments toward the sea. Once the rivers enter the sea, however, their swift motion is dissipated and the coastal currents carry on the horizontal transport of the material. In the nearshore environment, oscillatory tidal currents and surface waves act to keep the material in suspension so that the slower net flows can move it. (The term *net flow* refers to the progress that water makes in one direction above and beyond its oscillatory motion.) These currents, plus wave motion, sort the particles by keeping the smaller ones in suspension for a longer period of time than the larger ones. The latter settle out rapidly and are rolled around on the shallow sea bottom, gradually becoming ground down. The material kept in suspension is eventually affected by currents other than those of the coastal region and carried away from its source.

A particle suspended in water and being transported horizontally is continually sinking if (1) it is denser than water; (2) it is of sufficient size that it is not held in suspension permanently by Brownian motion, as are colloids; (3) the motion of the water is not so turbulent that random up-and-down velocities will render the speed of sinking inconsequential; and (4) the net upward flow of water is not equal to or greater than the sinking speed. Consider a particle delivered by a river to the edge of the sea. Initially kept from sinking by turbulent river flow, tidal current, and wave activity, it is then carried seaward by a slow-moving current that is not turbulent enough to maintain it in suspension; thus it will fall through the water toward the sea floor.

Experiments with quartz spheres have shown that the sinking rate of a small particle in still water is controlled by the difference in density between the particle and the surrounding water, the size of the particle (i.e., its diameter or radius), the force of gravity, and the viscosity of the fluid. The law governing the settling velocity for spheres is known as *Stokes' law:*

$$V = \frac{2}{9}\frac{g(\rho_1 - \rho_2)r^2}{\mu},$$

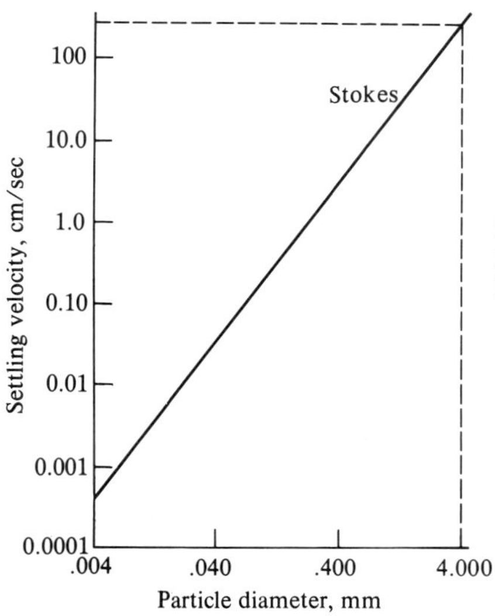

Fig. 5.1 Rate of sinking of quartz spheres in pure water at 20°C as a function of sphere diameter (Stokes' law). (From Ph. H. Kuenen, *Marine Geology*, Wiley, New York, 1950.)

where V is the terminal sinking velocity, g is the acceleration due to gravity, ρ_1 is the density of the sinking particle, ρ_2 is the density of the fluid, r is the radius of the particle, μ is the dynamic viscosity of the fluid, and $\frac{2}{9}$ is the proportionality coefficient for particles having a spherical shape.

Stokes' law is applicable to quartz spheres with radii between approximately 0.00001 and 0.0032 cm. The constant of proportionality, $\frac{2}{9}$, holds for spheres and may be termed a *shape factor*. If we were to test small cylinders or flat disks for settling rate, we would need a different shape factor or proportionality constant. Similarly, since their radii alone would not describe the size of the nonspherical particles, we would have to substitute another unit for r^2 to describe a characteristic size.

The sinking velocity equation may be simplified for average seawater conditions. If $g = 980$ cm/sec^2, ρ_1 of quartz $= 2.65$ g/cm^3, ρ_2 of seawater $=$ 1.03 g/cm^3, $\mu = 0.0135$ poise, and r is expressed in centimeters, we have for V in cm/sec:

$$V = 2.62 \times 10^4 r^2.$$

If we graph the sinking velocity versus the diameter of the particle, as shown in Fig. 5.1, it is apparent that smaller particles sink more slowly than larger particles. Relating sinking to horizontal transport by oceanic currents, we find that the larger particles remain in the moving water for only a short period of time before sinking to the bottom; therefore, they are transported horizontally only relatively short distances from their source. Since they sink slowly, the fine particles remain in the water column for a long period of time and are transported large horizontal distances by the current.

We now know why the size distribution of sediment particles runs from coarse near the source to fine as the distance from the source increases. Under natural conditions, many phenomena control the distribution of particle size and sediment type in any given location. Since there are many sources of sediment, coarse materials from one source may be added to the sea at a position marking the limit of fine material from another source.

Other mechanisms also transport terrigenous materials. Icebergs calved from glaciers float seaward with embedded rocks and materials of all sizes, carrying a great many large rock fragments considerable distances from their source. Even marine algae (kelp) attached by holdfasts to rocks in shallow water may be torn loose from the bottom by wind and waves and carried seaward, where they eventually die and drop their rock anchor. Marine mammals also transport material. The stomachs of sea lions, for example, may contain small pebbles which contribute to the sediment when the animals die.

Materials on the bottom close to shore may accumulate with such steep slopes that small seismic disturbances cause them to slide, creating submarine mudflows and turbidity currents which pour down the continental slopes to the sea floor. The flows transport all types of materials along the seabed and thoroughly disorganize any pattern of particle-size distribution. Winds may also transport material seaward, depositing it at random.

CLASSIFICATION OF SEDIMENT PARTICLES

The study of sediments of the sea floor is based in part on their size distribution, mineralogy, layering, fossils, relationship to source, and mechanical state (i.e., well-rounded or smooth or irregular in shape with sharp edges). Samples for analysis are obtained from the sea floor by devices such as corers, grabs, and dredges. Corers penetrate the sediment and extract a plug of material that shows the layering of sediments deposited over time. Subsamples are extracted from the core for size analysis and mineralogy tests to determine differences in the sediment with core length. Fossils are used to help date the layers and estimate the climatic conditions at the time of deposition. Samples from grabs and dredges are more useful when a large bulk of material from the top layer of sediment is desired and vertical variation in the sediment is not of interest.

Sediment samples are sorted according to size and quantity by first drying the loose material and then separating and passing it through sieves of specific screen sizes. The Wentworth size scale or phi size ($\phi = \log_2$) based on powers of two is in common use (Table 5.1). The screens are arranged in order of decreasing size. Each sieve retains particles too large to pass through it but too small to have been retained by the preceding screen. The particles in each size group are weighed to determine the percentage of the original sample falling into each size category.

We may separate the smallest particles (silt and clay) by suspending them in a tall cylinder of water. According to Stokes' law, the larger particles will sink fastest. Thus, if we use a pipet to sample an aliquot of the suspension at a fixed depth at time zero and at successive time intervals, we will find that particles of successively larger diameter will have settled out of the water

Table 5.1 Wentworth size scale

	Generic name	*Diameter in powers of 2, mm*	*Wentworth scale size range (φ)*	*Diameter, mm*
Gravel	Boulder	$>2^8$	>8	>256.0
	Cobble	2^6–2^8	6–8	65.0–256.0
	Pebble	2^2–2^6	2–6	4.0–64.0
	Granule	2^1–2^2	1–2	2.0–4.0
Sands	Very coarse	2^0–2^1	0–1	1.0–2.0
	Coarse	2^{-1}–2^0	−1–0	0.50–1.0
	Medium	2^{-2}–2^{-1}	−2–−1	0.25–0.50
	Fine	2^{-3}–2^{-2}	−3–−2	0.125–0.25
	Very fine	2^{-4}–2^{-3}	−4–−3	0.0625–0.125
Mud	Silt	2^{-8}–2^{-4}	−8–−4	0.0039–0.0625
	Clay	$<2^{-8}$	<-8	<0.0039

column above the sampling depth at each sampling time. With careful time selection, the samples may be taken so that particles above a certain size range are absent from the upper water column at each sampling. The sediment in each aliquot is then dried, weighed, and subtracted from the value of the previous sample to determine the mass of material settled out over each successive time increment. This procedure is known as *pipet analysis.*

Using the data obtained from screening and pipet analysis, we can graph the percent of mass of the original sample versus the particle-size range. We can then use the graph, equivalent to the signature of a sediment type, to trace trends in distribution and the effectiveness of sorting processes (Fig. 5.2).

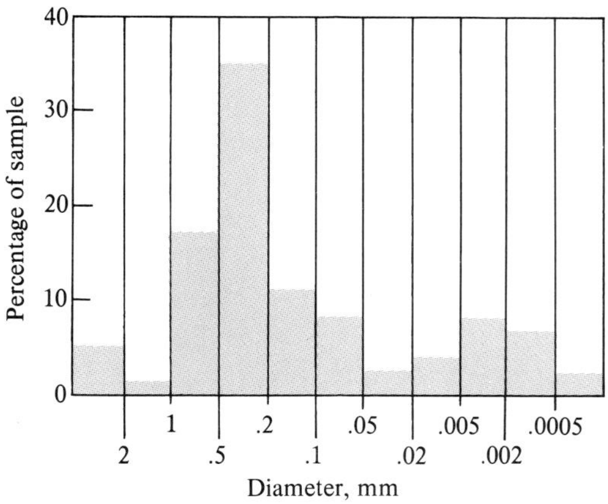

Fig. 5.2 A size fraction analysis diagram of a sample of *Globigerina* ooze showing good sorting of the sediments into both coarse and fine fractions, as denoted by the three peaks. (After Ph. H. Kuenen, *Marine Geology*, Wiley, New York, 1950.)

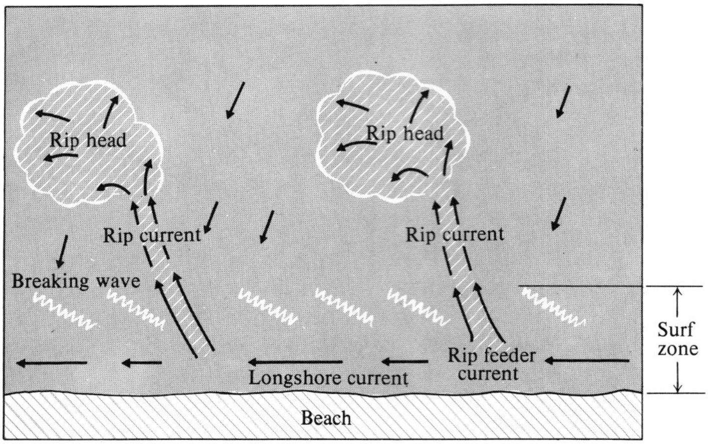

Fig. 5.3 Rip currents are seaward flows of surface water that return water carried shoreward by incoming waves and wind. The water accumulated at the beach moves parallel to the coast in a longshore current that is periodically disrupted by the seaward-flowing rip currents. The location of the rip current is not fixed but is determined by both the topography and the variations in the incoming wave trains. (After M. N. Hill, ed., *The Sea*, vol. 3, Interscience, New York, 1963.)

DEPOSITIONAL ENVIRONMENTS

In order to facilitate discussion of sediments, scientists customarily refer to specific horizontal and vertical areas of the ocean as *zones*. Although this zonation is appropriate primarily for biological investigation, certain terms will be introduced here for their convenience. In the horizontal dimension, the water area extending from the shore to the edge of the continental shelf is known as the *neritic zone*, while the region seaward of the shelf break is called the *pelagic* or *oceanic zone*. The former is greatly influenced by land; the latter is more oceanic in character. The sea floor and the water column are also classified according to depth and light availability. The sea bottom, underlying the neritic zone and extending from the wave splash zone to the sea floor at depths of less than 200 m, is called the *littoral zone*. It is subdivided into other zones, which we will discuss later. At greater depths, the sea floor under the pelagic region is divided into two zones, the uppermost of which is the *archibenthic zone* (200 to 1000 m deep) and the deeper of which is the *abyssal benthic zone* (greater than 1000 m). In both the neritic and oceanic regions the portion of the water column with sufficient sunlight energy to allow marine plants to photosynthesize is called the *euphotic zone*. Below it is the *aphotic zone*. (See Fig. 14.2).

Terrigenous materials contributing to the sediments of the deep sea must be deposited in or pass through the neritic or littoral zone. Some sediment materials, in fact, originate in this region, since waves working at the shore gradually wear away the continental margins to produce fine material, which is carried in suspension or along the sea floor by currents. Breakers impinging on the beach not only supply energy to grind up material but also create a net movement of water shoreward that must eventually be returned seaward. If the predominant wave direction is oblique to the shore, longshore currents may form that transport the surplus water and its suspended material along the beach. Occasionally rip currents form which carry material seaward into

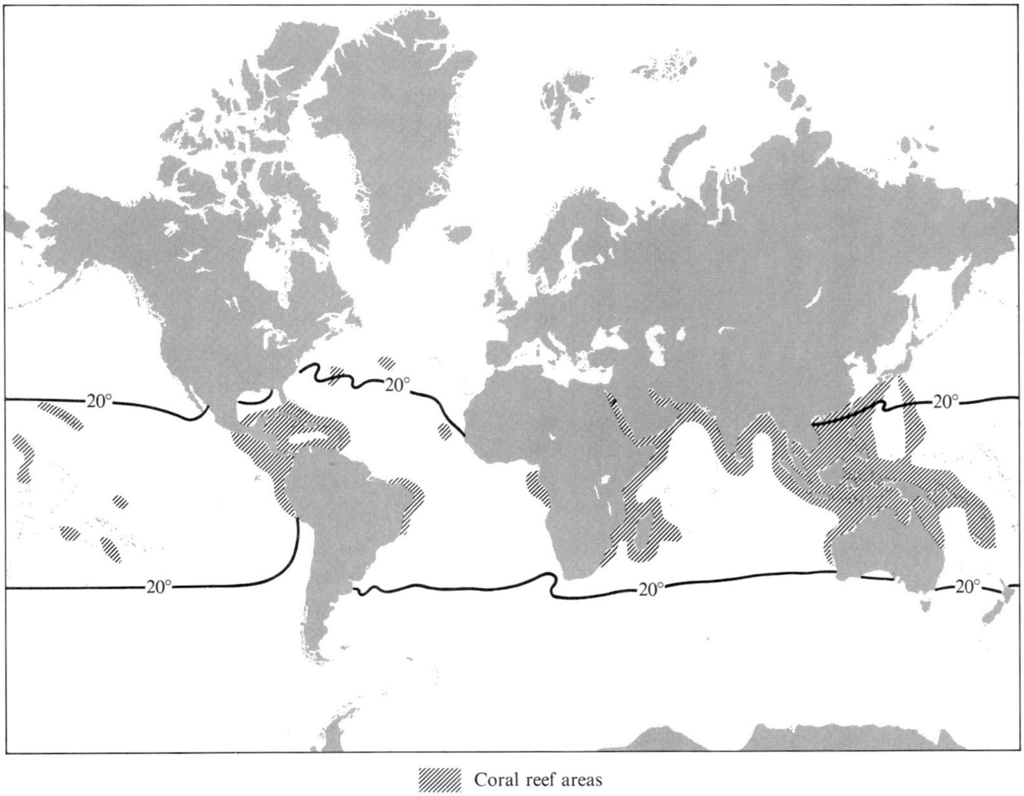

Fig. 5.4 Distribution of the major coral-reef areas of the world. These regions are limited in their latitudinal extent by water temperatures. Nearly all major coral-reef systems are found where water temperatures exceed 18 to 20°C.

deeper, quieter water (Fig. 5.3). Their location depends in part on the topography of the sea bottom, since they occur where the shoreward transport of water by waves is reduced.

Deposits of biogenic material are characteristic of the littoral zone. In coral structures, deposits of the skeletal parts of plants and animals constitute almost 100% of the sediment. Shells and shell fragments of marine mollusks are also common. Despite the proximity of these shallow sea areas to the land masses, the local contribution of terrigenous (inorganic and organic) materials in regions of little erosion may be small in comparison to the marine biogenic material.

The coral reef regions, with their white sand beaches and shallow lagoons, are the most familiar areas of shallow-water biogenous material. They are found in all major oceans, but are confined to shallow water and limited in their north and south extents by the average temperature of the water. Large calcium carbonate deposits associated with substantial coral reef structures are rarely found in areas where the average water temperature falls below 18°C (Fig. 5.4).

MAJOR DEEP-SEA SEDIMENT TYPES

Untold numbers of animals and plants exist in the water and on the bottom of both the neritic and pelagic zones of all oceans. The remains of these organisms also contribute to the sediments. In some oceanic areas, biogenic materials may be the primary components of sediments; in others the biogenic supply, though substantial, may be sufficiently diluted by another material to give the sediments a different quality. For this reason, we always use the major component of the sediment to identify its general type.

The free-floating plants and animals of the sea leave hard remains of either a calcereous or siliceous nature. The planktonic plants (diatoms) of the euphotic zone have siliceous *tests* (exoskeletons) which are relatively insoluble in water; in regions of the world remote from land influence where marine conditions favor the growth of diatom populations, the tests may be the principal constituents of the sediment. Although diatoms require sunlight and nutrients, they can survive in colder water than can some other forms, and thus their remains are the principal components of midoceanic sediments at mid-latitudes. Diatoms are also present in the equatorial belt, but in this warm-water region another planktonic animal, the *radiolarian*, is more prevalent. The skeletons of radiolarians, which are also siliceous, resist the dissolving powers of the sea as they slowly sink to the bottom and form the major biogenic component of the midoceanic sediments at these lower latitudes.

Calcareous deposits are found in the pelagic as well as neritic regions. In the open sea two principal planktonic animals contribute to the calcareous sediment type: *globigerina*, a pelagic *foraminifer*, and the *pteropod*, a pelagic *mollusk*. Since their calcareous remains are more soluble than the siliceous remains of diatoms and radiolarians, small fragments sinking to the sea floor with a velocity comparable to that defined by Stokes' law may dissolve before reaching the bottom. Therefore, although these animals populate large areas of the ocean, the distribution of their remains is somewhat restricted to the shallower regions of the pelagic zone.

Finally, there is the sedimentary material called *red clay*. Usually brown, it is formed from very small particles (diameter less than 1×10^{-3} cm) and has a low (less than $30\%_0$) content of biogenic material. It predominates in the deeper parts of the ocean basins under surface waters of low biological production. The source of red clay is open to dispute; some scientists have attributed its formation to wind-carried atmospheric dust from the land, others to fine material originating from land transported by currents. We will accept both theories here. Since red clay accumulates very slowly in comparison with coarser materials of terrigenous and biogenic origin, extraneous materials being added to red clay areas often appear uncovered on the surface of the sea floor. Sharks' teeth, ear bones from whales, meteoritic particles, and manganese nodules often litter the surface of red clay deposits. The manganese nodules, which apparently precipitate from the water and are therefore hydrogenic, commonly form around hard objects, which act as nuclei.

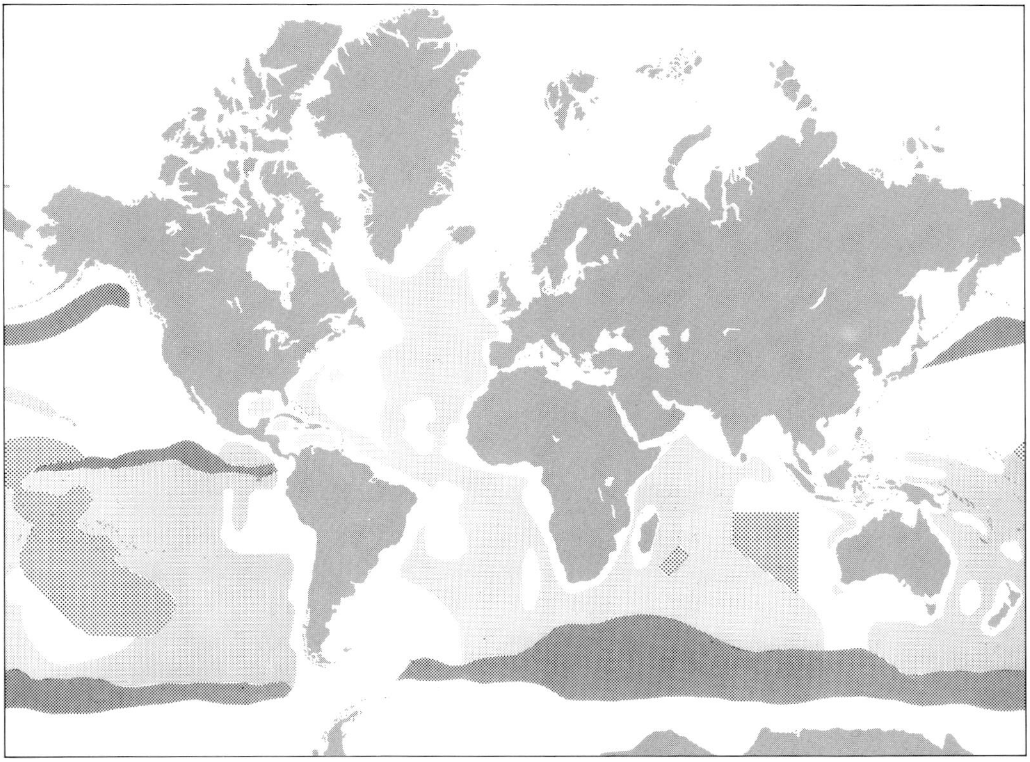

Calcareous mud | Lithogenous sediment | Hydrogenous sediment | Siliceous mud

Fig. 5.5 Distribution of the major sediment types on the deep sea floor. The sediments are classified according to the particular sediment type that is the major component.

Figure 5.5 shows the worldwide distribution of the deep-sea sediments discussed thus far. The areal distribution is evident, but its relationship to ocean depth is not. Early research into the distribution of these sediments in the Atlantic Ocean established maximum, minimum, and average deposit depths (Table 5.2); these depth values do not represent absolute limits but

Table 5.2 Depth zonation of pelagic sediments in meters

Sediment type	*Minimum depth*	*Maximum depth*	*Average depth*	*Texture*
Globigerina ooze	777	6006	3612	Coarse
Pteropod ooze	713	3519	2072	
Diatom ooze	1097	5733	3900	
Radiolarian ooze	4298	8184	5292	
Red clay	4060	8282	5407	Fine

After H. U. Sverdrup. M. W. Johnson, and R. H. Fleming, *The Oceans*, Prentice-Hall, Englewood Cliffs, N.J., 1942.

serve only as a guide to the range over which the sediments may be found. As the table shows, considerable overlap occurs in the various sediments. Although small in comparison with the total areas of the sediments, the boundary areas between each type are diffuse, indicating a gradual change from one to the next.

Sediments may also be classified according to their *mass properties*, which include color, water content, organic content, shear strength, particle-size distribution, and pore size (space between particles). Navigational charts indicate characteristics such as black or green mud, silt, clay, sand, or gravel as aids in describing the bottom. Although the color of sediments is usually obvious, it is difficult to distinguish between categories such as mud, silt, and clay. The Wentworth scale in Table 5.1 will help to separate these bottom types into discrete size grades.

THE SEDIMENT RECORD AND ITS DETECTION

The sediments of the deep sea may furnish the best record of past natural events on the earth. These materials, laid down year after year and maintained in a relatively undisturbed state, record variations in oceanic life, climate, and water conditions, as well as catastrophic events. Samples obtained by coring devices can be dated using modern isotope techniques and fossils. Changing mineral composition, abundance of organic materials, and the presence of foreign material from an unusual source create bands of varying colors and particle type in the sediments which furnish information on historical events against a measurable time scale. Variation in sedimentation rate may relate to major ice ages and climatic changes; fossils, which indicate that warm-water biotic forms existed in regions now occupied by cold water, aid in dating the sediments as well as understanding past climatic changes.

In some areas of the sea, cores appear homogeneous over their total length. However, isotope dating indicates that the material accumulated at reasonable rates, and therefore there should be evidence of seasonal events or other periodic or aperiodic changes. In some cases, the homogeneity of the sediments can be attributed to marine organisms living in the sediment which constantly work and rework the material to extract organic substance for food. This continual ingesting and plowing prevent the formation of discrete layers unless some unusual event supplies a large amound of material over a short time span.

Only a book devoted to all aspects of marine sedimentation can comprehensively present all the techniques employed by geochemists, geophysicists, geologists, and micropaleontologists to decipher the history and formation of marine deposits. However, we shall discuss how sedimentation processes are observed and analyzed in general terms. We have already mentioned the coring of the sea bottom and collection of material by dredge and grab as techniques for obtaining samples from the superficial upper layers of sediment. Deep-sea drilling projects currently in progress hold promise of obtaining samples from deep within the sedimentary layers. The difficulty and expense

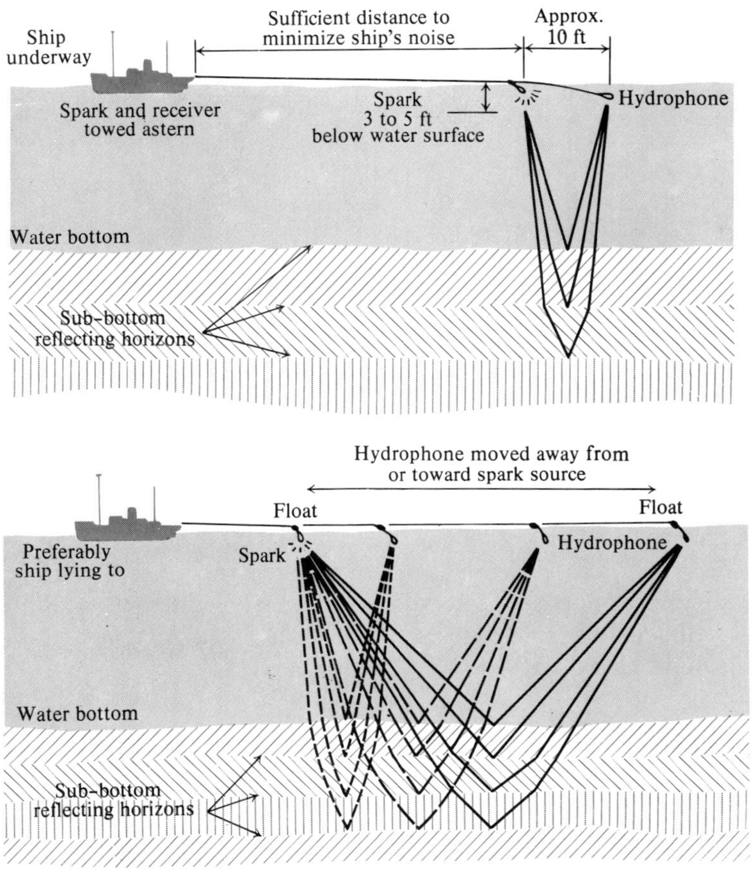

Fig. 5.6 Typical method of continuous seismic profiling, normal incidence reflection above and oblique reflection below, used to measure sub-bottom structure. (Adapted from M. N. Hill, ed., *The Sea*, vol. 3, Interscience, New York, 1963.)

of such projects, however, make it difficult to obtain samples that are representative of all the sedimentation conditions in the oceans. A technique using seismic profiling apparatus is currently used to study sediment thickness and layering over large areas.

In seismic profiling, a source of high energy density produces a shock pulse which travels through the water column and strikes the bottom. If the density level is high, the energy remaining after partial reflection at the sea floor penetrates the sediments and suffers partial reflections at each successive boundary. The bedrock under the sediments also reflects energy, marking the bottom of the sedimentary material. Since a reasonably large change must occur for the layers in the sediments to act as partial reflectors, subtle changes in layering are not readily detectable. The energy pulse, which is created by sound, electric sparks, explosives, or air-driven or magnetically driven oscillating plates, travels at a known speed in the water and in the sediments. The time interval between the outgoing pulse and the return of the reflected energy determines the distance between reflecting horizons (Fig. 5.6). Subbottom profiling yields information not only on sediments but also on the topography and structure of the oceanic crust supporting them. It can detect crustal faults and old channels formed by turbidity flows or streams at some lower stand of sea level or earlier period beneath the covering of sedimentary material.

SEDIMENTATION RATES

The thickness of sediments and rates of sedimentation are used to estimate the age of an oceanic basin. It is currently thought that the average length of time required for a particle with a diameter greater than 5×10^{-5} cm to reach a depth of 2900 m in the sea is 100 years. Particles one-fifth this size may require as long as 7800 years if Stokes' law describes their progress, but 200 to 600 years is considered the average time for settling. The very small particles apparently coagulate to form larger particles, which sink more rapidly.

For oceanic sediments, the accumulation rate of material in the sea floor may vary from about 5 cm/10^3 yr to 1 cm/10^5 yr, with the average being about 1 cm/10^3 yr. The slower accumulation occurs in deep regions remote from land having a low supply of biogenic material. Table 5.3 shows estimates of the recent pelagic sedimentation rates; these estimates use the last glacial age as a reference and assume that it occurred 20,000 years ago.

Table 5.3 Accumulation rates of recent pelagic sediments

	Region	*Blue mud*	*Globigerina ooze*	*Diatom ooze*	*Red clay*
Smallest, average, and highest rates respectively, cm/1000 years	Atlantic Ocean	0.9, 1.8, 3.3	0.5, 1.2, 2.1		< 0.5 < 0.9, 1.3
	Indian Ocean		0.3, 0.6, 1.0	0.3, 0.5, 0.7	

From Ph. H. Kuenen, *Marine Geology*, Wiley, New York, 1950.

In the littoral zone (and even at times in the deep sea), we find that accumulation rates as high as 20 to 500 cm/1000 yr are possible if terrigenous materials are supplied in great quantity. Under special circumstances, annual variations in organic and mineral sediment supply create varved sediments of alternate dark and light bands which record the yearly deposition cycle (much as tree rings indicate yearly growth) and provide an accurate time base against which the accumulation rate may be measured (Fig. 5.7). There may also be nonperiodic additions of large amounts of material to the deep-sea floor, e.g., the slumping of loose materials on the shelves and the resulting turbidity current may suddenly deliver a thick layer of sediment to a deeper area.

Fig. 5.7 Varved marine sediments. Top photo from M. G. Gross *et al.*, "Varved Marine Sediments in a Stagnant Fjord," *Science* **141** (3584), 918–919. Copyright 1963 by the American Association for the Advancement of Science. Bottom photo courtesy of Joe S. Creager.

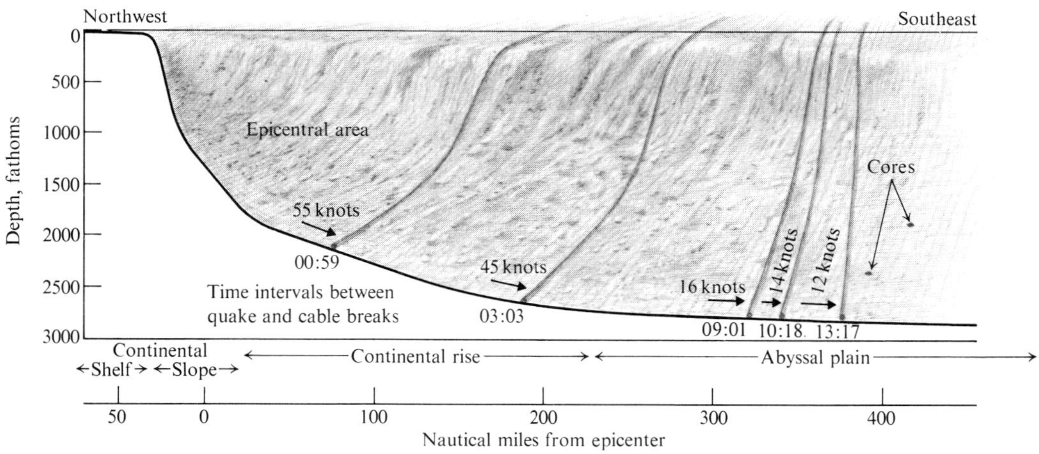

Fig. 5.8 Schematic view of slump and turbidity current associated with the Grand Banks turbidity flow. Speed of flow as determined from time of cable break and cable location is indicated.

A famous case history of a slump and turbidity current has been documented for the Atlantic Ocean. On Nov. 18, 1929, an earthquake occurred centered on the Grand Banks (Fig. 5.8). Following the earthquake telegraph cables lying on the sea floor broke one by one, in sequence from north to south. The time of each cable break and the position of the cable were recorded. It was subsequently surmised that slumping, which poured a mass of fluid mud downslope, caused the cables to snap. Scientists estimated the limit of the mud flow to be in excess of 833 km from the source, and from the placement and break time of the cables, the speed of the flow down the steepest slopes was calculated to be more than 25 m/sec. Figure 5.8 shows the features and extent of the flow. Cores taken at the positions indicated in Fig. 5.8 show a meter-thick layer of silt containing shallow-water fossils deposited by the flow. Thus this is a documented case of a spectacular rate of accumulation.

If the ocean basins are permanent features of the crust, it seems reasonable to estimate the earth's age from the average thicknesses of sediment and the average sedimentation rate. However, the accumulation rate today may not be applicable to that of earlier geologic times. Since the topography of the early land regions was probably not as extreme as it is today, the rate of supply of terrigenous material to the sea would have been much less, and the large inland seas would have caught much of the material as it moved seaward. Thus the present average accumulation rate is considered to be nearly twice as high as the average accumulation rate over geologic time. In addition to the change in sediment supply as a result of shifting topography, the biogenic substances that contribute material to today's sediments did not exist in the early stages of sedimentation, and sedimentary rocks on land having a marine origin indicate that recycling of sediments has occurred. Using 3 km as the average thickness of the present oceanic compacted sediments (equivalent to 9 km of noncompacted sediment, which still has interstitial water in its pore spaces) and an estimated sedimentation rate of about 0.5 cm/1000 yr, scientists have determined the geological age of the earth to be about 2×10^9 years.

There are several independent methods of estimating the total volume of Marine sediments within the earth's crust. The estimated accumulation rate, the estimated age of the earth, and the oceanic area, for example, give an approximation of sediment volume. We can also estimate the ratios in which the various sedimentary rocks are produced from weathered volcanic rocks as compared to the present ratios among sedimentary rocks. A third estimate is obtained by determining rates of denudation of the land and the contribution made to land by volcanics. These three methods yield values of 9×10^8 km^3, 8.5×10^8 km^3, and 12.5×10^8 km^3 respectively for the volume of marine sediments produced through geologic time. The last method yields the highest value because it does not compensate for the recycling of sedimentary material.

MINOR SEDIMENT SOURCES

The least important constituent of the sediments in bulk contribution is material of cosmic origin. In regions of the ocean having small deposition rates (red clay areas), small cosmic spherules called *tektites* (Fig. 5.9) have been found scattered on the sediment surface. In other areas these particles are found interspersed with the sedimentary material. The distribution of tektites is not uniform over the oceans, and their abundance in some areas seems to be indicative of localized meteoritic showers produced by disintegrating cosmic bodies.

Fig. 5.9 Tektites. (Photos courtesy of Virgil E. Barnes.)

Sedimentary processes are also influenced by airborne dust of cosmic or terrestrial origin smaller in diameter than spherules. Although it is difficult to evaluate the effect of the cosmic portion, that of the terrestrial portion can be locally important. The wind system blowing over the dry, dusty, and sandy areas of the earth can pick up and carry considerable material, as one can easily see from a dust or sand-storm. Figure 5.10 shows the regions of the oceans over which the air often carries large quantities of terrestrial particulate matter. We must remember that if the atmosphere is the transportation route for the fine particles forming the red clay deposits of the sea, airborne dust of terrestrial origin is not really a minor source on the global scale. Occasionally natural catastrophic events, as opposed to nuclear blasts, occur that release debris to the atmosphere for transport. In 1883 the volcano Krakatow erupted in a violent explosion that literally caused about four cubic miles of earth to disappear. The fine-particle material from the explosion was said to have been carried around the world and to have colored sunsets for over a year. The major portion of this material must have fallen on the sea, however, and eventually contributed to the sediments.

Volcanics supply material to the sediments by dispersing it directly into the sediments, the water, or the air. Lava flows interspersed among the layers of sediment from fissures can be detected by subbottom profiling, since the

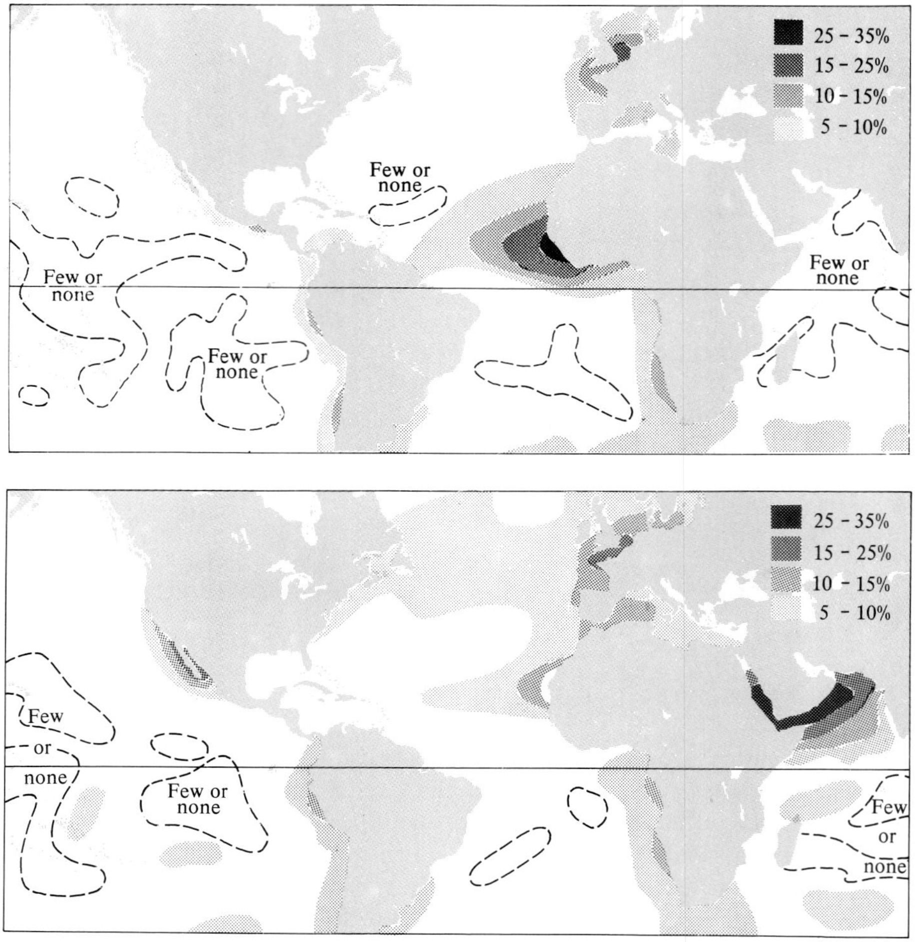

Fig. 5.10 Frequency of haze in percent of total observations due to airborne dust during Northern Hemisphere winter (above) and summer (below). (After M. N. Hill, ed., *The Sea*, vol. 3, Interscience, New York, 1963.)

layers form lenses of hard consolidated material which act as good reflecting horizons among the unconsolidated sedimentary material. Volcanoes that are submerged but break through the sea floor may also supply considerable quantities of molten material, solid fragments, and ash to the sea floor. The supply of volcanic products to the atmosphere by oceanic and land volcanoes and their subsequent dispersion in the air to settle out over the sea produce layers of unconsolidated material such as ash in the sediments. A volcanic island that appeared off Japan in the 1950's surfaced during active stages of eruption and then sank after the collapse of the soft ash cone when activity decreased. The Japanese Hydrographic Service was observing the activity of the cone when it suddenly erupted under the survey vessel, resulting in a total loss of ship and crew. A more recent contributor to the volcanic material

Fig. 5.11 Surtsey, viewed from the air on November 30, 1963. (Photo courtesy of S. Thorarinsson.)

on the sea floor is the volcanic isle of Surtsey, off Iceland. This peak made its sudden appearance on Nov. 14, 1963, and has remained active since (Fig. 5.11). Despite the frequency of volcanic activity on the earth, however, its direct contributions to the sediments on a global scale are relatively small.

The decomposable organic content of the pelagic deep-sea sediments is small. A long time is required for small-particled organic material to sink to the sea floor, and there is active bacterial decomposition of organic matter during sinking. In the oceanic zone the organic content is, on the average, less than 1% of the weight of the sedimentary material. As depth decreases toward shore, organic content increases to an average of about 2.5% composition by weight.

Once organic material reaches the bottom, decomposition continues in the presence of oxygen in the water. If considerable decomposition is occurring and circulation in the overlying water is so poor that oxygen cannot be supplied by currents at the rate it is consumed (as in isolated basins near shore with high organic content in the sediments), the environment may become depleted in oxygen, or *anoxic*. Under these conditions hydrogen sulfide is commonly found in the sediments and water. When oxygen is not depleted the organic content of sediments is controlled in part by the grain size of the sediments. Smaller particle size means less interstitial pore space in which oxygen-carrying water may circulate; this decreases the oxygen supply into the sediments and slows the rate of decomposition, all other factors being equal. The decomposable organic content of sediments is therefore determined by the interaction of the supply of organic material, the circulation of water, the grain size of sediments, the availability of oxygen, and the depth over which sinking occurs.

High organic content is found in the sediments of basins that are near land, have poor water circulation, and support an abundance of planktonic life in the surface waters. For example, the basins behind the Channel Islands off Santa Barbara, California, may have organic material making up as much as 10% of the sedimentary material by weight. In some basins, compaction and aging of sediments with high organic content over long periods of time forms oil, which flows under pressure between the sedimentary layers or along fractures until it becomes trapped in pockets within the lithified sediments. Oil is produced in the marine environment; its freshwater counterpart on land is coal, formed by organic material from the ancient fern forests. That oil companies recently paid over $600 million for oil leases in the Santa Barbara Basin area attests to its long history of organic accumulation and abundance of oil.

Hydrogenic materials, such as slabs, nodules of manganese, and other heavy minerals, are widely scattered over the ocean floor at moderate depths. In the regions of low deposition, they appear to litter the sea floor. Their abundance on the surface of the sediments makes them attractive as a source of heavy minerals when methods of gathering become economically feasible. Despite their economic potential, hydrogenic materials are only a minor constituent of the sediments; the real mineral wealth lies in the finer-particle sediments, which also contain heavy minerals.

Source, size, composition, color, and texture are all used to describe sediments. Near the borders of the continental land masses sources and supply rates may vary considerably over small distances, creating a wide range of sediment types. In the nearshore environment physical conditions such as waves, currents, and topography can also act to control the depositional environment, causing further variation. In the deeper regions of the open ocean fewer factors can influence deposition; thus a single sediment type may exist over a large area. It is in these areas that the principal sediment types are found.

STUDY QUESTIONS

1 Discuss why terrigenous sea-floor deposits near the shore are likely to have a greater variety of form than the deeper sea deposits farther from shore.

2 Why are large, hard-boned components of biogenic origin commonplace on the surface of sediments considered low in biogenic materials? What does this tell you about the sedimentation rate?

3 Study the distribution of biogenic sediments in the oceans and their relationship to depth of water and latitude. Discuss what clues this distribution can give you to understand the organic productiveness of the overlying water column.

4 A sample of fine-grain quartz sediment is subjected to pipet analysis. It is found that within 15 minutes, 25% of the sample has settled past the 10-cm mark in the settling flask. Thus 25% of the sample has a mean radius greater than what value? Use Stokes' law to find the answer.

5 If a suspended particle has a diameter of 0.02 cm, how far can it be transported by a horizontal current having a uniform speed of 25 cm/sec when the depth of water is 100 m and the density of the particle is 2.82 g/cm^3?

6 Why are siliceous biogenic deposits more likely to be found at midocean depths than are carbonate deposits?

7 Why does the mean particle size of oceanic sediments tend to decrease as we move seaward from land?

8 Why is a deep, semi-isolated nearshore basin likely to contain a large percentage of organic matter in the sediments?

RECOMMENDED READING

Fairbridge, R. W., ed., *The Encyclopedia of Oceanography*, Reinhold Publishing Corp., New York, 1966.

Hill, M. N., ed., *The Sea*, vol. 3, Interscience Publishers, New York, 1965.

Kuenen, Ph. H., *Marine Geology*, John Wiley & Sons, New York, 1950.

Shephard, F. P., *Submarine Geology*, 2nd ed., Harper and Row, New York, 1963.

Sverdrup, H. U., M. W. Johnson, and R. H. Fleming, *The Oceans*, Prentice-Hall, Inc., Englewood Cliffs, N.J., 1942.

CHAPTER 6

THE CHEMISTRY OF SEAWATER

INTRODUCTION

The title of this chapter could refer to the study of the organic, inorganic, or physical chemistry of the complex solution recognized as seawater. However, to consider only these aspects of seawater removes it from the context of its natural environment. In oceanography, it is as important if not more so to study seawater in its oceanic setting, where chemical changes and distributions occur in response to natural processes. Thus we shall not emphasize chemical techniques or the chemical properties of seawater, but rather present this complex solution as part of its oceanic environment.

The oceans have been receiving soluble materials from the earth since their formation. This supply has produced a seawater that contains a host of soluble solid materials and gases in both large and small amounts. Since the major constituents of the dissolved solids in the sea have a remarkably constant composition, seawater is a uniform and unique solution regardless of its total salt strength. Under natural conditions, however, it is continually changing; thus a sample of seawater that is isolated in a bottle loses some of its identity.

Although the proportions of the major dissolved materials are fairly constant in the sea, important but subtle variations in concentrations of the less abundant materials occur in response to local addition and extraction of chemicals and gases due to biological and chemical processes and the land supply of salts. To understand the sea as a chemical environment, we must be cognizant of these changes as well as the total salt concentration.

THE CHEMICAL UNIFORMITY OF SEAWATER

Since their formation, the seas have acted as catch basins not only for the particulate material that forms the sediments but also for the soluble material carried by the water that drains from the surrounding higher elevations of the earth. The water that carries this dissolved and suspended load of material to the sea is continuously recycled to the atmosphere, to the land, and back to the sea by rivers and ground seepage. Most of the soluble materials do not complete the hydrologic cycle, since the process of evaporation, which transfers water from the sea to the atmosphere, leaves them behind in the oceans. This flow of soluble material from the elevated surface of the earth is only one mechanism of supplying the chemicals found in the oceans. The rocks or sediments forming the basins holding the seas contribute soluble material through direct leaching by seawater, while volcanic processes add material directly from the earth's interior.

We commonly think of the dissolved solid materials in seawater as salts that have undergone a separation into their chemical components. When dissolved in water, sodium chloride (common table salt), for example, dissociates into sodium and chlorine, which then exist as free ions of sodium (Na^+) and chlorine (Cl^-) in solution. Other salts behave in the same way. Thus most of the chemicals supplied to the sea by soluble materials are present as free ions. Ionized salts, however, are not the only chemicals in seawater; dissolved atmospheric gases are also present, and are as necessary to support biological life in the sea as they are on land.

The oceans have both received and relinquished soluble materials over a considerable period of geologic time. As a result of continuous mixing and stirring by the currents, the sea now appears to have a homogeneous chemical content. We mentioned previously that atmospheric processes act to give some surface regions of the sea a high salt content and others a low salt content. Primarily, the processes that cause this variation only remove or add water; thus they do not significantly affect the chemical homogeneity of the dissolved solid materials.

This homogeneity concept implies that even though the total dissolved salt content per unit mass of seawater may vary over the oceans' breadth and depth because of the addition or subtraction of freshwater, the major ions (most abundant) in solution always appear in constant ratios to one another. However, processes in the sea and at its borders can upset the constancy of the chemical composition of seawater.

In restricted areas of the oceans the chemical composition of seawater fluctuates as materials are extracted from or liberated to their soluble states by biological processes. In general, chemicals extracted from the sea near its surface are used to build the body tissues of plants and animals, which are then decomposed at depth by processes of decay. The borrowed chemicals are returned to a soluble state or deposited on the sea floor as biogenic sediments. Under special conditions, the precipitation of certain chemicals can also alter the amount of material in solution, as in the formation of hydrogenic sediments. In coastal regions large rivers entering the sea may supply a chemical load sufficient to change the chemical composition of the adjacent sea and locally void the constancy of composition concept. For this reason, we shall consider these regions separately from our general discussion of the homogeneous chemical nature of the oceans.

THE DETERMINATION OF SALINITY

Since the chemical composition of seawater remains constant for open-ocean conditions, the chemical oceanographer may conveniently determine the dissolved salt content of the major ions by measuring only one chemical constituent. Constancy of composition requires that the salt content be such that the ratio of each major chemical ion to the total salt content, as well as the ratio of one dissolved major element to another, remains constant. *Salinity*, the term used to describe the salt content of seawater, is expressed as grams of salt per kilogram of seawater, or parts per thousand (‰). By definition, salinity is *the total amount in grams of solid material dissolved in 1 kg of seawater when all the carbonate has been converted to oxide, all the iodine and bromine have been replaced by chlorine, and all organic matter has been completely oxidized.*

If salinity is to be determined by measuring only one of the major ions, then an ion that is abundant and easy to measure precisely by chemical methods, e.g., the chloride ion, should be used. The preceding definition states that iodine and bromine must be replaced by chlorine before salinity can be determined. Chlorine, bromine, and iodine (halides) have similar chemical properties; if the chloride ion content of seawater is measured by chemical titration with silver nitrate, all three halides react with the silver ion and are measured as chlorine. Therefore, to determine salinity by chloride content, we must know the chlorine equivalent of bromine and iodine. The original empirical relationship between salinity and chloride content is

$$S‰ = 0.03 + 1.805\ Cl‰, \tag{6.1}$$

where Cl‰ is the *chlorinity* expressed in parts per thousand.

We can use the chemical titration of a seawater sample with a silver nitrate solution (precipitating out the halides) to determine salinity if we know the chlorine equivalent of the silver nitrate on a volume basis. To find the latter we must standardize the silver nitrate titer solution with a sample of known chlorinity. To enable marine chemists all over the world to use a universal method of standardization, the Hydrographic Laboratory of the International Council for the Exploration of the Sea in Copenhagen produces a special standard seawater called "normal water," the chlorinity of which is accurately controlled to the third decimal place. All oceanographers use this normal water as their primary standard.

The original empirical relationship between salinity and chlorinity was based on work by Knudsen, Forch, and Sørensen in 1901 and 1902. More recent physical chemical analyses based on a larger number of seawater samples have shown that the empirical relationship is more accurately described by

$$S‰ = 1.80655\ Cl‰. \tag{6.2}$$

This new relationship, which has gained universal acceptance, grew out of high-precision electrical conductivity measurements of seawater which yielded more accurate ratios between conductivity, chlorinity, and salinity than those obtained by chemical titration methods. In 1966 UNESCO and the National Institute of Oceanography of Great Britain jointly published the *International Oceanographic Tables*, which related electrical conductivity, salinity, and chlorinity. The tables have been endorsed by the major agencies and committees in the oceanographic field and are recommended for use.

Since high-precision electrical conductivity measurements on seawater are becoming more standard than chemical titrations, the salinity of a seawater sample has been defined as a function of the conductivity ratio of this sample to that of a standard seawater sample having a salinity of exactly 35‰. This conductivity ratio is denoted by R_t where t is the temperature of the seawater sample and the 35‰ salinity standard. The standard temperature for conductivity determinations is 15°C; the pressure is one standard atmosphere. The polynomial describing S‰ of a sample of seawater in terms of R_t, where t is 15°C, is

$$S‰ = -0.08996 + 28.29720\,R_{15} + 12.80832\,R_{15}^2 - 10.67869\,R_{15}^3 + 5.98624\,R_{15}^4 - 1.32311\,R_{15}^5. \tag{6.3}$$

For conductivity measurements, this expression serves as the modern empirical definition of the salinity of a seawater sample. It does not, however, convey the physical significance of salinity as does the older definition, which relates salinity to the materials dissolved in seawater. The student learning about the salt content and chemistry of seawater will find that he can use the older definitions to obtain a conception of what is meant by salinity and the new definition to find its numerical value. Since this book stresses the acquisition of concepts rather than precision techniques, we will emphasize the older definitions of the chemistry of seawater in the hope that they will promote a better understanding of seawater as a solution.

The salinity of any sample of seawater may now be determined from its chlorinity, if the rule of composition constancy holds for all the oceans. The percentages of other chemicals present in seawater can also be determined from their chlorinity ratios. In one kilogram of seawater having a chlorinity of 19‰ or a salinity of 34.32‰, the weights of the major and minor chemical constituents and their ratios to chlorine are as shown in Table 6.1.

Table 6.1 Dissolved chemical constituents of seawater (based on chlorinity = 19‰)

Material	*g/1 kg seawater*	*Element/Cl‰*	*Percent salt by weight*
Chloride, Cl^-	18.980	0.99894	55.04
Sodium, Na^+	10.556	0.5556	30.61
Sulfate, SO_4^{2-}	2.649	0.1394	7.68
Magnesium, Mg^{2+}	1.272	0.06695	3.69
Calcium, Ca^{2+}	0.400	0.02106	1.16
Potassium, K^+	0.380	0.02000	1.10
Bicarbonate, HCO_3^-	0.140	0.00735	0.41
Bromide, Br^-	0.065	0.00340	0.19
Boric acid, H_3BO_3	0.026	0.00137	0.07
Strontium, Sr^{2+}	0.013	0.00070	0.04
Fluoride, F^-	0.001	0.00007	0.00
Total	34.482‰		99.99%

From H. U. Sverdrup, M. W. Johnson, and R. H. Fleming, *The Oceans*, © 1942. Reprinted by permission of Prentice-Hall, Inc., Englewood Cliffs, N.J.

Note that this table shows that the total dissolved solids, 34.482 g, in a kilogram of seawater of 19‰ chlorinity are greater than the calculated salinity of 34.32‰. This discrepancy is explained by the fact that in the determination of salinity from the chloride content of a seawater sample all the halides act as chloride. For example, one atom of bromine acts as if it were one atom of chlorine and is counted as such. Thus we must multiply the mass of each halide by the ratio of its atomic weight to the atomic weight of chloride to determine its effective chloride mass. We then find the effective chloride content of the kilogram of seawater used in Table 6.1 by adding the values of chloride, 18.980, and bromide converted to chloride,

$$0.065 \times \frac{35.453}{79.909} = 0.029, \tag{6.4}$$

to yield an equivalent chloride content of 19.009 as opposed to a halide content of 19.045. The former value is still slightly more than the chlorinity of 19‰ given for this sample of water.

By definition, however, chlorinity is given as numerically equal to the atomic weight silver in grams necessary to precipitate the halogens in 0.3285233 kg of the seawater sample. In the chemical titration, one atom of silver combines with each chloride atom to form silver chloride (AgCl). Thus it is necessary to establish the ratio of chlorine to silver on a weight basis to determine the chlorine equivalent to the silver:

$$\frac{\mathrm{Cl}}{\mathrm{Ag}} = \frac{35.453}{107.870} = 0.3286641, \text{ the chlorine equivalent of silver.} \tag{6.5}$$

The ratio of chlorine equivalent to the chlorinity therefore becomes

$$\frac{0.3286641}{0.3285233} = 1.00043, \tag{6.6}$$

which implies that the chloride equivalent in the total dissolved solids is 1.00043 times the chlorinity. This requires that the chloride equivalent for a chlorinity of 19‰ be 19.009, which is also our calculated value.

If we convert the carbonate to oxide and replace bromine by chlorine, as stipulated in the definition of salinity, the adjusted total salt content of the seawater sample given in Table 6.1 is 34.324‰, which agrees well with the salinity determined from a chlorinity of 19‰. The total dissolved solids are therefore considered separate from the salinity and are expressed as

$$\text{total dissolved solids, ‰} = 0.073 + 1.8110\ \mathrm{Cl}‰. \tag{6.7}$$

In the open sea, away from the influence of land drainage, the ratio of dissolved materials to chlorinity is nearly constant; thus we can calculate the amount of dissolved material expressed as total solids or salinity from the chlorinity.

In titration procedures, it is often desirable to determine the chloride content of a given volume of sea water rather than of a given mass. In this case, the chloride content is expressed as chlorosity, Cl_0: grams of chloride per liter of seawater. Chlorinity and chlorosity are related by the density of the seawater sample at 20°C:

$$Cl‰ = Cl_0/\text{density of the sample at } 20^\circ C. \qquad (6.8)$$

THE SUPPLY OF SALT FROM LAND

We stated earlier that the ionic ratios can change near rivers such that the rule for constancy of composition does not hold. It has been estimated that the world's rivers contribute a total of 2.73×10^9 metric tons of dissolved materials to the sea each year (one metric ton equals 10^3 kg). This value represents a tremendous quantity of material, but it is still small when compared with the 5×10^{19} kg of sea salt currently dissolved in the oceans. It is possible to calculate the average chemical composition of river water on a worldwide basis and determine an average percentage composition of these river-borne dissolved materials (Table 6.2).

Table 6.2 Average composition of soluble material in rivers

Material	*Percent composition by weight*
Carbonate, CO_3^{2-}	35.15
Calcium, Ca^{2+}	20.39
Sulfate, SO_4^{2-}	12.14
Silica dioxide, SiO_2	11.67
Sodium, Na^+	5.79
Chloride, Cl^-	5.68
Magnesium, Mg^{2+}	3.41
Oxides $(Fe, Al)_2O_3$	2.75
Potassium, K^+	2.12
Nitrate, NO_3^-	0.90
Total	100.00%

After C. W. Wolfe *et al.*, *Earth and Space Science*, D. C. Heath, Boston, Mass., 1966.

If we compare the percent compositions of dissolved materials in seawater and river water (Tables 6.1 and 6.2), we see that the major ions which contribute most to the salt in seawater form a much smaller part of the soluble materials in average river water. This presents a puzzle if we consider river-supplied salts as the primary source of sea salts. The following solution to the problem has been proposed, but there is no way to establish its validity.

When water percolates through the land to leach soluble material which is then desposited in the sea, it first removes the most soluble substances. In early geologic time, therefore, the rivers should have had a preponderance of very soluble salts such as sodium chloride. After these materials were removed, slightly less soluble substances presumably made up the bulk of the dissolved materials in the rivers. Thus rivers today are leaching material from land that has already been stripped of its most soluble salts, and only the less soluble materials such as calcium carbonate are available in abundance. By this reasoning, the reversal in the chemical composition of river water compared to that of seawater is indicative of an old earth and verifies the assumption that once soluble salts are delivered to the sea they are not returned to the elevated land except in minor quantities on the global scale. Tectonic processes, which may cause upheavals of the sea floor to produce new land bodies or may isolate regions of the sea, creating landlocked bodies of salt water, are mechanisms that can return small amounts of soluble oceanic salts to the elevated land for recycling.

THE MINOR CONSTITUENTS OF SEAWATER

Table 6.1 presents only the major chemical constituents of seawater. If we could perform analyses sensitive to all known chemical elements in seawater, then we could detect nearly every element, even those present in minute concentrations. Chemical elements present in small quantities in the sea are called the *minor chemical constituents.* Some minor constituents, known as *trace elements*, are present in such weak concentrations that they can only be detected and not measured accurately. In some cases trace elements cannot be measured directly in seawater, but can be found in concentrated form in biological material. Even gold, one of the noble metals, is present in soluble form with a concentration of about 6×10^{-9} g per kilogram of seawater. However, the minor constituents of sea salt, which make up only about 0.01‰ by weight of the dissolved solids, more than compensate for their lack in bulk by their importance in biological processes. One need only consider the consequences suffered by land plants and animals that have been deprived of critical trace materials (minerals and vitamins). Sea animals and plants also need trace materials, in some cases in concentrations many times those found in seawater. For example, there are only minute quantities of iodine in the sea; yet some seaweeds can concentrate this element within their tissues to the point that we can harvest the plants to obtain commercial quantities of iodine for man's use.

THE SOURCES OF SEA SALT

The weathering of the earth's igneous crustal rocks over geologic time has been cited to account for the 5×10^{19} kg of the solid material dissolved in the sea. It is estimated that 600 g of igneous rock have been weathered for each kilogram of water in the oceans. Using the average chemical composition of crustal rock, we can calculate the amount of each chemical delivered to the sea from this weathered source. However, an analysis of chemicals currently found in seawater indicates that in general only a part of the material delivered to the sea has remained in solution; some of the material has been removed to the deep-sea sediments, while some of it played a part in the formation of the earth's sedimentary rocks in the early inland seas and never reached the oceans.

The chemicals in seawater may be placed in three categories:

1 Volatile materials present far in excess of the amount that could be supplied by weathered crustal rock. This group includes sulfur, chlorine, bromine, and boron, which possibly were abundant in the primeval atmosphere of the earth.

2 Calcium, sodium, potassium, magnesium, carbon, strontium, and iodine, which are present in amounts compatible with the estimated supply and represent about 1‰ of the earth's potential supply to the oceans.

3 The remaining materials, present only in very small amounts.

We have already stated that today's rivers supply materials in proportions differing from those of the sea. The chemical composition of igneous rocks is such that silicon, aluminum, and iron are the most abundant heavy minerals produced by their decomposition. Although these minerals are found in large amounts in river water, they are present in only small quantities in seawater. Therefore, the average chemical composition of existing crustal rocks may not be indicative of the chemical compositon of the crust in the early history of the earth and oceans. If this is the case, it may not be adequate to estimate the origin of the dissolved materials in seawater based on the chemical composition of modern-day crust. Even the source of the oceans' water presents a problem, for of 16.7×10^{20} kg, the total amount of water estimated to be present on the earth, only 0.13×10^{20} kg or 0.78‰ is thought to have come from the hydrogen and oxygen liberated from weathered rock. Thus the balance must have had some other source.

There are many unanswered questions concerning the source of both seawater and dissolved oceanic material. The liberation by volcanic processes of water and volatiles from deep within the earth and the changes occurring in magma as it ages have contributed materials to the seas at unknown rates. We cannot reach definite conclusions concerning the sources of dissolved materials in seawater until we have further knowledge of the earth's formation and early history. We do know, however, that the composition of seawater is not static, but that its change over relatively short time periods is so small that it is nearly immeasurable. We may also conclude that processes acting in the sea which tend to remove soluble material from the water by depositing it on the sea floor require an amount of material in excess of what is currently supplied.

CONSERVATIVE AND NONCONSERVATIVE CONSTITUENTS OF SEAWATER

In the preceding discussion, we assumed that once chemicals are delivered to the sea from the land they never return to the land or become a part of the solid earth except through occasional tectonic processes. This statement is not technically true, however, since many of the minor and even some of the major contituents of the sea can be extracted from solution and converted into nondissolved material by biological or chemical processes. Chemicals that can be altered in this manner are described as *nonconservative*, as opposed to *conservative*. To illustrate, assume that water containing a dissolved material flows into one end of a large basin at a constant rate. At the other end of the basin, it is allowed to run out at the same rate, but chemical tests show that it is devoid of the dissolved material. Thus the concentration of the material must increase in the basin at a rate equal to the influx—which is indeed the case if the material is conservative. Suppose, however, that an animal or plant population exists in the basin and that it extracts the material from the water and incorporates it into its body tissues at a rate equal to the rate of supply. In this situation the water shows no increase in concentration since the material is removed locally as fast as it is added; thus material removed from the soluble state is nonconservative. If the animals or plants were removed from the basin, the material would also be effectively removed. Plants, for example, use the soluble nutrients, phosphate, nitrates, and other fertilizers and minerals in their growth processes. Small *crustacea* that feed on the plants and fish that feed on the crustacea pass these materials up through the food chain.

Not all nonconservative materials are removed from the sea, however; indeed, the amount removed by man's harvesting of the biota is insignificant. In the oceans, nonconservative materials that are taken up by the biota in one region are usually liberated in either the same or another region as waste products of the food chain or as remains of organic material. In the basin example, the material carried into the tank would have been released into the water had the user population died, decomposed, and given up its body materials to soluble forms.

In some cases there is not a balance between the consumption and liberation of nonconservative materials by the biological population. Some of the material entering the body structures of the plants and animals is bound up in rather insoluble materials. Upon the death of the biota this material often sinks and is accumulated on the sea floor as the biogenic part of the marine sediments. Since the materials involved in biological cycling are primarily minor constituents and dissolved gases, it is only under extreme conditions of production or decay of organic material that the constancy of composition of seawater is significantly affected.

Animals and plants that produce body structures of calcium carbonate by removing carbonates from seawater may form sizable accumulations such as coral reefs which become part of man's habitable world and are not easily reclaimed by the ocean waters. Manganese nodules found on the sea floor are an example of a chemical deposition that converts soluble materials into a relatively insoluble form. The processes of chemical addition by rivers, volcanic activity, and the removal and redistribution of soluble materials by chemical and biological activity occur continuously at irregular rates. However, the volume of the oceans is so large, and mixing so effective, that changes in its average chemical composition are infinitesimal even over prolonged periods of time.

THE DISSOLVED GASES IN SEAWATER

One of the major boundaries of the sea is the air-sea interface through which the gases of the atmosphere freely enter and leave the sea surface. The turbulent motion of the surface waters, generated in currents and waves, continually exposes new water to the gases of the atmosphere and encourages a free exchange of gas. Table 6.3 shows the gas composition of normal dry atmosphere.

Table 6.3 Atmospheric gases

Gas	*Percent by volume*
Nitrogen	78.03
Oxygen	20.99
Argon	0.94
Carbon dioxide	0.03
Hydrogen, neon, and helium	0.01
Total	100.00%

Nitrogen, the most abundant gas in the atmosphere, is relatively unimportant in the sea since it is not required for most biological processes except by nitrogen-fixing bacteria. Argon, which is chemically inert, is also unimportant compared to oxygen and carbon dioxide. The latter two gases, which are nonconservative, play important roles in the marine environment. Both are produced or consumed in nearly every biological process, and oxygen must be present at all depths in order for animals to respire.

The amount of gas that seawater can hold in equilibrium at a specific salinity, temperature, and pressure is the *saturation value.* Since it decreases with increasing temperature and salinity and increases with pressure, the gas content of seawater can be below (undersaturated) or above (supersaturated) the saturation value. If water, either undersaturated or supersaturated with an atmospheric gas, is brought into contact with the atmosphere, an exchange of gas occurs that makes the gas in solution approach the saturation value. The saturation values of oxygen and carbon dioxide in seawater under standard atmospheric pressure are given in Table 6.4 as functions of the chlorinity and temperature of the water.

Table 6.4 Seawater saturation values of O_2 and CO_2 in milliliters per liter at 760 mm of mercury

Temperature, °C	*Chlorinity 0‰*		*Chlorinity 10‰*		*Chlorinity 20‰*	
	O_2	CO_2	O_2	CO_2	O_2	CO_2
0	9.91	1724.8	8.85	1581.4	7.78	1433.6
5	8.66	1456.0	7.74	1310.4	6.83	1191.6
10	7.64	1200.6	6.86	1104.3	6.07	1012.5
15	6.83	1046.1	6.15	945.3	5.47	860.1
20	6.19	882.5	5.58	817.6	4.98	754.9
25	5.68	766.1	5.13	712.3	4.58	658.6
30	5.27	669.7	4.75	627.2	4.24	582.4

From H. Barnes, *Apparatus and Methods of Oceanography*, part 1: Chemical. Interscience, New York, 1959.

When analyzed, water samples taken at sea (Cl‰ about 19) show oxygen values ranging from 0 to as high as 10 ml/l, indicating that water can range from being totally anoxic to supersaturated. Under the same conditions, carbon dioxide is usually found to be about 40ml/l and is never likely to reach saturation.

FACTORS INFLUENCING THE DISTRIBUTION OF DISSOLVED GASES

The plants of the sea, like those of the land, both photosynthesize and respire. The two processes occur simultaneously but not necessarily at the same rate. If sufficient radiant energy (sunlight) is available, the rate of photosynthesis surpasses that of respiration, with carbon dioxide, water, and energy producing simple sugars and oxygen:

$$6CO_2 + 6H_2O + \text{energy} \underset{\text{respiration}}{\overset{\text{photosynthesis}}{\rightleftharpoons}} C_6H_{12}O_6 + 6O_2. \tag{6.9}$$

Plants in the sea consume carbon dioxide and produce oxygen during periods of sufficient sunlight. When the light energy is no longer sufficient, the process is reversed and proceeds to the left, causing a net removal of oxygen from the water, a breakdown in sugars, and a release of carbon dioxide. Photosynthesis occurs not only right at the sea surface, but also at depths below the surface if sufficient light is available. The depth at which the available light energy can drive the reaction to the right at the same rate that respiration drives it to the left is called the *compensation depth* because photosynthesis and respiration are in balance. Above this level the plants produce oxygen at the expense of carbon dioxide; below it carbon dioxide is produced at the expense of oxygen.

The foregoing discussion considers only the role of plants in the production of carbon dioxide and oxygen. Animals, which do not photosynthesize but respire, consume oxygen and produce carbon dioxide. This process occurs at all depths in the sea; it is not limited to the upper sunlit (euphotic) zone. Carbon dioxide is produced at all levels in the sea as well as absorbed from the atmosphere at the sea surface. Oxygen, however, is produced in the sea only in the upper sunlit layers or is taken into the water at the surface through free exchange with the atmosphere.

In the open sea, therefore, the distributions of oxygen and carbon dioxide with depth appear as mirror images of each other. Oxygen found in high concentrations at the surface decreases with depth as respiration becomes more important than photosynthesis. We might expect the oxygen content to decrease continuously with depth since animals and bacteria exist at all levels and along the bottom, and the oxygen source is confined to the surface waters. However, in regions where surface waters sink at high latitudes and flow along gradually at depth, oxygen is supplied to replenish that consumed by biological processes. Though very slow, this oxygen-supplying mechanism is sufficient to satisfy the demands of the sparse population of deep-water animals and bacteria. It also maintains an elevated oxygen content in the deeper water; thus the oxygen minimum occurs at intermediate depths (200 to 1000 m) where biological demands are greater and supply rates are small. If there are deep trenches or basins that are well isolated by surrounding ridges, the circulation of deep water may be restricted and the oxygen supply reduced. When consumption of the available oxygen by biological processes exceeds the supply, near-anoxic conditions, as in the Black Sea, result. Anoxic regions can be found even in the unrestricted waters of the open ocean. If water is trapped in a cyclonic gyre or eddy for prolonged periods of time, the oxygen at depth can be stripped away by excessive decomposition processes which occur beneath the productive surface waters. The retention of the water in the gyre prevents it from gaining oxygen by exchange with adjacent oxygenated water. The gyre system known as the Costa Rica Dome, centered at 10° N Lat and 90° W Long, has such an anoxic region underlying the surface waters.

Early attempts to measure the carbon dioxide content of seawater led to the discovery that a strong acid must be added to the water sample in order to release all the carbon dioxide for analysis. Therefore, considerable CO_2 must be bound up in bicarbonate (HCO_3^-) and carbonate (CO_3^{2-}) within the seawater system. Pure water freely dissociates into hydrogen (H^+) and hydroxyl (OH^-) ions, but when CO_2 is added to water the gas combines with the hydrogen ions:

$$CO_2 + H_2O \rightleftharpoons H_2CO_3 \rightleftharpoons HCO_3^- + H^+ \rightleftharpoons CO_3^{2-} + 2H^+. \quad (6.10)$$

This expression presents the carbon dioxide in seawater as weak carbonic acid, bicarbonate and carbonate ions, and undissociated molecules of CO_2 gas, all in equilibrium with each other and with the hydrogen ions. In equilibrium, about 1% of the free CO_2 gas in seawater is present as carbonic acid. If CO_2 is removed from seawater by photosynthesis, the equilibrium is upset; bicarbonate ions convert to carbonate ions, reducing the hydrogen ion concentration of the water and raising the pH. If CO_2 is added to the system by respiration the reverse process takes place. Changes within this equilibrium occur rapidly if any one of the variables is modified; however, the movement of carbon dioxide into or out of the system at the air-sea interface takes place very slowly. The formation of loose and stable combinations of carbon dioxide with bases in seawater helps keep a reserve of CO_2 available for plant use and permits seawater to act as a buffer, preventing any sudden large-scale change in the pH, or alkalinity.

The pH is a measure of the hydrogen ion concentration. Pure water dissociates into

$$H_2O \rightleftharpoons H^+ + OH^-. \tag{6.11}$$

The concentrations of H^+ and OH^- in chemical equivalents per liter of pure water are essentially identical and are equal to 10^{-7} at 25°C. Since H^+ and OH^- are present in equal quantities the water is said to be neither acidic nor basic, but *neutral.* The pH of pure water is defined as

$$pH = \log \frac{1}{[H^+]} = \log \frac{1}{10^{-7}} = 7. \tag{6.12}$$

An *acidic* solution has a hydrogen ion concentration greater than 10^{-7} and a pH of less than 7, while a *basic* solution has a hydrogen ion concentration less than 10^{-7} and a pH greater than 7.

In the open sea, the CO_2 equilibrium controls the hydrogen ion concentration, causing the pH of the upper layers of the sea to vary over a very narrow range (8 to 8.3). In shallow waters and isolated tide pools, the pH range can be much greater, since photosynthesis raises it and respiration lowers it. However, the open sea system never turns acidic; under normal conditions the pH remains above 7.5.

The solubility of carbon dioxide in water is great compared to that of oxygen, since CO_2 reacts chemically with the water. As in the case of oxygen, seawater in contact with air will gain or lose CO_2 slowly depending on whether the partial pressure of the gas in the water is lower or higher than that in the atmosphere. Seawater can absorb considerably more carbon dioxide from the atmosphere if the partial pressure of the atmospheric carbon dioxide increases; if it decreases, the water liberates considerable quantities of carbon dioxide. Thus the sea acts as a CO_2 regulator for the atmosphere.

THE NUTRIENTS OF THE SEA

The inverse relationship between carbon dioxide and oxygen at great depths in the sea (i.e., one gas is generated at the expense of the other) is also reflected in distributions of other materials that are utilized in biological systems. These other materials are therefore nonconservative and are part of the minor dissolved constituents of seawater. For simplicity, we will classify these materials as *nutrients* since they act as dissolved fertilizers for phytoplankton, the microscopic sea plants. Soluble phosphates and nitrates are the most important nutrients, although other trace elements are also included. Dissolved silicate, though not a fertilizer, is also considered in this group, since it is required for the exoskeletons of the phytoplankton.

Nutrients are used by plants as they photosynthesize, grow, and reproduce; thus the nutrient supply is low in areas of increased photosynthesis, where oxygen is produced at the expense of carbon dioxide. When the plant cells die or sink below the photosynthetic zone to decompose, oxygen is consumed in the decomposition process or in respiration and carbon dioxide is produced. The same process liberates nutrients and returns them to the water in soluble form.

Chemists and biologists have ascertained that certain ratios exist between nitrogen, oxygen, phosphorus, and carbon in plant material produced in the sea and that constant ratios exist between the nitrate-nitrogen and phosphate-phosphorus content of seawater. Plants use the latter materials in the same proportions as they are found in seawater, and return them to the same proportions during decomposition.

The combining of carbon into organic material in the sea requires a certain amount of oxygen, nitrogen, and phosphorus. The ratios between these elements are

$$\begin{aligned} \mathrm{O:C:N:P} &= 212 : 106 : 16 : 1, \text{ by atoms,} \\ \mathrm{O:C:N:P} &= 109 : 41 : 7.2 : 1, \text{ by weight.} \end{aligned} \tag{6.13}$$

Both ratios are easy to interpret. By atoms, 212 atoms of oxygen and 16 atoms of nitrogen are used for every 1 atom of phosphate to produce 106 atoms of organic carbon, carbon bound up in organic material. Similarly, two atoms of oxygen are used to produce one atom of organic carbon, which follows from the ratio of carbon to oxygen in carbon dioxide, which acts as the source of carbon. The interpretation by weight follows the same pattern if we substitute gram weight for atoms. That is, 109 g of oxygen and 7.2 g of nitrogen are used with 1 g of phosphate to produce 41 g of carbon. Since the production of organic carbon requires the use of carbon, nitrogen, phosphorus, and oxygen in these ratios, any one of the elements may become a limiting factor in plant production.

The production of organic carbon in the surface waters and its decomposition at depth drain nutrients from the surface water and eventually limit production. In regions of the oceans where nutrient-laden deep water rises to the surface and continuously resupplies the nutrients, production is usually not limited chemically. The food supply is abundant in these areas, which are among the richest regions of the sea in terms of sea life if there is sufficient sunlight. These *upwelling* areas are discussed at length in another section of this book.

An area of chemical studies that is receiving increased attention from both chemists and biologists is that of organics in seawater. Organic materials from the excretory products of living material and the decomposition of dead organisms are present in the sea in both soluble and particulate form. Both types of material are thought to be usable as a food supply for living organisms.

Organics exist in seawater as carbohydrates, proteins, amino substances, lipids, enzymes, and vitamins—which are as basic to life in the oceans as they are on land. If these organics can be determined as the products produced during the breakdown and decomposition of known living material, then their presence in the sea and their alteration in time may be used to understand what is occurring in the living organisms using the reverse relationships.

Since organic materials are present in very dilute concentrations in seawater, which has a rather high concentration of salt, they are difficult to obtain for analysis. Despite their low concentrations, however, soluble organics greatly influence the life cycle of marine organisms and govern in part the succession and survival of species.

The study of organics in essential to an understanding of both the fundamental factors that control life in the sea and how organisms may affect each other by the production or consumption of materials that interfere with or are necessary to life. Examining the production, consumption, or alteration of organics leads to the determination of energy flow through the food chain, as well as the development of techniques that can enhance the production of living organic material to man's benefit.

THE SCOPE OF CHEMICAL OCEANOGRAPHY

Chemical oceanography is not limited to the study of solids and gases in solution, their distribution in oceanic space, and their role in biological phenomena. It also includes physicochemical, biochemical, and radiological chemistry, as well as chemical engineering problems associated with the sea. Desalination of seawater, which involves the retrieval of both dissolved solids in usable form and freshwater at reasonable cost, combines chemistry and engineering. Chemical oceanographers and geophysicists have recently given considerable attention to the mineral resources of the sea floor and the processes that form them, while physical chemists are studying the effects of pressure and temperature on solubility products of chemicals in complex and strong solutions such as seawater. Corrosion problems created by the marine environment, which incur great expense, are also being examined. Almost every chemical known on earth exists in the sea. The practical problem of concentrating them and extracting them economically is all that stands in the way of the chemist's claim to the sea's wealth.

The study of the chemistry of seawater has many facets. In its early days, chemical oceanography focused on the development of analytical techniques that would allow accurate determination of the constituents of seawater by classical chemical methods. A period of development of new methods followed which reduced the emphasis on classical analysis and introduced the measuring of trace materials in seawater.

Complicated electronic systems with electrodes can now determine the composition of seawater by measuring the electrical conductivity rather than the chloride content to determine salinity; they can also detect materials with concentration levels of a few parts per billion. Despite the invention and increasing use of complex instruments, the classical stoichiometric approach to chemistry is usually used to standardize the new methods and therefore remains important.

In essence, the chemistry of seawater is simply the chemistry of a complex solution. A chemist could devote his life to studying this chemical solution in the laboratory; however, the study of the rates of chemical alteration in natural seawater systems would greater enhance his comprehension of the oceanic environment. Indeed, the study of the natural oceanic environment using chemical methods is often the key to an understanding of the physical, geological, and biological processes of the sea. Observed chemical alterations in seawater, for example, can be used to establish rates of biological production and the rates at which water is replenished at depth. The chemist may also use isotopes of elements as labels to trace the body processes of marine animals and to establish dating techniques for the sediments. Or he may use isotope ratios to estimate past climatological events in the earth's history as they are recorded in the fossil remains buried in the sediment. In summary, an understanding of the chemical processes of the sea is necessary to the study of any branch or subarea of oceanography.

STUDY QUESTIONS

1 Why can the concentration of a major salt ion in sea water be calculated once the chlorinity is known?

2 Why is the chlorinity of a seawater sample higher than the chloride ion content?

3 Write a simple equation for salt concentration of a sample of seawater in grams of salt per cubic meter of seawater, assuming that the salinity is known.

4 If the chlorosity of a water sample is 18, what are the chlorinity and salinity?

5 Why is the total percentage of dissolved solids in a sample of seawater greater than the salinity?

6 What is meant by a *nonconservative* chemical constituent of seawater as opposed to a *conservative* constituent?

7 If samples of freshwater and seawater have the same temperature and pressure and if both are 100% saturated with dissolved oxygen, which has the highest concentration of dissolved oxygen?

8 Why are the vertical distribution of oxygen and carbon dioxide nearly mirror images of each other in the sea?

9 In what way does photosynthesis tend to change the pH of seawater?

10 If the rate of depletion of phosphate in seawater is 0.04 g/m^3 per day and is assumed to be due to utilization by plants, what is the estimated rate for production of organic carbon?

RECOMMENDED READING

Barnes, H., *Apparatus and Methods of Oceanography—Chemical*, Interscience Publishers, New York, 1959.

Fairbridge, R. W., ed., *The Encyclopedia of Oceanography*, Reinhold Publishing Corp., New York, 1966.

Harvey, H. W., *The Chemistry and Fertility of Seawaters*, 2nd ed., Cambridge University Press, New York, 1957.

Hill, M. N., ed., *The Sea*, vol. 2, Interscience Publishers, New York, 1965.

Knudsen, M., *Hydrographical Tables*, G. E. C. Gad, Copenhagen, 1901. Reprinted 1959, Tutein & Koch, Copenhagen.

Strickland, J. D. H., and T. R. Parsons, *A Manual of Seawater Analysis*, Fisheries Research Board of Canada Bulletin No. 125, Ottawa, 1960.

Sverdrup, H. U., M. W. Johnson, and R. H. Fleming, *The Oceans*, Prentice-Hall, Inc., Englewood Cliffs, N.J., 1942.

Photo by
Bruce Anderson ▶

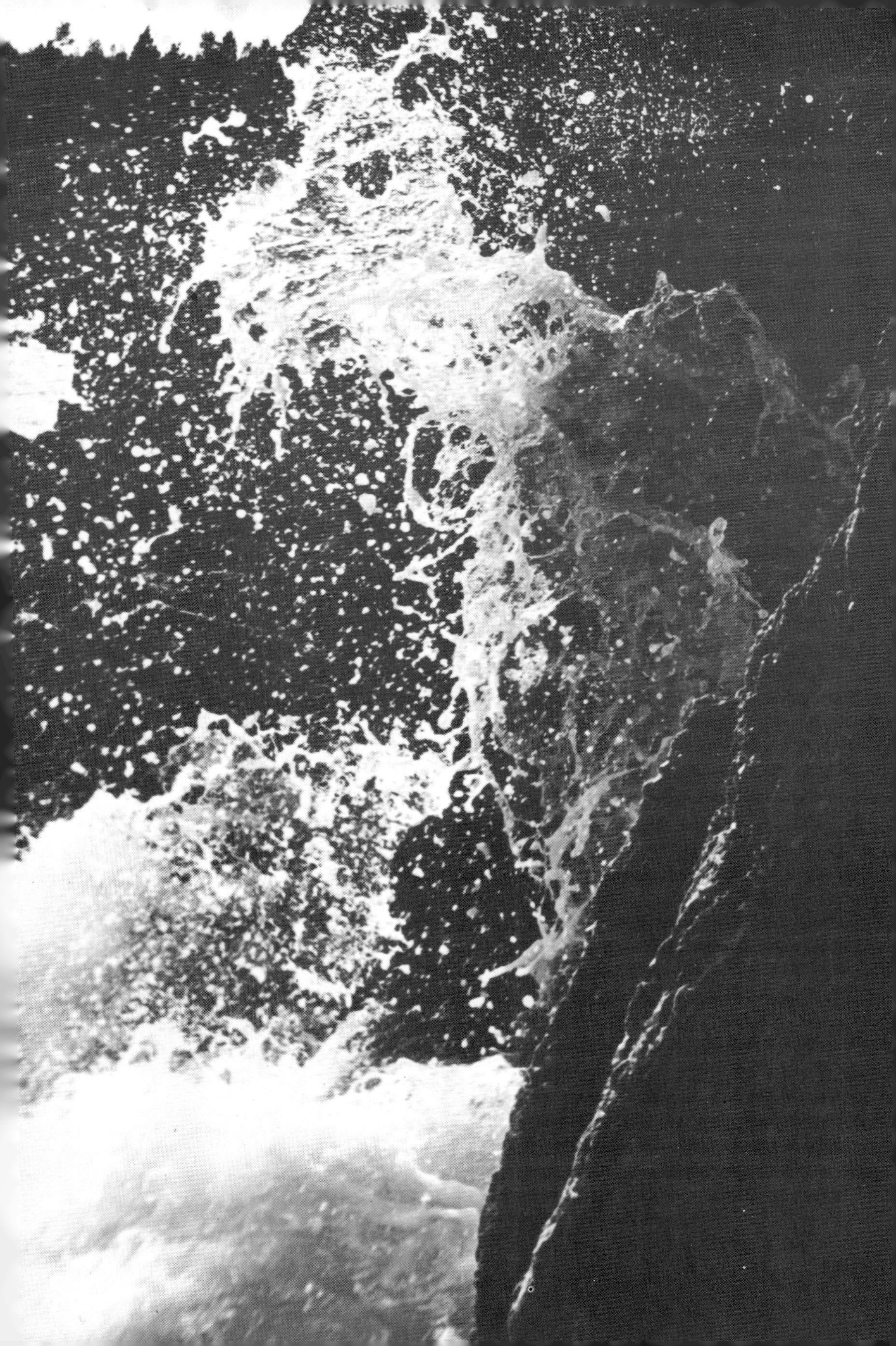

CHAPTER 7

PHYSICAL PROPERTIES OF SEAWATER

INTRODUCTION

Oceanographers must study the physical properties of seawater to understand not only the manner in which it responds to modifying factors but also the basic physical processes of the sea. The water at the oceans' surface is exposed to a variety of climatic regimes. For example, salinity increases in regions of excessive evaporation and decreases in regions of high precipitation; it is also affected by the formation and melting of sea ice. These processes, as well as temperature, alter the water's density. We must be aware of the importance of salinity and temperature in controlling density in order to understand why and where surface water sinks to depth to oxygenate the deep sea.

The physical properties of seawater enable the oceans to act as a thermal buffer for the earth, storing heat during the summer period of surplus and releasing it during the winter period of deficit. Thus, just as we must know the physical properties of rigid-body construction material to understand how it will react to environmental changes, we must know the properties of seawater to understand how the ocean behaves. One important additional factor must be considered in the latter case, however—namely that seawater is a mobile fluid and thus may be driven by gravitational force.

PROPERTIES OF WATER

Seawater is a complex solution of many chemical substances dissolved in pure water; 1 kg of standard seawater is composed of approximately 35 g of dissolved solids and 965 g of pure water. As used here, the term "pure water" is an undefined quantity; we shall stipulate only that it is water containing very few extraneous gases, solids, and other liquids. For practical purposes, we shall consider freshly distilled water as pure water. We shall assume that freshwater is the naturally occurring counterpart of pure water and has the same properties, even though it is actually slightly impure. Since seawater is mostly water and only partly salt, physically it behaves primarily like pure water. Salt affects its physical properties only slightly, but these slight modifications are extremely important in the oceanic system.

Water is a unique material—a universal solvent ideally suited for its role of chief component of living organisms. Its presence determines many of the characteristics of the land and atmosphere; in fact, without it life as we know it would not exist. Studies of compounds related to water indicate that pure water should theoretically freeze at $-150°C$ and boil at $-100°C$. Nothing in its chemical composition offers a logical explanation of why it freezes at 0°C

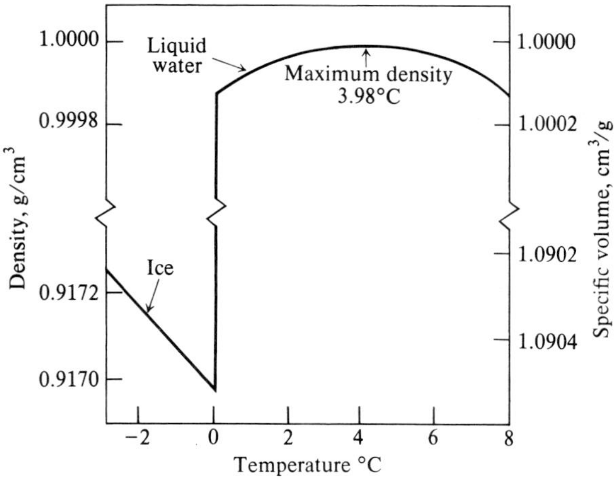

Fig. 7.1 Density of pure water near the ice point. Specific volume is the reciprocal of density. (Redrawn from M. G. Gross, *Oceanography*, Charles E. Merrill, Columbus, Ohio, 1967.)

and boils at 100°C. However, polymerization (the combination of groups of H_2O molecules) produces a physical state that does account for some of water's strange properties. Pure water occurring naturally on the earth or produced in the laboratory from natural earth waters is not identical to water made in the laboratory from mixtures of pure hydrogen and oxygen gases. In natural waters, isotopes of heavy hydrogen and oxygen are present in varying concentrations depending on the source of the water. Thus water produced in the laboratory cannot be expected to have the same isotopic composition as natural water.

The physical properties of pure water have been determined and are listed in Table 7.1. The dependence of these properties on temperature and pressure is obvious from a brief scanning of the data. In dealing with seawater we must consider an additional parameter: salinity.

In Chapter 6 we showed that salinity can be determined from chlorinity, which is measured chemically or by electrical conductivity. Since the physical behavior of seawater is dependent on temperature, pressure, and salinity, other methods may be used to determine salinity if (1) under conditions of constant pressure and temperature a small change in salinity produces a large change in physical behavior, and (2) the physical change can be accurately measured. Changes in physical properties such as electrical conductivity, refractive index, and density meet these requirements.

The *density* of pure water varies with temperature and pressure. At standard atmospheric pressure (760 mm of mercury) the density increases as the temperature decreases to 4°C. At about 4°C, pure water attains a *maximum density* of 1 g/cm³. Below 4°C the density decreases as the temperature decreases to 0°C (Fig. 7.1). Pure water is therefore most dense at approximately 4°C, which is above freezing; it reaches *maximum density* at 3.98°C. We can compare the density of any material to that of pure water at 4°C by dividing the former by the latter to obtain a ratio known as the *specific gravity*. Although this ratio has no units, it is numerically equal to the corresponding density in the cgs system because the density of pure water at 4°C is 1 g/cm³.

Table 7.1 Properties of water

Property (definitions, units)	*Comparison with other substances*	*Effects on the ocean*
Heat capacity: Quantity of heat required to raise the temperature of 1 g of a substance by 1°C. Common units: calories per gram per degree Celsius.	Highest of all commonly encountered solids and liquids.	Water's high heat capacity prevents extreme variations in temperature in the oceans and atmosphere.
Latent heat of fusion: Quantity of heat gained per unit mass by a substance changing from a solid to a liquid phase without an accompanying rise in temperature. Common units: calories per gram.	Highest of all common liquids and most solids.	The thermostatic effect at freezing point results from the release of latent heat during freezing and the absorption of heat during melting.
Latent heat of vaporization: Quantity of heat gained per unit mass by a substance changing from a liquid to a gas phase without an increase in temperature. Common units: calories per gram.	Highest of all common substances.	The release of heat during condensation and the absorption of heat during the vaporization of water is a major process affecting surface temperatures in the ocean and the transfer of heat to the atmosphere.
Density: Mass per unit volume. Common units: grams per cubic centimeter. *Specific gravity*: The ratio of the density of a substance relative to the density of pure water at 4°C. In the cgs system, the two terms can be used interchangeably because they have the same numerical value.	Density of seawater is controlled by (1) temperature, (2) salinity, (3) pressure. Temperature of maximum density for pure water is 4°C. For seawater it decreases with increasing salinity. For salinities greater than 24.70‰ the temperature of maximum density occurs below the initial freezing point.	Density is an important factor controlling the temperature distribution and vertical circulation in the oceans. *Substance* — *Density* (g/cm^3) Cork — 0.21 Ice (pure) — 0.917 Distilled water — 1.0 to 0.99 Seawater — 1.02 to 1.03 Mercury — 13.5
Dissolving ability: Ability to dissolve (act as a solvent for) gases, liquids, and solids.	Dissolves more substances, and in greater quantities than any other common liquid.	This property is of prime importance in physical, chemical, and biological processes.
Transparency: Ability of a medium to transmit (or not absorb) radiated energy such as light waves.	Relatively great for visible light.	Transparency is important in physical and biological phenomena (photosynthesis). Water is most transparent for blue-green light and less transparent in the infrared end of the visible light spectrum.
Quantity present: In the outer 5 km of the earth's surface.	Three times as abundant as all other substances combined; the only common inorganic liquid.	The abundance of water on the earth's surface is responsible for the presence of the oceans.

Table 7.1 (continued)

Property (definitions, units)	*Comparison with other substances*	*Effects on the ocean*
Physical state: Matter can occur in three states: as a gas (or vapor), liquid, or crystalline solid.	Water is the only substance which occurs naturally in all three states on the earth's surface.	This is an important factor in the transfer of heat in the oceans and the atmosphere (see latent heats of fusion and evaporation).
Surface tension	Highest of all common liquids.	Important in physiology of the cell. Controls certain surface phenomena and drop formation and behavior and the formation of bubbles.
Conduction of heat	Highest of all common liquids. (Mercury of course is higher.)	Although important on small scale, as in living cells, the molecular processes are far outweighed by eddy conduction (turbulent mixing).
Compressibility: A coefficient giving the proportional change in specific volume per unit pressure applied: per bar.	Seawater is only slightly compressible. A change in specific volume due to an increase of pressure of approximately 1 atmosphere is about 4×10^{-5} cc/g of material.	Density changes little with pressure. Descending water undergoes an adiabatic temperature rise in the deep ocean because of its compressibility.
Viscosity: The property of a fluid that makes it resist changes in shape: dyn-sec/cm^2.	Decreases with increasing temperature. Salt effect and pressure effect small. Water has relatively low viscosity for a liquid.	Low viscosity means that non-turbulent water motions can be considered friction-free in some cases. The low friction present can act as a dissipative mechanism to dampen out all motion in the sea. Turbulent motions have eddy viscosities associated with them that increase friction.
Refractive index: A ratio that gives the ability of a substance to bend light rays: no units.	Similar to that of distilled water but increasing with increasing amounts of ionized salts in seawater and decreasing with increasing temperature.	The refractive index can be used to determine salinity.
Sound transmission: The ability of the water to transmit compressional sound waves, sound velocity cm/sec.	Water transmits sound well in comparison with other fluids such as gases and plastics.	Allows rapid determination of water depth with good accuracy and the detection of subsurface vessels and fish.

Adapted from H. U. Sverdrup, M. W. Johnson, and R. H. Fleming, *The Oceans*, © 1942. By permission of Prentice-Hall, Inc., Englewood Cliffs, N.J.

DENSITY OF SEAWATER IN TERMS OF SALINITY, TEMPERATURE, AND PRESSURE

Density, ρ, is used in oceanography in several ways. The density of water may be given as $\rho_{s,t,p}$, where s is salinity, t is temperature, and p is pressure. It may also appear as σ_t (sigma-t), a shortened expression for $\rho_{s,t,0}$, when the pressure effect on density is considered zero:

$$\sigma_t = (\rho_{s,t,0} - 1) \times 10^3. \tag{7.1}$$

The reciprocal of density, $1/\rho_{s,t,p}$, is known as the *specific volume*, $\alpha_{s,t,p}$; the *specific volume anomaly*, δ, is defined as the deviation of the specific volume of a sample of water with salinity s, temperature t, and pressure p from a standard water with $s = 35‰$, $t = 0°C$, and pressure p:

$$\delta = \alpha_{s,t,p} - \alpha_{35,0,p}. \tag{7.2}$$

The numbers substituted for the subscripts s, t, and p should be interpreted as particular values of s, t, and p. When $p = 0$, the water is at the sea surface and not subject to pressure from a hydrostatic head; however, it is subject to atmospheric pressure, which is neglected.

It is extremely difficult to directly determine the density of seawater in place (*in situ*). Usually a sample of water is removed from the sea after the temperature has been measured *in situ*. After the salinity of the sample has been determined, we can calculate its density from the salinity, temperature, and depth at which it was taken. The depth is related to the pressure by the hydrostatic equation:

$$\text{Pressure} = \text{density} \times \text{acceleration due to gravity} \times \text{depth}.$$

We can further modify σ_t by stipulating that t is 0°C; thus σ_t becomes σ_0. Since p and t are constant, we can relate σ_0 to salinity or chlorinity and σ_t to σ_0:

$$\sigma_0 = -0.069 + 1.4708\ Cl - 0.001570\ Cl^2 + 0.0000398\ Cl^3, \tag{7.3}$$

$$\sigma_t = \sigma_0 - D. \tag{7.4}$$

In the latter equation σ_t is determined by calculating the effect of salinity or chlorinity on the density, σ_0. From this value we subtract D, the thermal expansion of seawater of a particular salinity expressed as a function of temperature; D is not constant, but varies with σ_0. Its values are given in Knudsen's hydrographic tables.

We must now consider the pressure effect in order to determine the dependence of seawater density on s, t, and p. The calculation of the *in situ* density $\rho_{s,t,p}$ is an extremely cumbersome process: we must know how the density is affected by (1) salinity as t and p change; (2), temperature as s and p change; and (3), pressure as s and t change. The empirical equation which we must use is cubic in each of its base parameters and contains about 47 terms. Tables can be constructed that simplify the task considerably by solving for $\alpha_{s,t,p}$ rather than $\rho_{s,t,p}$. We showed earlier that $\alpha_{s,t,p}$ is related by δ to the specific volume of a particular water sample, $\alpha_{35,0,p}$. The values of $\alpha_{35,0,p}$ need be calculated only once as a function of pressure (Table 7.2); we then need only determine δ to find $\alpha_{s,t,p}$. Since a column of seawater 10 m in height produces a hydrostatic pressure of about 1 bar (10^6 dyn/cm^2), the pressure expressed in decibars (one-tenth of a bar) is nearly numerically equal to the depth expressed in meters.

Table 7.2 Specific volume of seawater under pressure when $s = 35‰$ and $t = 0°C$

Depth, m (approximate values)	*Pressure, decibars*	$\alpha_{35,0,p}$ *in cm³/g*
0	0	0.97264
500	500	0.97040
1000	1000	0.96819
1500	1500	0.96602
2000	2000	0.96388
2500	2500	0.96177
3000	3000	0.95970
4000	4000	0.95566

Material for Tables 7.2 through 7.5 from SP-68, *Handbook of Oceanographic Tables,* 1966. Courtesy of the U.S. Naval Oceanographic Office.

The specific volume anomaly, δ, which separates $\alpha_{s,t,p}$ from the tabulated values of $\alpha_{35,0,p}$, is a function of s, t, and p. The magnitude of δ is dependent on the departure of s from 35‰ and t from 0°C:

$$\delta = \delta_s + \delta_t + \delta_{s,t} + \delta_{s,p} + \delta_{t,p} + \delta_{s,t,p}. \tag{7.5}$$

Each term on the right-hand side of this equation represents a fraction of the total specific volume anomaly, which is a result of both the deviation of s and t from 35‰ and 0°C respectively and the effect of s and t on the compressibility of water.

Table 7.2 shows how seawater of 35‰ salinity at 0°C is compressed by increasing pressure; that is, the volume occupied by each gram of water decreases as the pressure increases. If a comparable tabulation is made for seawater of different salinity and temperature, the specific volume reduction for each increase in pressure will not be the same. This reduction in the specific volume with pressure is equal to $\delta_{s,p} + \delta_{t,p} + \delta_{s,t,p}$. The equation has no δ_p term because it is not dependent on the deviation of s and t from the standard values 35‰ and 0°C, and is therefore already incorporated in the tabulation of $\alpha_{35,0,p}$ as a function of pressure. The first three terms on the right-hand side of the equation, δ_s, δ_t, $\delta_{s,t}$, are corrections involving only salinity and temperature. They are often combined to form the *thermosteric anomaly*, $\Delta_{s,t}$ (Table 7.3). The

Table 7.3 $\Delta_{s,t} \times 10^5$ as a function of salinity and temperature in cm^3/g

Temperature, °C	*Salinity ‰* 30	32	34	35	36	38
−1	379.4	225.6	72.2	−4.4	−80.8	−233.5
0	382.3	229.1	76.3	0.0	−76.2	−228.4
1	386.6	233.9	81.7	5.7	−70.3	−221.9
2	392.1	240.1	88.3	12.6	−63.1	−214.2
4	407.1	256.1	105.4	30.2	−45.0	−195.2
6	426.9	276.8	127.1	52.3	−22.4	−171.6
8	451.2	302.0	153.2	78.8	4.5	−143.9
10	479.8	331.4	183.4	109.4	35.5	−112.1
15	568.7	422.0	275.6	202.5	129.4	−16.6
20	680.5	535.1	390.0	317.5	245.1	100.3
30	966.8	822.9	679.1	607.3	535.6	392.1

Table 7.4 $\delta_{s,p} \times 10^5$ for salinity and pressure in cm^3/g

Pressure, decibars	*Salinity ‰* 30	32	34	35	36	38	40
0	0.0	0.0	0.0	0	0.0	0.0	0.0
500	−3.8	−2.3	−0.8	0	0.8	2.2	3.7
1000	−7.5	−4.5	−1.5	0	1.5	4.4	7.3
1500	−11.0	−6.6	−2.2	0	2.2	6.5	10.8
2000		−8.7	−2.9	0	2.9	8.6	14.3
2500			−3.6	0	3.6	10.7	17.7
3000			−4.3	0	4.2	12.7	21.0
3500			−4.9	0	4.9	14.6	24.2
4000			−5.6	0	5.5	16.5	27.4

Table 7.5 $\delta_{t,p} \times 10^5$ for temperature and pressure in cm^3/g

Pressure, decibars	*Temperature, °C* −2	−1	0	1	2	4	6	8	10	15	20	30
0	0.0	0.0	0	0.0	0.0	0.0	0.0	0.0	0.0	0.0	0.0	0.0
500	−2.8	−1.4	0	1.4	2.6	5.0	7.1	9.0	10.8	14.6	17.4	20.9
1000	−5.5	−2.8	0	2.7	5.3	9.8	14.0	17.7	21.3	28.7	34.2	41.1
1500	−8.1	−4.1	0	4.0	7.8	14.4	20.5	26.2	31.5	42.3	50.5	
2000	−10.6	−5.3	0	5.2	10.1	19.0	26.9	34.4	41.3	55.5	66.2	
2500	−13.1	−6.5	0	6.3	12.3	23.4	33.1	42.2	50.7	68.2		
3000	−15.4	−7.6	0	7.4	14.4	27.6	39.1	49.9	59.8	80.4		
3500	−17.7	−8.7	0	8.4	16.4	31.5	44.9	57.3	68.6	92.2		
4000	−19.9	−9.9	0	9.5	18.4	35.2	50.5	64.4	77.1			

combined salinity-pressure and temperature-pressure corrections, $\delta_{s,p}$ and $\delta_{t,p}$, are given separately in Tables 7.4 and 7.5. The last term, $\delta_{s,t,p}$, is so small that it is neglected. The physical significance of the thermosteric anomaly is given by

$$\alpha_{s,t,0} = \alpha_{35,0,0} + \Delta_{s,t} = 0.97264 + \Delta_{s,t}. \quad (7.6)$$

The calculation of the $\alpha_{s,t,0}$ from $\alpha_{35,0,0}$ and $\Delta_{s,t}$ leads directly to the determination of $\rho_{s,t,0}$, where $\rho_{s,t,0} = 1/\alpha_{s,t,0}$, and σ_t, where $\sigma_t = (\rho_{s,t,0} - 1) \times 10^3$.

Oceanographers go to great lengths to make accurate density determinations of seawater, carrying their calculations to the fifth decimal place. Indeed, if accurate temperature measurements and salinity determination could be routinely achieved and if the effect of temperature and salinity on the density of seawater were known more exactly, oceanographers would gladly calculate density to six decimal places. What may appear to be an excessive preoccupation with accuracy is in reality very important, since minute changes in density caused by surface processes and mixing govern the vertical and in part the horizontal *thermohaline circulation* of the oceans. Accurate measurements of density are also necessary to determine variations in the speed of sound waves and their refraction, as well as the integrated horizontal currents of the ocean.

Although we are taught that water is an incompressible fluid, scientists must occasionally consider seawater as slightly compressible. In many cases the accuracy of measurements or oceanographic theories is such that the incompressibility assumption is valid; in the determination of density, however, we must take compressibility into account. Table 7.2 shows that the volume occupied by 1 g of seawater of 35‰ salinity and 0°C at the sea surface is 0.97264 cm³, while at 4000 m depth it is 0.95566 cm³. This change, which is brought about by an increase in pressure by 4000 decibars (0.01698 cm³/g), is very small, but since each gram of water is affected, the total decrease over a column of water is quite significant. The average depth of the oceans is about 4000 m; the average pressure over a column of this depth is therefore 2000 decibars, or the pressure at about 2000 m depth. The average change in volume of each gram of water over the 0- to 4000-m column is $\frac{1}{2} \times 0.01698$ cm³/g, or 0.00849 cm³/g; thus the column is compressed slightly. The total compression is the sum of all the reductions in volume over the column, or

$$(4 \times 10^3 \text{ m}) \times (0.00849 \text{ cm}^3 \text{ g}^{-1}) \times (1/0.96388 \text{ cm}^3 \text{ g}^{-1}) \simeq 35 \text{ m}. \quad (7.7)$$

The product, ~35 m, would be the rise in sea level if seawater were truly incompressible and the oceans' area did not increase. It is an impressive distance if one considers that the mean elevation of land on the earth is only 840 m. The term $1/0.96388 \text{ cm}^3 \text{ g}^{-1}$ is the average density of the compressed water column and is used to convert the average change in volume per unit mass due to pressure, to the average change in volume per unit volume of seawater.

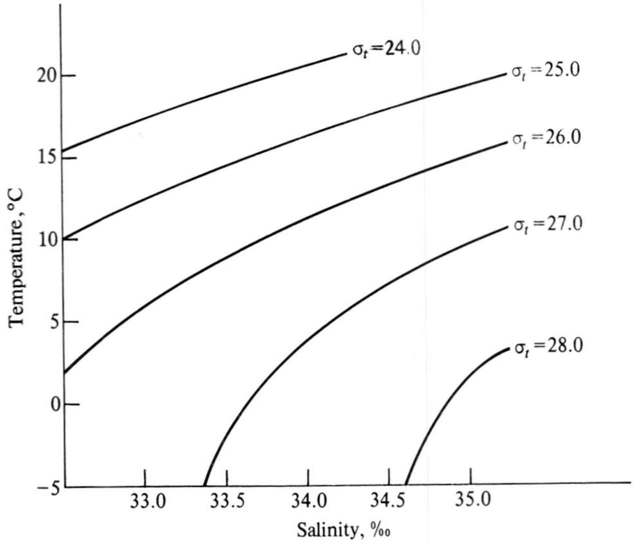

Fig. 7.2 A temperature versus salinity diagram (*T-S* diagram), showing how σ_t varies as a function of these two variables.

A variety of salinity and temperature values can be combined to produce the same density, as shown in Eq. (7.6) and the tabulated values of $\Delta_{s,t}$ in Table 7.2. Whenever $\Delta_{s,t}$ is zero, $\alpha_{s,t,0}$ is equal to $\alpha_{35,0,0}$, which requires that their reciprocals, the densities, also be equal. Table 7.3 shows that $\Delta_{s,t}$ has zero values at $s = 35‰$, $t = 0°C$; $s = 36‰$, $t \sim 7.7°C$; $s = 38‰$, $t \sim 15.9°C$. We might also expect zero values of $\Delta_{s,t}$ at minus temperatures and low salinities; however, they are unlikely to occur in the oceans except in arctic regions during ice melt.

Lines of constant value σ_t or $\rho_{s,t,0}$ can be shown on a *T-S* (temperature versus salinity) diagram (Fig. 7.2). Table 7.6 gives the pairs of *s* and *t* values that produce these lines. The use of a *T-S* graph as a powerful diagnostic tool in oceanography will become apparent when we discuss the physical behavior of ocean waters.

Table 7.6 Values of *s* and *t* for *T-S* graph (Fig. 7.2)

$\sigma_t = 25.0$		$\sigma_t = 26.0$		$\sigma_t = 27.0$	
$s‰$	t, °C	$s‰$	t, °C	$s‰$	t, °C
33.0	12.3	33.0	6.0	33.0	(off scale)
33.5	14.2	33.5	8.8	33.5	−2.0
34.0	16.0	34.0	11.1	34.0	4.1
34.5	17.6	34.5	13.1	34.5	7.3
35.0	19.1	35.0	14.9	35.0	9.8

It should be noted that the lines of constant density are curved rather than straight, especially at low temperatures and high salinities. At high temperatures and low salinities, the lines of constant σ_t indicate that both s and t exert a marked influence on the density, and that a unit change (1‰ or (1°C) in either produces a reasonably large change in σ_t. At low temperatures and high salinities, a unit change in s or t has a less marked effect on the density of seawater; however, in this region of the T-S diagram the density change caused by a unit change in s is more important than that caused by a change in t (Fig. 7.2).

We have presented the data needed to calculate the density of seawater in its many forms and approximations, provided that salinity or chlorinity, temperature, and depth are known. The small increments of temperature and salinity which have not been tabulated may be found by interpolation between tabulated values.

DIFFUSIVITY IN SEAWATER

Water can conduct heat by *molecular diffusion*. Water heated at its upper surface will gradually conduct heat downward by molecular activity; pure water at 15°C has a coefficient of thermal conductivity of 1.39×10^{-3}. If water is heated from below, the resulting decrease in density at the bottom will cause free convective overturn of the water. This process distributes heat throughout the water at a rate many times that produced by molecular conduction. In nature, however, the sun's radiant energy heats the water at the upper surface of the oceans rather than at the bottom; thus the surface water becomes less dense and thermally driven convective overturn does not occur. This method of warming seawater is relatively inefficient. However, the transmission of heat to depth is not governed by molecular processes alone, but is aided by the turbulence in the surface waters caused by waves, tidal currents, and wind-driven currents.

The downward flux of heat into the oceans is given by an equation form

$$\frac{\Delta Q}{\Delta t} = -\gamma \frac{\Delta \theta}{\Delta Z}, \tag{7.8}$$

where $\Delta Q/\Delta t$ is the heat carried through a unit surface area per unit time, i.e., a *flux of heat*, and $\Delta\theta/\Delta Z$ is the vertical *thermal gradient*, equal to the change in temperature per unit vertical distance. (Since t has been used to denote time, θ indicates temperature.) The thermal gradient determines the direction that the heat moves, in this case up or down. The flux of heat always goes from a region of higher temperature to a region of lower temperature. The *coefficient of thermal conductivity*, γ, is a constant of proportionality relating the heat flux to the gradient. For purely molecular transfer processes, this coefficient is 1.39×10^{-3} in cgs units. If other mechanisms produce turbulence, so that warmer water is displaced downward and mechanically mixed with cooler water, the coefficient becomes much larger and a larger flux of heat occurs for the same gradient value. In this case the coefficient is a *turbulent* or *eddy* coefficient and depends on the extent and type of turbulence.

The expression relating the flux of heat to the gradient is called a *diffusion* equation; it is equally applicable to the diffusion of chemicals, momentum, or dissolved gases in water. The movement of these materials by diffusion can occur in any direction, but it always goes from regions of high concentration to regions of low concentration. As in the case of heat, the coefficient of diffusion can have a value related to molecular or turbulent processes. In the diffusion of matter or energy, the flux through a unit of surface area per unit of time is always proportional to the gradient, the change in the concentration with distance; the surface through which the flux occurs is always at right angles to the gradient direction.

Packets of water, with their characteristic temperatures and dissolved chemical and gases, have a larger mass and momentum and move a greater distance due to turbulence than do single molecules undergoing Brownian movement. Thus eddy diffusion coefficients can be as large as 10^7 horizontally and 10^4 vertically. The smaller vertical value arises from the tendency of stability (density increasing with depth) to inhibit turbulence; this is not usually a factor in the horizontal direction. Under conditions of strong density stratification, the vertical coefficient can be as small as 10. Eddy coefficients of this magnitude cause diffusion transfers which are 10^4 to 10^7 times greater vertically and 10^{10} times greater horizontally than those caused by molecular processes. Molecular diffusion is therefore insignificant compared to turbulent processes in the sea, except in instances (usually at depth) in which naturally occurring turbulence is slight or in which soluble material is transferred through semipermeable cell walls. The latter occurs when the plants of the sea absorb nutrients and gases from the water to produce body materials.

THERMAL PROPERTIES OF SEAWATER

The *specific heat* of seawater at constant pressure, C_p, varies slightly with temperature and salinity. The specific heat of any material is the heat capacity of the material divided by the heat capacity of pure water; it is therefore a ratio and is never given in units. In the cgs system, thermal capacity is expressed in gram-calories per gram of material per degrees Celsius. Since the heat capacity of pure water is 1 cal g^{-1} $°C^{-1}$, the specific heat and heat capacity of a material are numerically equal. In working problems, you will find it often desirable to use heat capacity rather than specific heat to facilitate the balancing of units and detection of errors.

The specific heat of seawater changes slightly with both salinity and temperature. At 17°C it changes from 1.00 at 0‰ to 0.926 at 40‰, while at about 35‰ it changes from 0.942 at −2°C to 0.932 at 20°C. Since the deviation of these values from unity is small, we can neglect it in most heat problems involving seawater. The assumptions made in many calculations account for larger unknown errors than the assumption that $C_p = 1$ cal g^{-1} °C^{-1}.

In certain calculations, such as that of the rate of sound propagation, we must consider the relationship of C_v, the specific heat at constant volume, to C_p. For example, the ratio C_p/C_v at 34.85‰ and at atmospheric pressure is 1.004 at 0°C and 1.0207 at 30°C. Pressure is also an important factor in modifying C_v. Water at 0°C and 1000 m (or 1000 decibars) pressure has a C_p/C_v ratio of 1.0009; at 10,000 m (or 10,000 decibars) pressure C_p/C_v becomes 1.0126.

A considerable amount of heat is transmitted from the sea to the atmosphere by processes of evaporation at the sea surface. Evaporation occurs at all temperatures found at the earth's surface; even ice can evaporate without melting. Each gram of water that is removed from the sea by evaporation carries with it the *latent heat of vaporization*, the heat required to convert the water from its liquid or solid to its gaseous phase. This heat is usually provided by the seas' surface layers; thus evaporation causes the sea to cool. The latent heat of vaporization of pure water is 540 cal g^{-1} of water at the boiling point, or the heat required to convert 1 g of water at 100°C to 1 g of water vapor at 100°C. Evaporation processes in the sea as well as those on land normally occur at temperatures well below 100°C. Therefore, the latent heat of vaporization must be adjusted to the lower temperatures and to seawater rather than pure water. The latent heat, L, of seawater is

$$L = 596 - 0.52\,\theta \text{ cal g}^{-1},$$

where θ is the temperature in degrees Celsius at which evaporation occurs.

To accomplish the change in physical state, the melting of ice requires the *latent heat of fusion*, which is a function of the temperature and salinity of the ice. If the ice is pure (i.e., contains no salt) then the latent heat of fusion at 0° to -1° is 80 cal g^{-1} and 81 cal g^{-1} at -2°C. If the ice contains trapped salt, such as that found in sea ice, the latent heat of fusion decreases as the salt content of the ice increases. A sea-ice salinity at 10‰ decreases the latent heat of fusion at -1°C to 37 cal g^{-1} and 39 cal g^{-1} at -2°C. Thus the latent heat of fusion for sea ice decreases with increasing salt content and increasing melting-point temperature.

At this point we might consider the importance of evaporation in extracting heat from the oceans. In our discussion of the hydrologic cycle, we mentioned that 97 cm of water evaporates from the world's oceans annually. In volume, this amounts to about 350×10^{12} m^3, or about 360×10^{18} g of seawater. If we assume the average ocean surface temperature to be 17.5°C and the latent heat of vaporization to be supplied by the sea, the evaporation of 360×10^{18} g of seawater at 17.5°C removes 211×10^{21} cal from the sea annually. This loss, roughly equivalent to the energy in about 10^{18} instant breakfast drinks, must of course be balanced by an energy input, as we shall learn later.

The processes of heat conduction and diffusion in the oceans also have some interesting facets. We have mentioned that heating the sea at the surface is inefficient; however, what about cooling at the sea surface? Surface cooling increases water density and causes convective overturn, which carries the cooled water to depth. The deeper warmer water then rises to the surface and is cooled in turn. If the atmosphere is very cold relative to the water, then there is a rapid flux of heat from the sea to the atmosphere and the transfer of heat upward in the water column is turbulent and convective. When the sea surface is cooled to its initial freezing point, as it is in arctic regions, ice can form which acts as a barrier through which heat from the water must flow to reach the atmosphere in order for more ice to be produced. Heat cannot be conducted through the solid ice except by molecular processes. Therefore, if heat is continuously extracted from the water through the ice, the ice will increase in thickness and act as a greater barrier to the transfer. As a heat-transfer barrier, the ice serves as an insulator. If the arctic air temperature is -40°C and the water is -2°C, the temperature gradient across the ice is 38°C per thickness of the ice. An increase in ice thickness decreases the gradient and thus decreases the heat flux. Since the formation of sea ice in the arctic soon reaches a point of diminishing return, the undisturbed ice formed in one winter season reaches a thickness of only about 2 m. Snow accumulated on the ice surface further restricts the extraction of heat from the underlying water.

When a salt is dissolved in water, the boiling and freezing points of the resulting solution are not the same as those of pure water. The deviation from these temperatures is controlled by the amount of salt added. Thus the lowering of the freezing point and the elevation of the boiling point of seawater are functions of its salt content. The latter is of little consequence, since seawater does not normally reach such high temperatures, but the temperature and density of the water at the freezing point are significant.

Figure 7.3 shows the relationship of the temperature of maximum density and the initial freezing point of seawater to salinity. It is important to note from the figure that at chlorinities less than 13.92‰, $s = 24.70$‰, the temperature of maximum density is reached during cooling before initial freezing occurs. Therefore, dilute sea water (less than 24.7‰ salinity) behaves like freshwater, attaining maximum density prior to freezing. The temperature of maximum density of pure water is 4°C; freezing occurs at 0°C. Seawater with salinities higher than 24.7‰ does not attain its maximum density upon cooling, since maximum density occurs below the initial freezing point.

We have emphasized the word "initial" in discussing the freezing point. When seawater freezes and forms sea ice, the initial freezing process forms ice from the freshwater component of seawater and leaves the salts in solution. This selective process immediately increases the salt content of the remaining seawater and further depresses the ice point. Recently formed sea ice is so low in salt content that it is nearly potable. The freezing process in sea-ice formation is usually not uniform; pockets of seawater in the ice become very salty as more water is extracted by further freezing. These brine pockets are also centers of melting when the ice warms, giving the melting sea ice a spongy appearance.

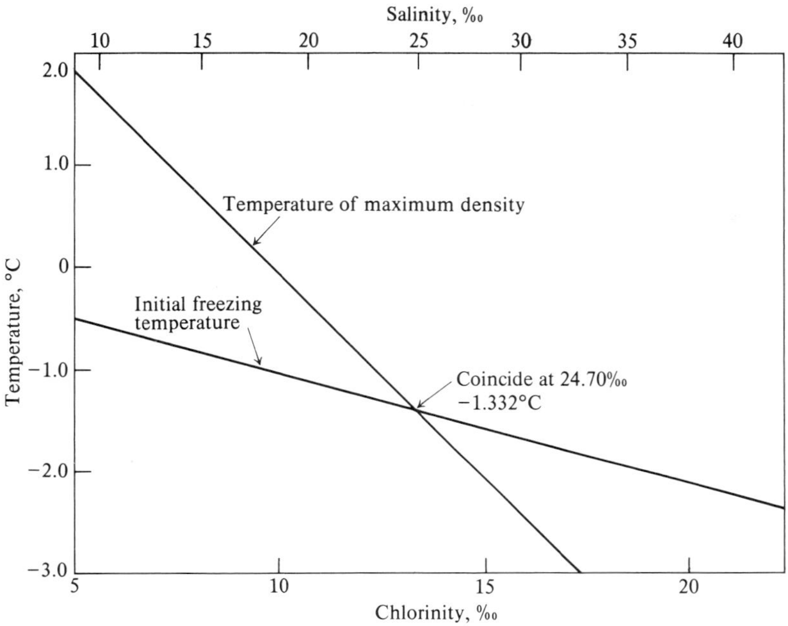

Fig. 7.3 The depression of the initial freezing point and temperature of maximum density of seawater with increasing salinity.

Like seawater, sea ice has a specific heat which is a function of its temperature and salt content. Thus to calculate the heat required to melt sea ice, we must know its specific heat, since it dictates the number of calories required to raise the temperature of a mass of ice of a given salt content to its melting point. The specific heat of sea ice increases with its salt content and decreases as the temperature of the ice falls below freezing. If the salt content of the ice is zero, its specific heat at its melting temperature is 0.48.

ADIABATIC PROCESSES IN SEAWATER

As we have stated, seawater is slightly compressible. When a gas is compressed its temperature rises; when it expands, its temperature drops. Liquids behave in the same manner if they are subjected to pressures great enough to compress them. In the deep ocean trenches at depths greater than 10,000 m, there are corresponding pressures in excess of 10,000 decibars or 1000 atm. If we put water from these depths in a flexible container, insulated so that heat can flow neither into nor out of it, the water will register a lower temperature than it had at depth due to its expansion during retrieval. The container allows expansion to occur *adiabatically*, i.e., without heat gain or loss. If we return the water to the deep sea, its temperature will increase adiabatically because of compression.

Table 7.7 Adiabatic temperature rise (Δt) $\times$ 10^2 in °C as a function of temperature and salinity of water lowered to various pressure levels from the surface

	Pressure, decibars, or depths, m	*Initial temperature, °C*							*Salinity*
		−2	*0*	*2*	*4*	*6*	*8*	*10*	‰
Salinity constant	1000	2.6	4.4	6.2	7.8	9.5	11.0	12.4	34.85
	2000	7.2	10.7	14.1	17.2	20.4	23.3	26.2	34.85
	4000	21.7	28.4	34.7	40.6	46.3	51.9	57.2	34.85
	6000	42.8	52.2	61.1	69.4				34.85
	8000		81.5	92.5	102.7				34.85
	10,000		115.7	128.3	14.2				34.85
Pressure constant	1000		3.5	5.3	7.0	8.7	10.3	11.8	30.00
	1000		3.9	5.7	7.3	9.0	10.6	12.1	32.00
	1000		4.3	6.0	7.7	9.4	10.9	12.4	34.00
	1000		4.7	6.4	8.1	9.7	11.2	12.7	36.00
	1000		5.1	6.8	8.4	10.0	11.6	13.0	38.00

Adapted from H. U. Sverdrup, M. W. Johnson, and R. H. Fleming, *The Oceans*, © 1942. By permission of Prentice-Hall, Inc., Englewood Cliffs, N.J.

Adiabatic expansion and contraction occur in nature in two ways. If a material is either compressed or allowed to expand very rapidly, there is no time for heat to flow into or out of it, even though a temperature gradient exists. Thus insulation is not necessary. Since the vertical motions of the sea are not very rapid, this situation does not occur in the oceans; however, it exists occasionally in the atmosphere. Adiabatic processes can also occur where the pressure changes slowly and there is appropriate insulation. It is under these conditions that dense water sinks adiabatically to great depths. In regions of sinking, the characteristic values of salinity and temperature of large volumes of water are determined by conditions at the sea surface. If all the particles of dense water formed at the surface are considered identical and move downward at the same rate, they are all compressed at the same rate and suffer the same rise in temperature per increase in pressure. Thus, since there is no temperature gradient to allow heat to pass from one adjacent particle to another, each particle is insulated by its neighboring particles. This insulating process also occurs in the atmosphere.

If derived from a single source, water in the deep ocean basins will show an increase in temperature but no significant change in salinity with depth. It is difficult to relate this deep water to its surface source if we do not know the value of the adiabatic temperature rise; if we do, however, we can correct the temperature *in situ* by subtracting the adiabatic temperature rise from it, obtaining a temperature at depth that is essentially the same as that of the source water.

At great depths, measured temperature increases adiabatically with depth while salinity remains uniform, producing decreasing σ_t values with depth. If we use the decrease in σ_t as a basis for determining density structure, it seems to create an unstable distribution—dense water above less dense water. In reality, however, this does not occur, since the decrease in σ_t disappears if we subtract the adiabatic temperature rise from the measured temperature. If we determine the density of the deep water (derived from a single source) as a function of salinity, observed temperature, and pressure, the density increases with depth because the effect of pressure is greater than that of the adiabatic temperature rise.

We can easily eliminate the appearance of an unstable distribution by determining σ_t from the measured salinity and the *potential temperature*, θ. The *in situ* temperature, t_m, minus Δt, the adiabatic temperature rise, equals θ. Typical values for Δt as a function of salinity and source temperature are given in Table 7.7.

The value of 34.85‰, $\sigma_0 = 28.0$ is used as the standard salinity in most tables listing the physical properties of water. It is characteristic of the mid-depth to deep open ocean waters.

SURFACE TENSION

The *surface tension* of water, which gives the air-sea boundary its elastic quality, varies with both temperature and salinity. In Chapter 16, we shall show that wind stress stretches the water surface to produce tiny capillary waves, which are restored by surface tension. These waves in turn roughen the water surface, strengthening the wind's "grip" and creating waves of ever-increasing size. In the oceans, however, natural oils from the biota decrease the surface tension. We can determine surface tension in cgs units from

$$\text{surface tension (dyn/cm}^2) = 75.64 - 0.144\, t(^\circ\text{C}) + 0.399\, Cl‰$$

neglecting interference from organics or a surface oil film.

The expression "to pour oil on troubled waters" refers to the calming effect of oil on turbulence in a region of wind generation. A thin layer of light oil cannot remove waves that have already been generated, but it does mitigate the locally created wind chop which rides on the backs of the larger waves. One does not have to look far through the literature of oceanic sailing adventures to find an example of a boat running before a storm under bare poles and trickling oil to quiet the surface waves in the upwind direction. Even today, oil and an oil bag are common items in the emergency stores of most lifeboats and small vessels that sail the open sea.

DETERMINATION OF SALINITY BASED ON PHYSICAL PROPERTIES

We have suggested that the salt content of seawater can be determined by measuring physical properties of water that are functions of salinity as well as by stoichiometric chemical procedures. In practice, three physical techniques can be used. The least accurate uses a *hydrometer* to measure density at fixed temperature and pressure. If we hold temperature and pressure constant,

Table 7.8 Conductivity of seawater in millimhos

Salinity ‰	Temperature, °C 0	2	4	6	8	10	12	16	18
2	2.00	2.14	2.27	2.39	2.53	2.67	2.81	3.08	3.23
5	4.80	5.09	5.40	5.71	6.02	6.35	6.68	7.35	7.70
10	9.15	9.74	10.32	10.89	11.49	12.09	12.70	13.97	14.63
15	13.34	14.15	15.01	15.88	16.75	17.63	18.52	20.34	21.28
20	17.39	18.46	19.55	20.66	21.79	22.93	24.08	26.46	27.68
25	21.35	22.66	23.98	25.34	26.71	28.09	29.50	32.42	33.89
30	25.24	26.77	28.33	29.92	30.66	32.24	33.86	38.22	39.96
35	29.03	30.78	32.58	34.39	36.20	38.07	39.98	43.87	45.86
40	32.73	34.73	36.73	38.73	40.78	42.86	44.97	49.32	51.58

From B. D. Thomas *et al.*, "The Electrical Conductivity of Seawater," *Journal du Conseil*, **9** (1), 28–34 (1934).

the stem of the hydrometer can be calibrated in units of salinity, since under these conditions it alone determines density. Hydrometers can be constructed that cover narrow ranges of salinity (or density) with an accuracy of about ±0.5‰. We can make more precise density determinations of salinity with a total-immersion hydrometer, which eliminates surface tension and meniscus effects on the stem.

The refractive index of seawater (i.e., its capacity to bend a light beam) can also be used as an indicator of salt content, since at constant temperature the relationship of the index to chlorinity is linear. This technique is useful because, although the refractive index does not change greatly with salt content, it can be determined to a high degree of accuracy. A typical range of index values at 20°C, for example, is 1334.8 at 5‰ *Cl* to 1339.7 at 20‰ *Cl*.

The most successful physical method of determining salt content is that based on the measurement of electrical conductivity. Special electrical bridge systems are used to measure the resistance (in ohms per cubic centimeter) of a precisely controlled volume of water between two electrodes of fixed surface area. The conductance of water, which is the reciprocal of resistance, is then determined in mhos (Table 7.8). With good conductivity equipment, a trained operator can routinely determine salinity with a higher degree of accuracy than is usually possible by precision chemical titration. The ease with which conductivity bridges can be used at sea and their high accuracy have led to the adoption of this technique for measuring salinity at most well equipped oceanographic facilities.

The electrical conductivity of seawater is highly temperature-sensitive, as the conductivity values in Table 7.8. indicate. If conductivity is used to determine salinity only, it is desirable to rule out the temperature effect by placing a water sample from the sea in a conductivity cell that is surrounded by a temperature-controlled bath. The sample is then heated or cooled by the bath to fix the sample temperature to a constant value. If all samples are analyzed at this precise temperature, conductivity becomes a function of salinity only.

If the salinity is to be measured *in situ*, a conductivity cell accompanied by a temperature sensor must be lowered into the sea. The sensor is usually a resistance thermometer constructed to have a resistance change with temperature that approximately compensates for the change in resistance of seawater with temperature. Thus the temperature effect is canceled out and we can obtain direct measurements of *in situ* salinity.

An electrodeless conductivity sensor has recently come into use that produces a toroidal magnetic field around the sensing unit. The field strength in the center of the cell is a function of the conductivity of the seawater, and the measurement is based on induction rather than resistance. The inductive salinometers are quite accurate and eliminate the problem of changing cell calibration in exposed electrode-type cells, which are affected by use.

In conclusion, we have seen that the physical properties of seawater affect both its response to environmental changes and the nature of many oceanic processes. Thus the properties of seawater are of primary importance in determining the nature of the marine environment. We have yet to consider two of the more important physical properties of seawater: the transmission of light and of sound. The next chapter is entirely devoted to these topics in order to allow a lengthy discussion of their characteristics and ramifications.

STUDY QUESTIONS

1 Produce a table of S and T values that combine to produce σ_t = a constant. Plot the curve of σ_t = constant on a T-S graph.

2 If seawater were truly incompressible, sea level would be higher than it is at present. Calculate the elevation in level that the oceans would undergo given that seawater is of 35‰ salinity at 0°C, the oceans' area does not change, and the mean depth of the oceans is 3870 m.

3 Determine an equation that relates σ_t to the thermosteric anomaly, $\Delta_{s,t}$. Use relationships between $\alpha_{s,t,0}$, $\alpha_{35,0,0}$, $\Delta_{s,t}$, $\rho_{s,t,0}$, and σ_t.

4 Calculate the adiabatic temperature rise for a parcel of water having a source temperature of 2°C at 3000 m depth that is lowered to 7000 m. Assume the salinity to be 35‰.

5 Water of S_1‰, T_1°C, and σ_t = 27.0 is mixed in equal proportions with water of S_2‰, T_2°C, and σ_t = 27.0. Show on a T-S diagram why the resulting water mixture has a σ_t greater than 27.0.

6 Determine the electrical conductivity of seawater having a salinity of 34‰ at 8.5°C.

7 Discuss the importance of the temperature of maximum density being higher than the initial freezing point of seawater when the salinity is less than 24.7‰. Emphasize the role of *thermally* driven overturn in the sea during both cooling and warming cycles.

8 If the initial freezing point of seawater is passed in the cooling cycle, what additional process must we consider to understand vertical convection in the water column?

9 What amount of heat must be extracted from 1 m^3 of seawater of salinity 33‰ to convert it to sea ice if the initial temperature of the seawater is 9°C?

RECOMMENDED READING

Defant, A., *Physical Oceanography*, Pergamon Press, New York, 1961.

Hill, M. N., ed., *The Sea*, vol. 1, Interscience Publishers, New York, 1965.

Knudsen, M., *Hydrographic Tables*, G. E. C. Gad, Copenhagen, 1901. Reprinted 1959, Tutein & Koch, Copenhagen.

Neumann, G., and W. Pierson, Jr., *Principles of Physical Oceanography*, Prentice-Hall, Inc., Englewood Cliffs, N.J., 1966.

Sverdrup, H. U., M. W. Johnson, and R. H. Fleming, *The Oceans*, Prentice-Hall, Inc., Englewood Cliffs, N.J., 1942.

U.S. Naval Hydrographic Office, *Processing Oceanographic Data*, H. O. Publication No. 614, Washington, D.C., 1951.

CHAPTER 8

TRANSMISSION OF SOUND AND LIGHT

INTRODUCTION

The fact that seawater can transmit sound and light is of particular interest to the oceanographer. The transmission of sound with low attenuation enables scientists to use acoustical methods to determine depth, sea-bottom structure, and locate submerged submarines, as well as to communicate over considerable distances in a medium that is nearly opaque to most electromagnetic radiation. This property of seawater, however, has little influence on the natural oceanic environment; in fact, it is important only in that marine animals can use sound as a communicating device.

The transmission of light (electromagnetic radiation in the visible range) is much more important in controlling the natural environment of the oceans. Because of its attenuation, light penetrates the sea surface for only short distances to allow photosynthesis to occur in the upper layers of the sea and supply the primary energy needed to support the biological population. It also supplies heat to the surface waters, determining in part the thermal structure of the oceans.

THE PROPAGATION OF SOUND IN SEAWATER

Both light and sound travel through water as waves, but of differing types. Light propagates as a transverse wave, in the form of electromagnetic radiation; its oscillation amplitude is at right angles to the direction of propagation. Sound is transmitted as a longitudinal wave in the form of pressure disturbance, with compression and rarefaction occurring alternately in the direction of propagation. Both sound and light are forms of energy, but there the similarity ceases.

Water transmits sound more effectively than the atmosphere and at a faster rate. The velocity of sound waves is determined by

$$V = \sqrt{\text{elasticity/density}} \tag{8.1}$$

where the elasticity is $C_p/C_v \times 1/K$. The ratio of specific heat at constant pressure to specific heat at constant volume (C_p/C_v) is usually called gamma, γ; K is the compressibility of the transmitting medium. Since elasticity is related to $1/K$, it is evident that an easily compressible material has a small elasticity and thus a low speed of sound propagation. A material that is only slightly compressible has a large elasticity and conducts sound rapidly. In the sea the speed of sound increases with increasing salinity, temperature, and pressure. Near the sea surface in the open ocean, changing temperature with depth exerts the

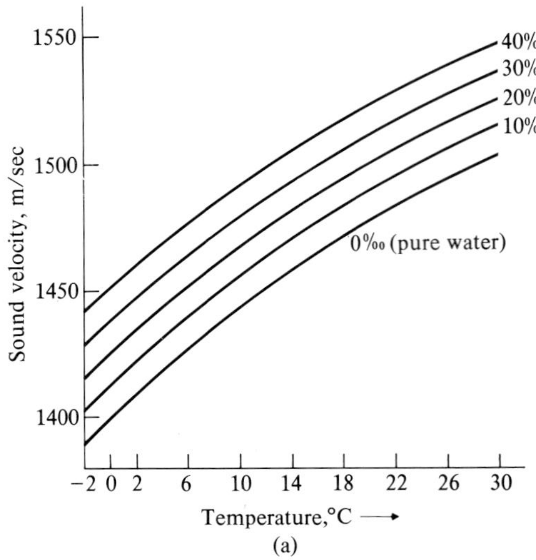

Fig. 8.1a Sound velocity in water as a function of salt content and temperature at atmospheric pressure.

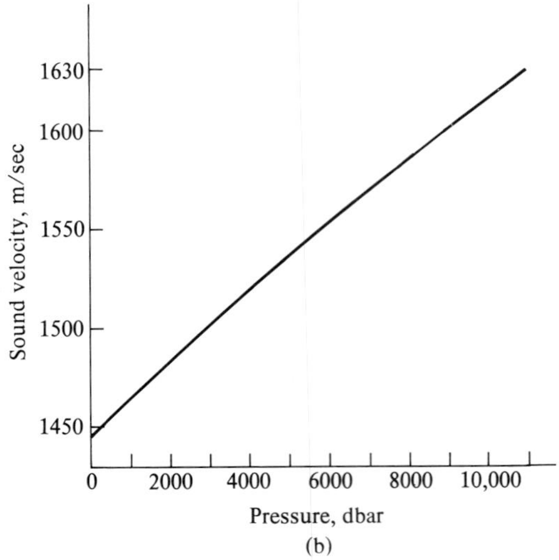

Fig. 8.1b Sound velocity in seawater of 0°C and 35‰ as a function of hydrostatic pressure. (After G. Neumann and W. J. Pierson, *Principles of Physical Oceanography*, © 1966. By permission of Prentice-Hall, Inc., Englewood Cliffs, N.J.)

greatest influence on the speed of sound propagation, while at moderate to great depths, where salinity and temperature change very little, pressure is the dominant controlling factor (Fig. 8.1).

Sound waves traveling vertically pass through zones of water having different values of salinity, temperature, and pressure. These zones have different speeds of sound propagation, but in general are oriented horizontally so that the vertical sound beam strikes the zones at a right angle. Thus little or no refraction of the sound beam occurs. Since sound travels at different speeds in each depth zone, we must use a weighted average sound speed for the particular water column and depth. This average can then be used to determine the depth of water. If a sound pulse is generated at the sea surface and directed downward, the water depth equals the average speed $\times$ $T/2$, where T is the time required for the sound pulse to reach the bottom, be reflected, and return to the generating source. For seawater, the average speed used in sonic sounding is about 1500 m/sec. A sonic depth-sounding device need only determine the time increment between the outgoing pulse and its reflected return, since we can convert this measurement to depth quite easily when the sounding speed is constant. If very accurate depth sounding is required, we can make corrections based on the actual sound speed distribution and the depth of the location being studied.

Sound waves beamed obliquely through the water strike the zones of differing propagation speed at an angle and are bent or refracted according

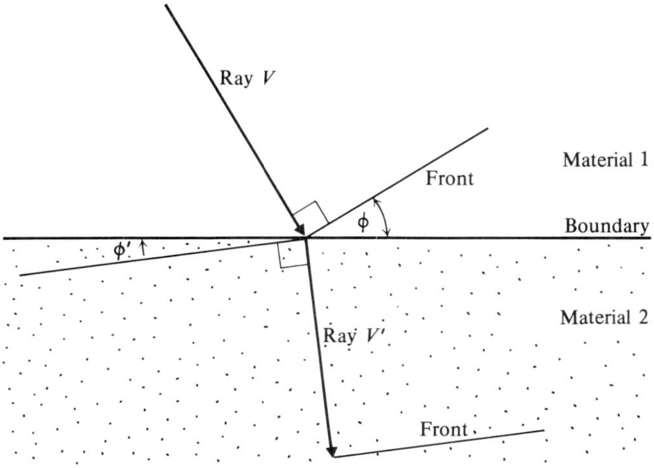

Fig. 8.2 A wave progressing through a material where it has one speed of propagation into another material where its propagation speed changes will be refracted according to Snell's law.

to Snell's law (Fig. 8.2). If a cone of sound passes through water of one uniform sound speed and then strikes a zone in which the sound speed is smaller, the speed of the sound wave at the lower edge of the cone is reduced before that of the upper edge and the cone is bent downward (*refracted*). In the same manner, if the cone passes from one zone into another with a higher sound speed, it is refracted upward as the lower edge of the sound wave starts to move faster than the upper.

Snell's law governing refraction, which gives the relationship between the velocities in two substances and their effect on creating curvature, is

$$V' \sin \phi = V \sin \phi'.$$

In the example shown here V is greater than V', so downward bending results (Fig. 8.3). The wave fronts are perpendicular to the rays, which show the direction of propagation. Snell's law also applies to optics and the study of water waves. In some cases, if angle ϕ is large enough, the ray does not penetrate the boundary and total reflection occurs; thus ϕ is equal to or greater than the critical angle. Most of us have witnessed the significance of a critical angle in causing total reflection of light. When submerged in a swimming pool with a quiet water surface, you need only open your eyes and look through the surface from below. If you look at a right angle through the surface, you can see what is above the water; if you look obliquely, at a small angle, the surface appears as a mirror and you cannot see above it. When this occurs your line of sight is form-

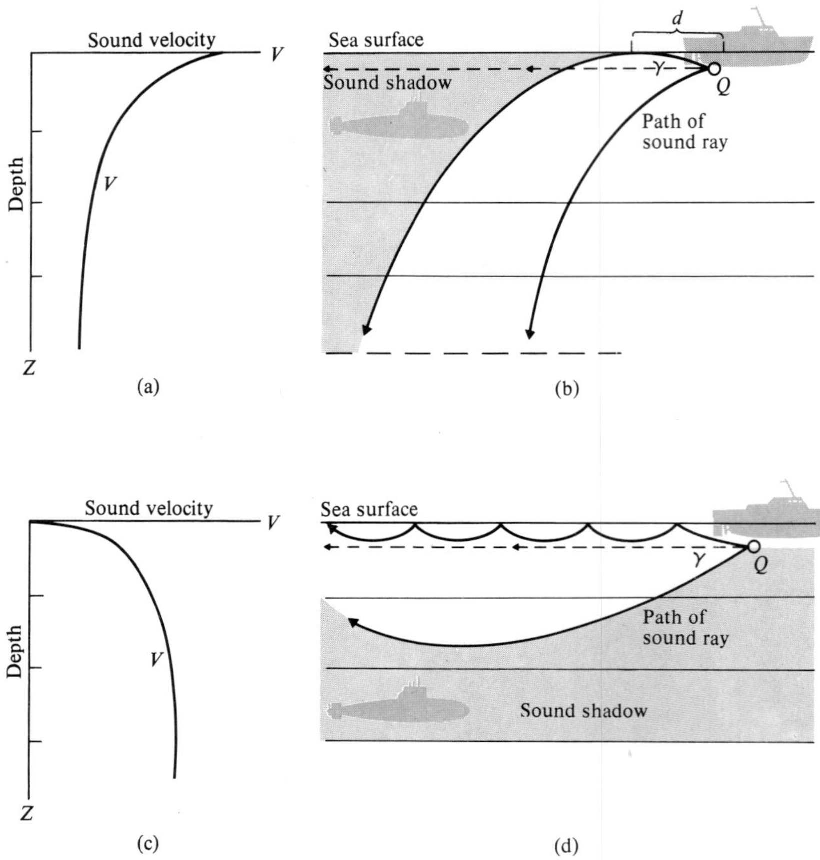

Fig. 8.3 In near-surface waters the temperature structure and salt distribution can cause the sound velocity to decrease or increase with depth to drastically alter the path of the sound rays. (After G. Neumann and W. J. Pierson, *Principles of Physical Oceanography*, © 1966. By permission of Prentice-Hall, Inc., Englewood Cliffs, N.J.)

ing an angle of incidence with the surface that is greater than the critical angle and total reflection is occurring.

The distribution of sound speed with depth in the sea is such that variable-direction sound-beam generators used for sound ranging and the detection of undersea craft (SONAR) often send out rays that meet zones of changes in sound speed at critical angles. Thus there are regions in the sea that cannot be penetrated by sound beams if the sound is directed toward them at small angles (Fig. 8.4). Regions that exclude sound in this manner are called *shadow zones.* A subsurface vessel can maintain a position of proper depth and distance relative to a searching sound-beam source and remain hidden in a shadow zone, a fact of considerable importance in military situations. Even if the vessel is not in a shadow zone, its location is difficult to ascertain. A typical sound transducer instrument tells you only that an underwater target is so many

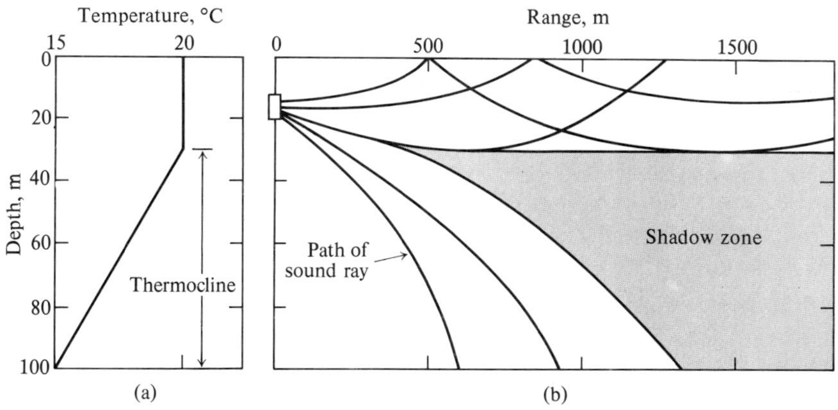

Fig. 8.4 Near the surface of the sea the temperature structure (a) will act to produce a sound shadow zone (b) if the salinity is constant. (After L. E. Kinsler and A. R. Frey, *Fundamentals of Acoustics*, 2d ed., Wiley, New York, 1962.)

degrees bearing relative to the ship's heading, at a downward inclination of so many degrees, and at a given distance. Although the depth of the target can then be determined from the downward angle of the sound source and target distance, considerable error results from the fact that the sound beam follows a curved rather than straight path. We must make additional measurements of the water characteristics or direct measurements of the distribution of sound velocity with depth to find the actual path of the sound beam and location of the target.

The combined effect of salinity, temperature, and pressure produces a middepth minimum in sound velocity in the oceans. The depth zone centered about this minimum is called the *SOFAR channel* (Fig. 8.5). A sound generated in this zone escapes if it passes out at a large angle to the horizontal; however, if it strikes out from the source at low angles it is refracted back and

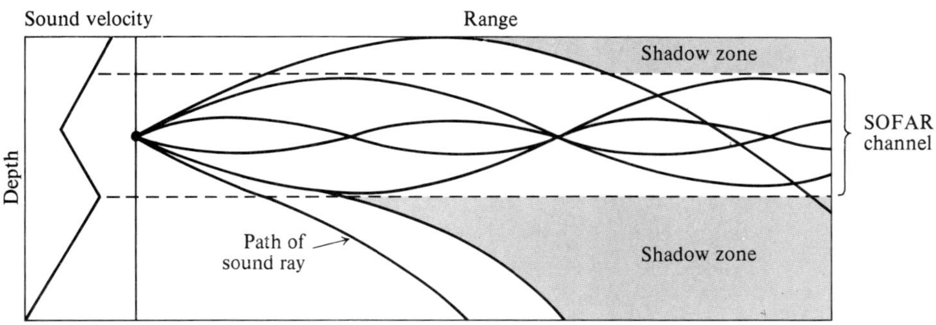

Fig. 8.5 The temperature, salinity, and pressure variation with depth combine to produce a minimum sound velocity at about 1000m. Sound generated at this depth can be trapped in a layer known as the SOFAR channel. (After L. E. Kinsler and A. R. Frey, *Fundamentals of Acoustics*, 2d ed., Wiley, New York, 1962.)

forth along the center of the channel. Sound held captive in this manner is not dispersed spherically through the entire water body, but radiates outward from the source in a relatively thin layer. Even moderate amounts of sound energy can traverse great horizontal distances when spherical divergence is prevented. In the SOFAR channel signals have been transmitted over distances as great as from Australia to Bermuda.

THE ATTENUATION OF SOUND IN THE SEA

Sound, like light, is attenuated in seawater; its intensity issuing from a point source decreases with the square of the distance from its source due to spherical spreading, i.e., the cross-sectional area of the cone increases with the square of the distance. Sound energy is kinetic, and since water is viscous sound is absorbed and dissipated as heat in the sea. The attenuation of sound due to absorption is exponential in character:

$$I_{x\lambda} = I_{0\lambda}\, e^{-2vx}, \tag{8.2}$$

where $I_{x\lambda}$ is the intensity of a specific frequency of sound of wavelength λ after the sound has traveled x meters from its source; $I_{0\lambda}$ is the intensity of the same frequency at the source, e is the Napierian e which forms the base of natural logarithms, and v is the *absorption coefficient* for sound, given by

$$v = 8\pi^2\mu/3\lambda^2\rho V. \tag{8.3}$$

In this expression, μ is the viscosity of the water, λ is the length of the sound wave, ρ is the density of the water, and V is the speed of sound. Although V controls λ, it is not dependent on it because sound waves are nondispersive. Since V is dictated by water conditions and equals λ/T, where T is the period of the generated sound signal, high-frequency sound, having short periods, must also have small wavelengths. Similarly, sound waves of longer periods have longer wavelengths. All other factors being constant, small values of λ produce large values of v; therefore, high-frequency sound is absorbed more rapidly than low-frequency sound, which has large T and λ. Frequencies of about 2000 cycles per second are usually satisfactory for good sound transmission.

Absorption and spreading over ever-increasing areas combine to attenuate sound. Reflection of sound by particulate matter, fish, etc., also contributes to attenuation in a fixed direction, since discrete bodies deflect the sound beams to a new direction. This process is called *scattering*. Exponential absorption of sound and scattering occur in the SOFAR channel, but the refractive funneling of the sound beams within the channel significantly reduces the loss due to spreading.

Sound or light issuing uniformly in all directions from a point source passes through larger and larger concentric spheres as it moves away from the source. If the intensity of the sound or light energy at the surface of a sphere with a radius of 1 cm is I/cm^2, then the total intensity over the unit sphere is

$4\pi R^2 \times I/\text{cm}^2$, where $R = 1$ cm. The total intensity over the area of a larger sphere ($R = 2$ cm) is the same as that for $R = 1$ cm, but the intensity per unit area on this new sphere is $(I/4)/\text{cm}^2$. Thus the intensity per unit surface area depends on the ratio R_1^2/R_2^2, that is, $I_2/I_1 = R_1^2/R_2^2$, which is true as long as the energy continues outward in a constant direction. When refraction and scattering occur the energy does not diverge uniformly as a function of the square of the distance from the source.

THE PROPAGATION OF LIGHT IN THE SEA

Light, the other form of wave energy transmitted in the sea, represents a very small part of the spectrum of electromagnetic radiation, i.e., the visible part (Fig. 8.6). Water is essentially opaque to all electromagnetic radiation except the narrow band called *visible light*, which is also the only band that human eyes can see. This is not surprising if we recall that the human body is composed primarily of a saline solution not unlike seawater. If animals needed sight for survival and if life developed in the sea, sight would be limited to using electromagnetic energy transmitted by seawater for visual perception.

The transmission of light in the sea is extremely important, since the sun is the primary source of energy for all biological phenomena. This energy is required for photosynthesis, the process by which land and sea plants produce the pasturage that supports the higher levels of herbivores and carnivores. In the sea, because visible light penetrates to depth, plant production is possible over a depth range rather than being limited to the surface film. Thus enough plant material is available to support the entire animal population of the sea, including that portion that lives below the lighted depths in the dark zones which extend to the ocean bottom.

Since solar light comes from both the sun and diffuse sky light, we can consider its rays as nearly parallel and thus nondivergent (i.e. not issuing from a point source, as does sound) over large areas of the sea. For this reason, natural light rays do not approach layers of water (in which their speed of

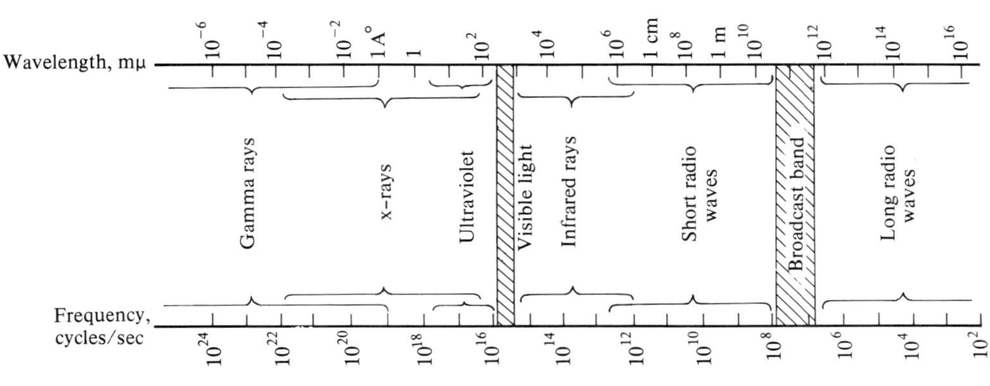

Fig. 8.6 Electromagnetic spectrum.

propagation may vary) from a variety of angles as do sound waves from a point source. Thus the formation of shadow zones or regions of excess light due to divergence and convergence of light rays is not as important as in the case of sound. We shall not discuss the refraction of solar light in the sea; however, it is obvious from the swimming pool example and the displacement of a stick as it passes through the air-water boundary that refraction is a common phenomenon. Light, like sound, obeys Snell's law.

Because light and sound in the oceans obey the same physical laws, it is convenient to treat them together. Both are refracted and attenuated; however, the refraction of sound is more important than its attenuation, while the reverse is true for light.

The attenuation, or propagation loss, of light, like that of sound, is due to spreading, scattering, and absorption. Light will also be considered non-dispersive, i.e., all its wavelengths have the same speed. This assumption is made only for convenience's sake, as we all know that different wavelengths of light travel at different speeds when passing through a medium other than a vacuum. As is that of sound, its absorption rate is dependent on frequency and wavelength:

$$I_{x\lambda} = I_{0\lambda}\, e^{-\alpha_\lambda x}, \tag{8.4}$$

where $I_{x\lambda}$ and $I_{0\lambda}$ are the same as those in Eq. (8.2) and α_λ is the *absorption coefficient* of light with wavelength λ. Table 8.1 gives the absorption coefficient of light as a function of wavelength for pure water. Deviation from these values is negligible in clean seawater, indicating that salt has little effect on absorption. It is obvious from Eq. (8.4) that if α_λ is small, the intensity decreases slowly with distance x, whereas if α_λ is large, the intensity decreases rapidly. The absorption coefficients for blue to green wavelengths are small for water (Table 8.1). This means that if *white light* (all components of the visible spectrum) passes through a column of clean water, its spectral quality as well as its quantity will change as it issues from the opposite end of the column. Very little of the red light component of white light is transmitted because of the high absorption of these wavelengths (large α_λ at the red end of the visible spectrum); the light emanating from the end of the column opposite the light source will be blue to green because these are the wavelengths transmitted with minimum loss.

If suspended particles are placed in the water column, they will further affect the light intensity as well as the spectral quality of the transmitted light; i.e., the light will be scattered not only by the water molecules but also by the suspended particles, with the result that more light will be deflected out of the column at an angle to the original light-beam path and the intensity will be further decreased at the point of observation. The available light shining on the particles will expose their color, adding an additional tint to the water. If the particles in the water are semitransparent, so that they pass light as well as scatter it, they may also alter the quality and quantity of light transmitted through them.

Table 8.1 Absorption coefficients of light for pure water, per meter

Color	*Wavelength of light, μ*	*Absorption coefficient per meter*
Ultraviolet	0.30	0.800
Blue	0.40	0.072
Green	0.50	0.016
Yellow	0.60	0.125
Orange	0.70	0.840
Red	0.80	2.400
	0.90	6.550
	1.00	39.700
Infrared	1.10	20.300
	1.20	123.200
	1.30	150.000
	1.40	1600.000
	1.50	1940.000

Adapted from H. U. Sverdrup, M. W. Johnson, and R. H. Fleming, *The Oceans*, © 1942. By permission of Prentice-Hall, Inc., Englewood Cliffs. N.J.

All the effects just discussed occur in natural waters. Absorption of light takes place because of molecular absorption and the presence of organic compounds in soluble or insoluble form in the water. Light is scattered by water molecules and by both organic and inorganic suspended particles. The contribution made by molecular scattering to the attenuation of light is considered part of the absorption registered in clean water. In the sea, the wavelengths of light that penetrate the farthest with minimum absorption are also those that are most available to be scattered. If we look at the ocean from above the sea surface, the predominant color is blue, since blue-light (short wavelength relative to molecular radius) scattering predominates. However, the presence of large quantities of soluble organic materials, small plants and animals, suspended silt, and other organic debris tends to make large areas of the sea appear more green than blue. Regions of the oceans that appear as very clear and blue have low organic production and therefore may be considered the biological deserts of the seas. The greenish-brown turbid coastal regions usually contain large quantities of biological material suspended in the water.

It is possible to treat the absorption and scattering of light as one process, since both lead to reduction or attenuation of light intensity. Equation (8.4) then becomes an equation for light attenuation, where α_λ is the attenuation coefficient, a. We can determine the attenuation coefficient for a specific wavelength of light by passing the monochromatic light as a nondiverging beam over a fixed distance in the sea. If we know the intensity of the light at its source and at a light-sensitive receptor a fixed distance away, we can calculate the attenuation coefficient from the exponential equation using x as the distance between light source and receptor. To find a for specific depths a hydrophotometer consisting of a light source and a photocell receptor mounted on a rigid bar with a fixed distance between the light source and the cell can be lowered into the sea. Often, if we do not need the attenuation coefficient for a particular wavelength of monochromatic light, we can use white light as a basis of measurement; the attenuation coefficient is then applicable to the entire light spectrum available from the source. In some cases, the determination of a as a function of depth and horizontal location yields information that can be interpreted to estimate the productivity of a region of the sea; i.e., the attenuation coefficient increases in areas in which the concentration of microscopic plants and animals increases (or, in general, where the quantity of particulate matter increases).

It is not necessary, however, to obtain the distribution of a with space to evaluate the abundance of particulate matter in the oceans. Since the radiant energy entering the sea comes from direct sunlight and diffuse sky light, we need only determine the depth to which it penetrates in sufficient quantity to support life. The depth at which the light reaches 10% of the surface value is of course a measure of the optical clarity of the water and thus is related to amount of suspended material. The decrease in solar light intensity with depth determines the *extinction rate* of solar light. In regions of low biological production, the extinction rate is small, the water is clear, and light penetrates to greater depths than it does in regions of high production.

Oceanographers measure the extinction of solar light in the sea by lowering a *Secchi disk*, a white disk about 35 cm in diameter, to a depth at which it is just visible. At this depth, solar light penetrating the sea is reflected off the surface of the disk in a quantity just sufficient to come back through the water and reach the observer's eye, which can then distinguish the disk from the scattered background light. The depth at which this occurs, D, determines the *extinction coefficient*, κ, which is equal to $1.7/D$. The extinction coefficient can be used in the exponential equation

$$I_z = I_0 e^{-\kappa z} \tag{8.5}$$

which has the same form as the absorption equation (8.4) and attenuation equation (8.2), where α_λ and a respectively are used in place of κ.

A clear distinction should be made between these three coefficients as they are used here. The absorption coefficient, α_λ, is the rate at which light of a given wavelength or white light (omitting the subscript λ) passing through a given thickness or layer of clean (no suspended particles) seawater is attenuated over distance due to molecular absorption and molecular scattering. The attenuation coefficient, a, is the rate at which light of a given wavelength or white light passing through a given thickness or layer of seawater is attenuated over distance due to both absorption and scattering by water molecules and by naturally occurring suspended material. The first coefficient, then, is applicable to clean, filtered, uncontaminated seawater; the second is applicable to natural

Table 8.2 Typical seasonal variation of κ in coastal waters

Month	*Depth, m, of Secchi disk*	κ, m^{-1}
January	8.5	0.20
February	7.1	0.24
March	6.0	0.28
April	5.7	0.30
May	7.7	0.22
June	9.4	0.18
July	10.6	0.16
August	11.3	0.15
September	9.4	0.18
October	12.2	0.14
November	10.6	0.16
December	9.4	0.18

seawater that is not filtered or treated to remove the contaminants. The instrumentation and techniques used in determining these coefficients permit us to evaluate them at discrete points in space so that we can determine their variation with depth as well as horizontal space.

Since $\kappa = 1.7/D$, it is applicable only to total available solar light, natural oceanic conditions, and the upper sunlit regions of the sea. This coefficient cannot be evaluated at discrete depths, but rather represents an integrated value applicable over a vertical water column of depth D. Because the immersion hydrophotometer used to measure the attenuation coefficient has an independent light source and receptor, it may be hung at any depth in the sea with the bar connecting the light source and receptor at any attitude, provided that solar light is excluded from contributing to the light beam.

In oceanic coastal areas at midlatitudes of the Northern Hemisphere, the extinction coefficient varies seasonally (Table 8.2). From the table we see that the extinction coefficient is greater in spring and late summer due to cyclic increases in the numbers of floating unicellular plants and other particles suspended in water.

Not all the solar energy striking the sea surface enters the water; some is reflected. However, only the energy that enters the sea to be attenuated contributes to the illumination and heating of the oceans. Figure 8.7 shows the distribution of radiant energy with wavelengths of light for light entering the surface and at differing depths for average oceanic water. It also shows that infrared radiation is almost totally absorbed in the first few centimeters of water, thereby adding to the heat content of the seas. Since the red wavelengths of visible light are quickly removed, the energy peak centers over the blue to green light wavelengths (0.4 to 0.6 μ) as depth increases. Within 100 m, the total radiant energy available at the surface has been reduced to about 3% of its surface value. If we construct a similar energy spectrum portraying

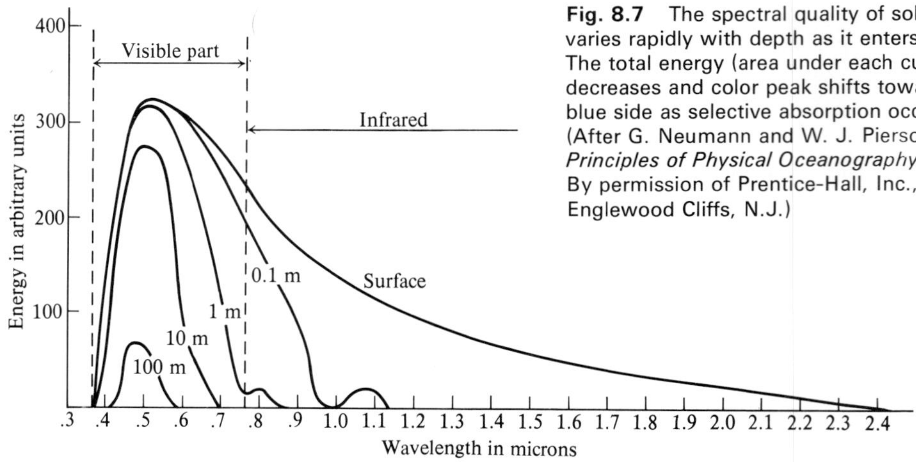

Fig. 8.7 The spectral quality of solar light varies rapidly with depth as it enters the sea. The total energy (area under each curve) decreases and color peak shifts toward the blue side as selective absorption occurs. (After G. Neumann and W. J. Pierson, *Principles of Physical Oceanography*, © 1966. By permission of Prentice-Hall, Inc., Englewood Cliffs, N.J.)

the distribution of light in turbid coastal water, we find an increase in light attenuation which decreases the energy more rapidly with depth and a shift in the spectral peak toward the yellow light (0.55μ). Thus coastal waters appear greenish.

In oceanography, as in any science, the loose usage of terms often creates misunderstanding. Any immersible instrument providing a light beam that traverses a path of fixed length through water between a light source and a receptor can be used to determine the light loss due to absorption and scattering for either monochromatic light or white light. Note, however, that coefficients determined in this manner *in situ* are *attenuation* coefficients, and are not to be confused with *extinction* coefficients, which are determined from Secchi disks. When confronted with coefficients cited in the literature, we must therefore know the techniques used to determine them, although they are often unspecified.

A common technique to determine the rate at which solar radiation decreases with depth in the sea uses a photoreceptor that points upward and can detect either total solar illumination or, when appropriate light filters are used over the light sensitive cell, one component wavelength of sunlight. This cell is lowered meter by meter and the intensity of solar illumination is recorded at each unit of depth. One can thus use successive values of illumination to calculate the attenuation coefficient for each 1-m depth increment, as well as to determine at what depth the illumination is reduced to a given percentage of the surface value. The coefficient determined for each increment of depth is an attenuation coefficient based on available sunlight as the light source. Needless to say, if we are measuring total solar radiation, the quality (wavelength composition of the light) is changing with depth and the calculated coefficient is sensitive to this change.

The foregoing discussion does not imply that ours is the only correct terminology; rather, we have tried to point out the differences in the coefficients and their applicability to the oceans. Care must be exercised to ensure the use of the correct coefficient at the right time for the right purpose.

The color of the sea is as varied as observers' descriptions of it. When seen under a cloudless sky and bright sun, the sea is blue unless sufficient suspended material is present to change it to green or even yellow-brown. As we shall discuss in Chapter 15, there are planktonic organisms which when present in abundance give the sea a red color; under storm clouds, the sea seems gray and forbidding. Homer saw the sea as "wine-dark" under the vermilion cast of the sunset. Waves also affect color, since light may pass obliquely through their steep crests and foamy whitecaps form part of the color background.

As we have shown, the sea gains its color from solar light; however, not all solar light is the same. The altitude (angle) of the sun, the water-vapor content of the atmosphere, and the presence of clouds and dust in the atmosphere all act to continually change both the quantity and quality of direct and diffuse light entering the oceans. The color of the sea therefore changes with the available sunlight. The direction from which the sea is viewed is also critical; when looking toward the sun we observe reflected sunlight from the sea surface, as well as scattered light from beneath the surface.

Because of these variables, including variation in the sensitivity of the human eye to color, a set of procedures must be followed to establish the color of the sea. The standard color scale, the *Forel* scale, consists of vials of colored water ranging from blue through yellow-brown to red. These colors are created chemically and each is given a number that corresponds to a fixed solution. To determine sea color, an observer lowers a Secchi disk from the shady side of the ship to a depth of 1 m and matches the color in the vials as observed against a white card to the color of the water against the Secchi disk. Though subjective, this method does somewhat standardize the determination of observed water color.

STUDY QUESTIONS

1 Both light and sound waves obey Snell's law and *Beers-Bouguer's* law (the exponential attenuation equation). For the sea, state which law can be neglected (a) in the case of light wave propagation and (b) in the case of sound wave propagation.

2 If the average speed of sound over a vertical water column is 1500 m/sec and it requires 4 sec for the sound to issue from the source at the sea surface to reflect off the bottom and return, what is the depth of water? If the accuracy of the 4-sec time increment is $\pm$ 10 μsec, how accurately is the depth known?

3 Why does seawater in midocean appear blue to the observer and more greenish in the coastal region?

4 According to Snell's law, sound beams that issue from a point source located at the depth at which the speed of sound is minimum are concentrated in a depth layer centered upon this minimum speed zone. Explain why.

5 Why would you expect the attenuation coefficient for light to decrease with depth from 20 to 1000 m?

6 The attenuation coefficient of sunlight determined for each meter of depth from natural light intensities measured at 1-m intervals of depth differs from that determined for each meter of depth using a device having a full spectrum white light source and a photocell to measure the intensity at a distance of 1 m from the light source. Why do the two attenuations differ?

7 Calculate the depth at which solar radiation decreases to 10% of the surface value given that the attenuation coefficient for solar light is 0.05/m and invariant with depth.

8 List at least five different techniques for measuring attenuation coefficients. State why each method yields a different coefficient.

RECOMMENDED READING

Albers, V. M., *Underwater Acoustics Handbook*, The Pennsylvania State University Press, University Park, Pa., 1960.

Defant, A., *Physical Oceanography*, vol. 1, Pergamon Press, New York, 1961.

Fairbridge, R. W., ed., *The Encyclopedia of Oceanography*, Reinhold Publishing Corp., New York, 1966.

Hill, M. N., ed., *The Sea*, vol. 1, Interscience Publishers, New York, 1965.

Jerlov, N. G., *Optical Oceanography*, Elsevier Publishing Co., New York, 1968.

Neumann, G., and W. Pierson, Jr., *Principles of Physical Oceanography*, Prentice-Hall, Inc., Englewood Cliffs, N.J., 1966.

Sverdrup, H. U., M. W. Johnson, and R. H. Fleming, *The Oceans*, Prentice-Hall, Inc., Englewood Cliffs, N.J., 1942.

CHAPTER 9

CONSEQUENCES OF THE PHYSICAL AND CHEMICAL PROPERTIES OF SEAWATER

INTRODUCTION

In previous chapters we discussed some of the chemical and physical properties of seawater. We stated that seawater contains dissolved solid material and gases; that its density is controlled by temperature, salinity, and pressure; that its physical properties are unique; and that it has a large capacity to absorb heat with little temperature change.

In reality, nearly every event in the sea is affected by and/or affects the physics and chemistry of the ocean water; therefore it is impossible to cite every example of this reciprocity. In this chapter, we shall discuss a few processes that illustrate the interrelation between the physical and chemical properties of seawater and the environmental conditions found in the oceans. Interestingly enough, many oceanic processes are related to the response of seawater density to environmental changes and the vertical circulation of water caused by changes in density.

THE SEA AS A CONTROLLING AGENT OF CARBON DIOXIDE

We stated earlier that the gas carbon dioxide (CO_2) is very soluble in seawater, and that when it is added to or removed from the sea a shift in the carbonate-bicarbonate system occurs to maintain the pH of the water. Carbon dioxide also enters into the photosynthetic processes of marine plants. The freedom with which it circulates in the atmosphere, passes back and forth through the air-sea boundary, is bound up in organic matter or liberated by decomposition, and is carried about by the oceanic currents makes CO_2 one of the most important media for the transfer and circulation of carbon on the earth.

In the earth-sea-air system, carbon dioxide is found in the *lithosphere*, the rock and solid portions of the earth; the *hydrosphere*, the liquid portions of the earth; the *biosphere*, plant and animal material; and the *atmosphere*, the gaseous envelope surrounding the earth. If we consider the carbon dioxide content of the atmosphere by mass as one unit, then the CO_2 content of the lithosphere is 38,000 units, of the biosphere 2.78 units, and of the hydrosphere 58.0 units. The atmosphere is therefore the smallest carbon dioxide reservoir and is most easily affected by alteration of the rate at which CO_2 transfers from one reservoir to another.

Man's effect on the carbon dioxide cycle is increasing. More and more he removes the plants that through photosynthesis take carbon dioxide from the atmosphere and transfer it to the biosphere, while at the same time he is burning fossil fuels at an increasing rate, thus pumping carbon dioxide from the lithosphere and biosphere into the atmosphere. Unless the transfer of carbon dioxide to the hydrosphere from the atmosphere can keep pace, the atmosphere will gain in carbon dioxide. Measurements of atmospheric carbon dioxide taken in a worldwide survey in January 1960 indicated that the average concentration of atmospheric CO_2 was $314.5 \times 10^{-4}\%$ and that it was increasing at a rate of about $0.72 \times 10^{-4}\%$ per year. The production of CO_2 by the burning of fossil fuels, petroleum products and coal, was estimated to supply $1.3 \times 10^{-4}\%$ CO_2 to the atmosphere, nearly double the observed increase. Thus not all the CO_2 produced by the burning of fossil fuels stayed in the atmosphere; some was consumed by land and sea plants and transferred to other reservoirs. However, man's removal of greenery and the vegetation mat of the forest, as well as his ever-increasing production of CO_2, will no doubt reduce the ability of plant life to keep pace and result in an even higher rate of CO_2 accumulation in the atmosphere.

The oceans and their plant life are an important factor in controlling the CO_2 content of the atmosphere. In Chapter 6, we found that carbon dioxide reacts with seawater to form carbonic acid, which dissociates into carbonate and bicarbonate. As CO_2 is removed from the water by photosynthesis, the ionic balance changes, allowing the water to absorb more CO_2 from the atmosphere. Organic plant material sinks upon dying to decompose at depth, where the CO_2 is again liberated. Calcium carbonate formed by plants and animals near the surface also acts to transfer CO_2 downward, where it is then dissolved or deposited as carbonate sediments. This transfer of CO_2 to deeper depths helps in extracting from the atmosphere the surplus CO_2 produced by man. This buffering capacity of the oceans cannot continue indefinitely, for the ocean waters gradually circulate and sooner or later the deeper water rises to the surface and releases its excess CO_2 to the atmosphere.

What will happen to the earth if there are large increases or decreases of CO_2 in the atmosphere? This question has interested scientists for some time. The CO_2 content of the atmosphere has been linked with the cyclic ice ages of the earth, because CO_2, which is transparent to short-wave radiation from the sun but absorbs long-wave infrared radiation, affects the heat budget of the earth. Thus, if the atmosphere has a high CO_2 content, the earth tends to warm as incoming short-wave radiation passes through the CO_2 barrier, while long-wave radiation being given off by the earth is absorbed by the CO_2 barrier, heating the atmosphere and being reradiated back to earth. This is the *greenhouse effect*, produced by both CO_2 and water vapor in the atmosphere.

If the CO_2 content of the atmosphere is decreased, the earth cools because of the increased heat loss from long-wave radiation directed out into space. It is estimated that a gross removal of atmospheric CO_2 by photosynthesis and its confinement in the lithosphere as peat and coal could start another ice age. The resulting change of seawater to ice caps would cause the remaining ocean water to become supersaturated with CO_2. This CO_2 would then be released to the atmosphere, causing a warming trend and a return of meltwater, undersaturated in CO_2, to the oceans. The undersaturated ocean water would gradually absorb the atmospheric CO_2, slowly starting the earth on its way to another period of icing.

The CO_2 content of the atmosphere is not the only factor affecting the earth's heat balance. Slight changes in cloudiness may alter the radiation balance and surface temperature of the earth sufficiently to upset climatic stability; a few clouds give rise to more clouds, increasing precipitation on the land as the temperature drops. Scientists are still far from reaching a complete understanding of the complex atmospheric moisture, radiation balance, and climate of the earth.

THE EFFECT OF CHANGING SOLUBILITY OF GASES IN SEAWATER

The concentrations of dissolved gases in the oceans may be determined on a weight or volume basis or as percentage saturation. We stated previously that the solubility of gases in seawater is a function of temperature, pressure, and salt content of the water. We also showed that oxygen and carbon dioxide are nonconservative in that they may be extracted from or added to the water by the biopopulation of the oceans.

If we could study a parcel of water in the oceans as it is carried along by the currents, an observed increase in CO_2 and a decrease in O_2 could indicate that respiration (the opposite of photosynthesis) or decomposition was exceeding photosynthesis. If only one of the gases were monitored, this interpretation might not be correct. Consider oceanic surface water with an initial high oxygen content at midnorthern latitudes that is being carried southward toward the equator. If this water is monitored as it proceeds southward, the oxygen content will be seen to drop gradually, even though the surface water is constantly exposed to oxygen exchange with the atmosphere. What is depleting oxygen in this case? It could, of course, be respiration, but we must also consider other factors. The surface water undergoes a temperature change during the journey; since the water warms as it moves toward the equator, the oxygen content must drop because warm water can hold less oxygen than cold water. The oxygen content of the water is also affected by the change in salinity, caused by the variation in precipitation and evaporation with latitude. Thus we must evaluate and compensate for all of these factors before we can assume that the oxygen depletion is due to respiration alone. It is possible that the percentage of oxygen saturation will remain constant while the oxygen content on a weight or volume basis decreases.

If both oxygen and carbon dioxide are monitored in the above example, both will be seen to decrease on a per-weight or -volume basis. However, the solubilities of oxygen and carbon dioxide are not affected equally by changes in water temperature and salinity. We may misinterpret this differential between the two concentrations if we assume that it is indicative of changes in respiration or photosynthesis. Although the changes in the solubility of gases in the surface waters caused by variations in temperature and salinity are subtle, we must take them into consideration before assuming the observed change is due to other factors.

DENSITY-DRIVEN VERTICAL CIRCULATION AND THE AERATION OF THE OCEANS

In the deeper portions of the oceans below the sunlit euphotic zone, the oxygen content is low and carbon dioxide content is high due to respiration. The oxygen at depth is supplied by water that once lay near the surface but was modified by surface processes to become dense, sink, and flow along slowly at depth. Once this water is removed from the surface, its oxygen supply can only be depleted, not increased. In the sea, oxygen is at a minimum at depths ranging from 200 to 800 m, since in this region considerable amounts of it are consumed by respiration and decomposition of the biota, while there is little replenishment by advection. At greater depths, the respiration demand on oxygen is decreased because there are fewer biota and the meager supply due to the advection of dense water is enough to maintain the oxygen content at a higher level; at the sea surface, photosynthesis and exchange with the atmosphere provide an elevated oxygen concentration, which leads to minimum oxygen content at middepth.

If the circulation of the deeper oxygen-bearing water is restricted so that the rate of oxygen supply is very small—especially in the depth range in which demands on the oxygen are high—oxygen consumption can cause total anoxia. Basins on the sea floor that are well isolated by surrounding ridges may suffer a reduction in their circulation; if they lie under regions of high surface productivity, the supply of organic material to depth and the accompanying decomposition may deplete their oxygen, producing an anoxic sulfide environment. Since once the oxygen is depleted the supply of organic material is greater than the decomposition rate, large amounts of organic material may accumulate on the basin floors.

The isolation of water below the sill depth of a basin need not be total to produce anoxic or near-anoxic conditions. In some cases, seasonal changes in temperature and salinity will modify the water outside a basin so that it is dense enough at sufficiently shallow depths to spill over the sill of the confining ridges and bodily displace the less dense oxygen-depleted water within the basin. Thus the basin is periodically flushed; it gradually returns to an anoxic state, however, unless the flushing occurs frequently enough to supply oxygen at its consumption rate.

The flushing of isolated basins depends on the capacity of water of greater density to displace water that is less dense. If the basin is in a region of considerable freshwater drainage (a fjord) and the water column is stably stratified, flushing occurs if temperature and salinity combine to increase the density of the water outside the sill at the fjord entrance above that of the fjord's bottom water, and at shallow enough depths to make this denser water spill over the sill into the fjord. If the basin is in a region of high evaporation and moderate temperatures, the density of the surface water will increase and the surface water will sink, displacing the bottom water and forcing it over the sill (as in the Mediterranean Sea). In the latter case, flushing is more likely and the bottom waters seldom become anoxic. The effect of salinity and temperature on sea-water density is very important in the control of vertical circulation and thus of the oceanic environment.

WATER-TYPE MIXTURES AND THEIR EFFECT ON DENSITY

The density of water is controlled by its temperature, salinity, and pressure. If the water is unaffected by pressure or is always at sea-level pressure, salinity and temperature alone control density. We have pointed out that in the latter case we can denote the density by σ_t and construct a salinity versus temperature diagram with a curve of constant σ_t.

Assume that surface waters from two different geographical areas with different values of temperature and salinity are mixed with each other. Further assume that the two waters have the same value of σ_t. Their complete mixing therefore produces a third water type with salinity and temperature values between those of the original two types. If the mixture is composed of one part of each of the original waters, its salinity and temperature values will be the average of those of the two original types. A straight line drawn between the point on the *T-S* diagram representing one water type and the point representing the second type with the same σ_t describes the possible salinities and temperatures of the resulting mixture. Thus a point midway between the two original types describes a mixture of equal parts, while a point one-fourth of the way from one of the original types describes a mixture obtained from three-fourths of the type closest to the point and one-fourth of the type farthest from the point.

This straight-line relationship for conservative mixtures always produces a water type that does not lie on the same σ_t curve as the parent waters do (Fig. 9.1). The mixture is also always denser than either parent water type and must sink. The formation and sinking of the denser water mixture is called *caballing*; it is believed to be a significant mechanism in causing vertical circulation (overturn) at high northern latitudes, where changes in the vertical density structure are slight.

When dynamic forces in the oceans bring together water types of different salinity, temperature, and σ_t values, a *T-S* diagram will show the resulting mixture as a straight line between the two parent water types. However, unlike the mixture of water types with the same σ_t value, this latter mixture is denser than one parent but less dense than the other.

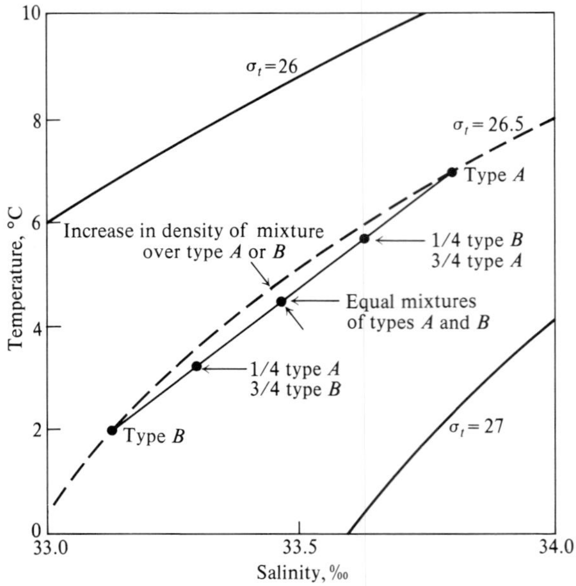

Fig. 9.1 The mixing of two waters having the same density but dissimilar $S‰$ and T°C values will produce a mixed water of $S‰$ and T°C values intermediate to the parent waters and with a higher density.

Because the density of water changes with temperature, pressure, and salinity, and since both temperature and salinity tend to decrease at the greater depths prevailing in the open ocean, a source of usable energy may be available in the sea. If a parcel of water at depth could be moved upward, conserving its lower salt content but expanding with the pressure decrease and equilibrating its temperature (but not its salt content) to that of the surrounding water at lesser depth, it would have the same temperature but a lower salinity than the surrounding water. Its density would be lower than its surroundings, and it would continue to rise.

Assume that we can construct a vertical pipe in the sea that chemically isolates the water on the inside from that on the outside but allows adequate thermal exchange between the two. If the water is started on its upward journey in the pipe by a pump, the flow will perpetuate itself by drawing heat from the surrounding water. This has been called the *salt-fountain effect* and can be demonstrated with care. If the flow is reversed, the warm saline surface water will start down the pipe, where it is cooled by heat exchange through the pipe wall and becomes denser than the outside water because of its higher salinity. The water in the pipe must keep sinking, drawing in additional quantities of warm saline surface water. Vertical motion sustained in this way may one day aid in the lifting of mineral-laden sediment from the sea floor to surface processing units, or aid in supplying nutrients to the photic zone.

WATER AS HEAT SOURCE AND HEAT SUMP

The abundance of water on the earth's surface and its specific heat also affect the oceanic environment. If water had a low specific heat, as does land, large seasonal changes in sea-surface temperatures would produce anomalous seasonal winds and major changes in water salinity, as well as extremes in weather and climatic patterns over land. Not only can water absorb large quantities of heat without changing its temperature drastically, but as a fluid it can also transfer this heat over a considerable length of the water column. Heat stored in the oceans during the summer cycle of heating is released to the atmosphere during the winter cycle, helping to temper the influence of the cold land surface on the atmosphere. The temperature differences between air, water, and land and between various types of land, plus the variation in these differences throughout the year, are partly responsible for keeping the atmosphere in motion, preventing it from stagnating, and enabling it to transfer both water and heat over the earth.

We can observe significant differences in climate and temperature between regions of the same latitude depending on whether the air masses above them have been influenced by a marine or continental environment. Neither low- nor high-latitude oceanic areas show a great change in surface temperature over the year; seasonal changes in midoceanic surface temperatures at midlatitudes average about 8°C, whereas there are 50°C changes in midcontinental areas at the same latitudes.

OVERTURN PROCESSES IN NATURAL WATERS

When the surface water temperature of isolated bodies of water undergoes large seasonal changes because of the influence of the surrounding land masses and seasonal radiation changes, the vertical density structure may be significantly affected. As we shall show, the vertical circulation of lakes and inland or adjacent seas at mid- to moderately high latitudes is driven by density changes resulting from this seasonal cycling.

A lake in the midlatitudes is subject to surface temperatures ranging from approximately 0° to 20°C. (Recall that the density of freshwater changes throughout this range, reaching its maximum at about 4°C.) If we study the vertical thermal and density structures of such a lake during the latter part of the summer cycle, we shall find warm, less dense water overlying cooler, more dense water. In a lake of 40 m depth, the bottom temperature in the summer may be typically about 10°C, while the surface temperature is about 20°C (Fig. 9.2). When the fall cooling period starts, the surface water is cooled below 20°C and sinks, because of its increase in density, to the depth at which its temperature is equal to that of the late-summer thermal distribution. As cooling continues, the surface water sinks to ever-increasing depths, until at 10°C it penetrates to the bottom of the lake, having gradually displaced water in

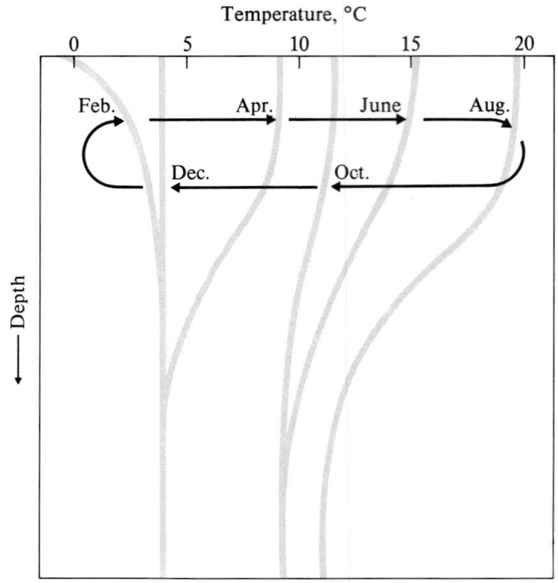

Fig. 9.2 The vertical thermal structure in a lake varies seasonally in response to thermally driven convective overturn during the cooling cycle and stability imposed by surface heating during the warming cycle.

the column to the surface to be cooled and sink in turn. This sinking and replacement of surface water constitute an *overturn*, which allows water in contact with the air to carry its dissolved gases to depth and distribute the liberated nutrients throughout the water column. Overturn is a free convection process and is thermally induced. If the lake is cooled no further, surface temperature will increase in the spring and the lake will immediately become stably stratified, with less dense water once more at its surface. Under these conditions, one free thermally driven convective overturn occurs annually during the cooling cycle.

However, if the lake is subjected to additional surface cooling, as indicated by the stipulated temperature range of 0° to 20°C, overturn continues as a free convective process until the temperature of the entire lake has been lowered to 4°C, the maximum density level of freshwater. Further cooling below 4°C makes the surface less dense than the underlying 4°C water and overturn ceases. At this point the lake becomes stably stratified. The surface water may now cool to the ice point, 0°C, while the deeper water remains at 4°C. Under severe winter conditions, ice may form at the surface and the temperature of the water underneath may drop below 4°C for some depth. This temperature drop is the result of the conduction of heat upward from the warmer water through the ice to the atmosphere.

At the end of the winter period, the ice melts and surface water warms slightly to temperatures above 0°C but below 4°C. Since the surface water is denser than the cooler water directly below it, convective overturn takes place again until the lake once more becomes isothermal at 4°C. The convective overturn need not occur all the way to the bottom of the lake if the deep water has remained at 4°C throughout the winter. However, when the lake becomes isothermal at 4°C at the beginning of the warming trend, as well as during the cooling period, the water column is of neutral stability (constant density over depth); thus any external disturbing force, such as wind, can mix the lake thoroughly to great depth.

Further warming produces stable stratification in the lake as the surface water is warmed above 4°C. We should stress that if heating or cooling causes the density of the surface water to increase with time, thermally driven *overturn occurs*. If heating or cooling causes the density of the surface water to decrease with time, *overturn ceases* and the water column becomes stable.

In the foregoing discussion, we stated that 10°C is an acceptable summertime bottom water temperature for a lake of 40 m depth at midlatitude. In discussions of overturn in a lake, we must specify not only its latitude and the temperature range to which it is exposed, but also its depth, because depth can be important in determining the bottom temperature and extent of overturn. For example, consider a very deep lake in which the volume of water is great compared to the surface area through which heat may be exchanged with the atmosphere. Winter cooling may cause convective overturn to reach only middepth, since the deep water is likely to be always at its maximum density, 4°C, and not subject to displacement unless there is severe wind mixing. If the surface waters drop below 4°C during the winter, then upon spring warming the free convective overturn will again penetrate only to middepth and hence affect only the water near the surface that has cooled below 4°C.

In this deep lake diffusion cannot extend surface heating to great depths during the summer. In the shallow lake, sufficient heat can work downward by diffusion to warm the water to 10°C, but the water at the bottom of the deep lake is almost always at 4°C.

In very shallow lakes, the effects of summer heating may extend to the bottom; the temperature of the bottom water will then be well above 4°C. In winter, if the extraction of heat from depth is sufficient to depress the bottom temperature below 4°C, two free convective overturns from top to bottom are assured.

To gain some idea of the possibilities in lake overturn, consider what happens in deep lakes and in shallow lakes when the cooling period is intense but short compared to the heating period. Try to determine the amount of heat that must be extracted from or added to a deep or a shallow lake through a constant surface area if the lake is to overturn and ice is to form at the surface. Why do shallow lakes ice over more rapidly than deep lakes? The mass of water that must be cooled or heated varies considerably in the two cases. We have described overturn in lakes first because only one variable, temperature, significantly affects the density of freshwater. The importance of overturn in aerating and redistributing nutrients in the water is obvious and applies to the oceans as well. But two variables, temperature and salinity, affect the density of seawater; both must be considered in studying oceanic convective overturn.

Figure 7.3 indicates the manner in which the temperature of maximum density and the initial freezing point of seawater decrease as salinity increases. The two curves intersect at a salinity of 24.7‰. At salinities higher than 24.7‰, the water never reaches its maximum density as it cools, since it reaches the initial freezing point first; thus ice is formed and the salinity of the remaining water increases slightly to lower the initial freezing point. Further cooling produces more ice and a more saline residual water.

We will divide our discussion of the overturn of seawater as caused by climatic conditions that affect the surface density into two parts. First, let us consider the overturn problem when the salinity is below 24.7‰, as in a shallow sea area on the fringe of land masses which is subject to extremes in air temperature and diluted by land drainage. The Baltic Sea, shallow inlets in Alaska, and regions near the Kamchatka Peninsula are examples. During the cooling part of the cycle, thermally driven overturn occurs in the water column just as it does in a lake, since with sufficient cooling the water passes through the maximum density point before ice is formed. In the short period between the passing of the maximum density point and the formation of ice, the water column is stably stratified and free thermally driven convective overturn ceases. When ice forms, the salinity of the water under the ice increases as the temperature slowly drops. Only if its salinity increases sufficiently will the water become denser than the underlying water and cause a chemically driven overturn to continue during ice formation. When the warming cycle commences, the meltwater returns to its original salinity by wind mixing and its temperature approaches that of maximum density, allowing thermally driven overturn to occur again. Thus, as in a freshwater lake, two periods of convective overturn can occur in dilute regions of the ocean.

In the second case, when the salinity is greater than 24.7‰, the cooling cycle produces thermally driven free convective overturn which continues until ice forms. At that point, overturn continues but is driven by the density increase due to increasing salinity under the ice and the gradual drop in temperature. When the warming trend starts, the water column immediately stratifies stably and free convective thermally driven overturn cannot occur. Thus only one period of extended overturn occurs in which cold saline water is sent to the bottom, since the density maximum is never exceeded. The wintertime formation of sea ice under salinity conditions of more than 24.7‰ and the resulting overturn combine to produce much of the dense water found in the deeper portions of the oceans. The Weddell Sea in Antarctica is usually cited as the classic example of the formation of dense bottom water under winter conditions. The seasonally formed dense water from this area flows virtually unimpeded down the slope of the shallow shelf to fill the deeper portions of the southern Atlantic, aerating the deep ocean and displacing water which must eventually fill the void left at shallower depths.

Conditions forming sea ice produce dense bottom water all around the fringes of Antarctica, but in the Pacific Ocean, dense bottom water is not detectable as a discrete water type. The deep basin in the southern Pacific Ocean is so large compared to that along the Antarctic continent that the quantity of dense water produced is insufficient to permit identification of the water type. In the Atlantic Ocean, the ratio of production area to basin area is large and dense water maintains its identity despite mixing processes.

Convective overturn brings deep water to the surface, where heat must be given off to continue the overturn and cooling of the water column. Thus a water system releases a large quantity of heat to the atmosphere during the seasonal cooling cycle and hence acts as a heat source. During the warmer cycle, water bodies becomes stably stratified and do not readily increase their temperatures. Compared with land, they remain cool and act as a heat sink. Thus large bodies of water are effective agents in tempering the annual temperature fluctuation of adjacent land masses and their overlying air masses.

Not all the consequences of the physical and chemical properties of seawater can be readily covered here, nor is it appropriate to do so. However, it is hoped that the reader has begun to discover from the few examples cited that the processes occurring in the natural oceanic state are governed by rules. We can usually explain some part of the governing forces and rules of nature if we know the properties of the substances in question.

STUDY QUESTIONS

1 What is likely to happen on a night when the water vapor content and CO_2 content of the atmosphere are high compared to a night when the air is dry and little CO_2 is present?

2 Why and how can the gas content of water change if the gas always remains at the 100% saturation level?

3 Why can deeper seawater that is semi-isolated approach anoxic conditions if the overlying surface waters are highly productive?

4 How does the thin-walled pipe in the salt-fountain problem promote the vertical movement of water?

5 Why is thermally driven density overturn important to a lake?

6 What general comments can you make about a lake's ability to support biological life throughout its entire depth range, given that is located (a) at low elevation in the tropics or (b) at high temperate latitudes?

7 How does the thermally driven seasonal overturn sequence of an arm of the sea compare to the overturn system of a lake if the salinity of the seawater is (a) less or (b) greater than 24.7‰?

8 When does chemically driven density overturn occur in seawater that is cooled below its initial freezing point? How important is this overturn compared to thermally driven overturn at low temperatures?

9 The thermal structure of a freshwater lake is shown below. On what day will the water at a depth of 50 m be reoxygenated by the cooling surface water? Assume that there is thermally driven convective overturn and no wind mixing.

Thermal structure of lake in September

Depth, m	*Temperature, °C*
0	18
10	12
15	9
20	7
30	6
50	5
80	4

Surface temperatures

Date	*Temperature, °C*
Sept. 1	18.
Oct. 5	12
Nov. 10	8
Dec. 3	6
Jan. 6	4.5

RECOMMENDED READING

(See also texts cited in the reference sections of Chapters 6, 7, and 8.)

Coker, R. E., *This Great and Wide Sea*, University of North Carolina Press, Chapel Hill, N.C., 1954.

Cowen, R. C., *Frontiers of the Sea*, Doubleday and Co., New York, 1960.

Von Arx, W. S., *Introduction to Physical Oceanography*, Addison-Wesley, Reading, Mass., 1962.

CHAPTER 10

THE HEAT BUDGET

INTRODUCTION

We have already pointed out that solar energy is absorbed by water, thus decreasing the available sunlight with depth. This absorption causes the water's temperature to rise. The sun also plays a very important role in controlling evaporation and precipitation; it furnishes enough energy to sustain both the motions of the earth's atmosphere and a large portion of the ocean current systems.

In this chapter, we shall consider the supply of thermal energy to the earth as well as its redistribution. We can state to a first approximation that, neglecting annual changes in the sea's and earth's temperature, the total heat content of the oceans and the earth is not changing with time. Therefore the heat added by the sun must be lost by some mechanism. This balance between the incoming and outgoing heat is referred to as the *heat budget* of the earth and oceans.

The heat budget of the oceans must include all major sources of supply and removal of heat. In this sense it is a balance of heat fluxes—units of heat per units of time. We may evaluate the flux of heat for the entire ocean system or for ocean regions; in the former case, variations in heat flux over the ocean are not considered. A flux of heat is usually given as a quantity of heat per unit time for each unit surface area of the ocean through which it passes. The products of fluxes given in this manner and the area of the earth to which they apply then determine the total flux. The mechanisms by which heat may be transferred to create a flux are *conduction*, *convection*, and *radiation*. All three are at work in the heat budget of the earth and oceans.

THE RADIATION BUDGET OF THE EARTH

Our most obvious source of heat is solar radiation, which reaches the earth in nearly parallel rays of energy. The intensity of solar radiation flux at a unit surface on the earth, oriented perpendicular to the sun's rays when the earth is at its mean distance from the sun and when atmospheric effects do not diminish the radiation, is referred to as the *solar constant*. The value of this flux is currently considered to be 2.0 g-cal cm^{-2} min^{-1}. If we assume that there is no absorption of solar energy by the atmosphere and that the earth's radius is small in comparison with the mean distance between the earth and sun, then the total solar energy delivered to the earth in one day is the product of the solar constant times one day times the cross-sectional area of the earth. The spherical shape of the earth and its rotation on its axis require that this total daily solar energy supply be distributed over the entire surface of the earth. Since the surface area of the spherical earth is four times the cross-

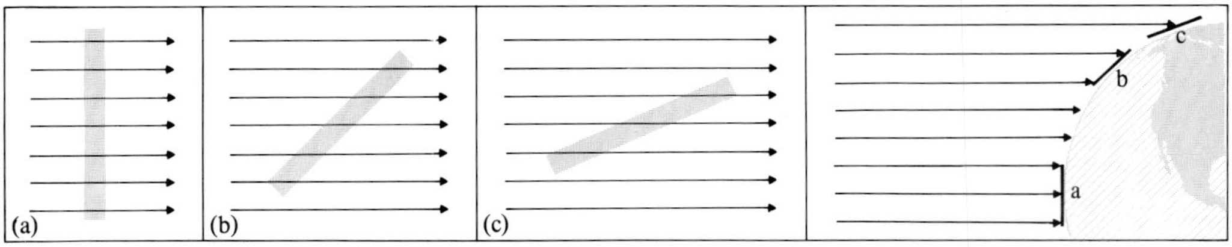

Fig. 10.1 Unit surface areas on the earth receive less radiant energy as they become more oblique to the energy source because of the curvature of the earth's surface. (After Earth Science Curriculum Project, *Investigating the Earth*, Houghton Mifflin, Boston, 1967.)

sectional area ($4\pi R^2$ compared to πR^2), the average solar flux level for the total earth is one-fourth the solar constant, or 0.5 g-cal cm^{-2} min^{-1}.

The spherical shape of the earth causes not only a fourfold reduction in the average energy flux from the sun, but also an unequal distribution of the flux over the earth's surface. The tilt of the earth's axis allows the sun's rays to strike the earth at right angles only at points between latitudes $23\frac{1}{2}^\circ$N and $23\frac{1}{2}^\circ$S (Fig. 10.1). At high latitudes the earth's surface is always oblique to the sun's parallel rays, whereas at low latitudes (between $23\frac{1}{2}^\circ$N and $23\frac{1}{2}^\circ$S) a point on the earth's surface is perpendicular to the rays at some time during the annual cycle. Thus high latitudes always receive a smaller radiation flux than low latitudes.

The earth, continuously exposed to solar energy, would get extremely hot if there were no heat loss. Since there is no significant rise in temperature, there must be a heat loss comparable to the heat gain. If we consider that the average heat flux from the sun over the earth's surface (0.5 g-cal cm^{-2} min^{-1}) is 100% of the available energy to the earth, we may assign percentages to the various mechanisms by which the earth loses heat to space. Of the 100% incident solar radiation available to the earth and its atmosphere, 7% is scattered from the stratosphere back to space, 4% is reflected from the earth's surface to space, 24% is reflected from the earth's cloud cover, 5.5% is reradiated from the earth's surface as blackbody* longwave radiation to space, 56.5% is reradiated as longwave radiation from the lower atmosphere, which contains moisture, to space, and 3% is longwave radiation from the stratosphere to space.

As we can see, the incident radiation is strongly affected by the atmospheric envelope of the earth. The average amount of incident direct shortwave radiation that reaches the surface of the earth after passing through the atmosphere is 26.5%. Of this, 4% is reflected back to space and 22.5% is

* A blackbody is an ideal body or surface which absorbs all of the incident energy falling on it from outside. Since a good absorber is a good emitter, a blackbody, the best absorber, will be the best emitter (or complete radiator).

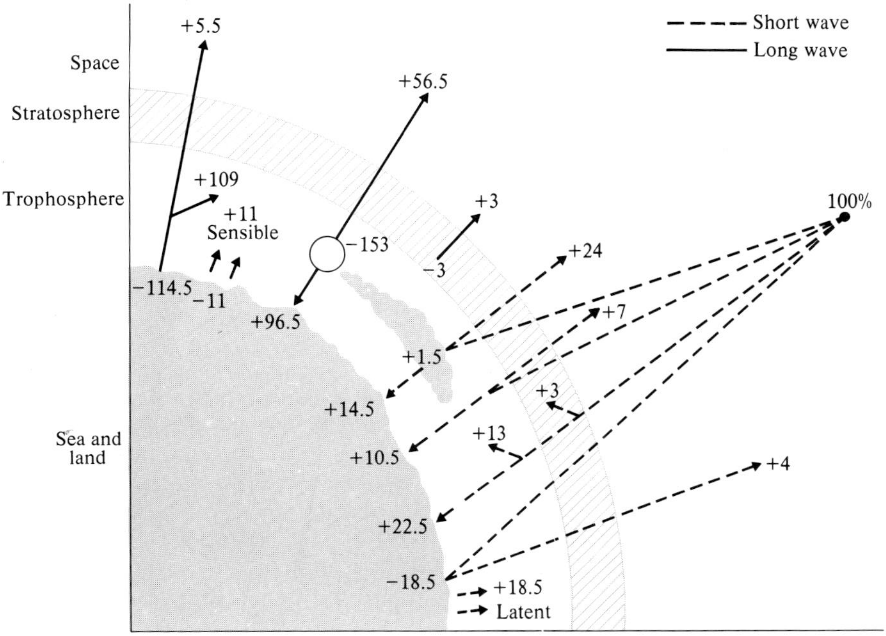

Fig. 10.2 The annual radiant-energy budget of the earth and atmosphere, showing how 100% of one unit of available solar energy is absorbed, redistributed, and reradiated back to space. (After G. Neumann and W. J. Pierson, *Principles of Physical Oceanography*, © 1966. By permission of Prentice-Hall, Inc., Englewood Cliffs, N.J.)

absorbed at the earth's surface. In addition, the earth's surface receives 10.5% of the solar radiation as diffuse scattered skylight and 14.5% as scattered light energy from clouds. All of this energy is shortwave. The earth then receives 96.5% of the longwave energy from the troposphere and reradiates 114.5% of it to the atmosphere and space. Additional losses of 11% and 18.5% as sensible and latent heat from the earth to the atmosphere account for all radiant energy fluxes (Fig. 10.2).

THE DISTRIBUTION OF SOLAR ENERGY ON THE EARTH

In addition to the effect of the spherical rotating earth on radiation flux levels, we must also consider the variation in time of exposure to radiation. The changing declination of the sun exposes the polar areas to long periods of alternate illumination and darkness in the annual cycle. During a polar summer, the long period of light can produce a significant energy flux over the period of one day. Thus time of exposure as well as the portion of the curved earth's surface that is perpendicular to the incident solar radiation determines the amount of solar radiation the earth receives.

We stated previously that 47.5% of the total incident shortwave solar radiation is passed through the atmosphere and absorbed by the earth; of this

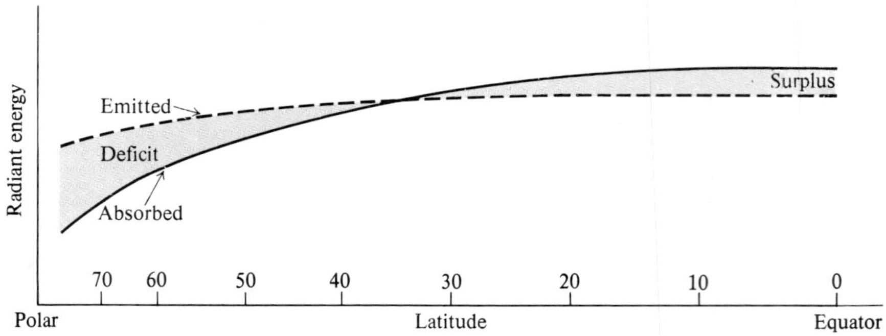

Fig. 10.3 Incoming radiant energy per unit area absorbed by the earth at the lower latitudes exceeds the loss of energy per unit area due to back radiation to space. The reverse is true at higher latitudes, so that between the poles and the equator no net gain or loss of heat is evident. Deficit equals surplus. (After Earth Science Curriculum Project, *Investigating the Earth*, Houghton Mifflin, Boston, 1967.)

amount 22.5% is direct solar radiation and 25% is scattered radiation from the sky and clouds. Thus during the equinox a unit of the earth's surface on the equator perpendicular to the incident shortwave solar flux receives a flux that is about 47.5% of the solar constant (0.95 g-cal cm^{-2} min^{-1}). A point at the poles receives 47.5% $\times$ cos 90° $\times$ the solar constant, or zero, and a point at 45° N or S receives about 47.5% $\times$ cos 45° $\times$ the solar constant. These values are theoretical; although they reflect the distribution of the maximum flux levels, they are not necessarily indicative of radiation levels at actual earth positions.

Sometimes measurements are made in a mountain basin surrounded by snow-covered peaks. Solar radiation can be reflected off the peaks and focused at the basin center; thus the recorded flux is the sum of the direct and reflected solar radiation. The same phenomenon can occur when a bank of cumulus clouds reflects solar radiation from its edge to a position on the ground that is free from cloud cover. High values of radiation flux measured under these circumstances require that another area have a lower than normal value.

Solar radiation varies seasonally with latitude because of the tilt of the earth to the ecliptic plane, the translation of the earth about the sun, and the annual change in the mean distance between the earth and sun. The tilt of the earth and the earth's passage around the sun cause the earth-sun line of centers to move from $23\frac{1}{2}$° N to $23\frac{1}{2}$° S, crossing the equator twice in the annual cycle. Thus over the annual cycle the low-latitude belts receive more radiant energy than the high-latitude belts (Fig. 10.3). The nearness of the earth to the sun during the Southern Hemisphere summer causes that hemisphere to receive slightly more radiation than the Northern Hemisphere.

If there were no redistribution of heat over the earth, the average temperature of the equatorial regions would increase and that of the polar

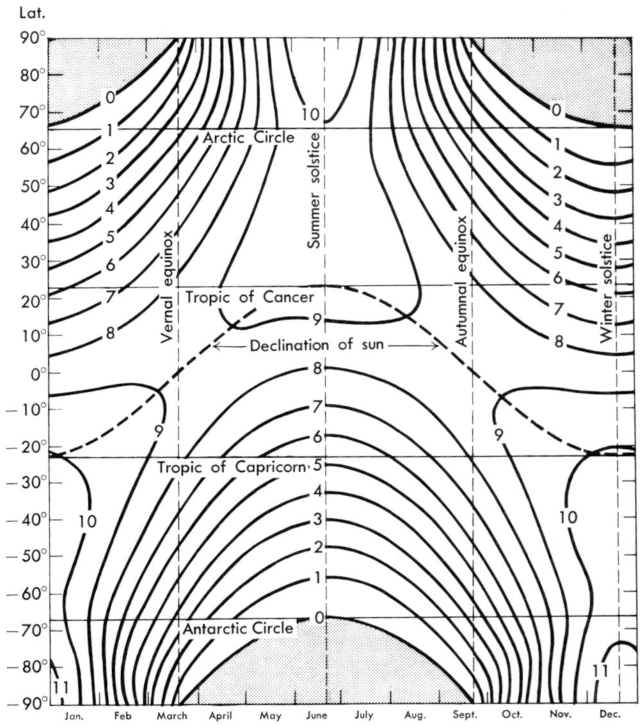

Fig. 10.4 Solar radiation on a horizontal surface outside the earth's shadow in hundreds of g cal cm^{-2} day^{-1}. The north-south seasonal migration of the sun and its effect of changing the radiation from a surface horizontal to the sun's rays are evident. (From W. S. von Arx, *Introduction to Physical Oceanography*, Addison-Wesley, Reading, Mass., 1962.)

regions decrease until the back radiations of the two areas equaled their incoming radiations. Back radiation is like blackbody radiation: the flux is proportional to the absolute temperature of the body raised to the fourth power. Since longwave back radiation radiates outward as an energy flux and is everywhere perpendicular to the earth's surface, it does not spread over increasing surface areas as does incoming radiation. Thus back radiation changes little with latitude; it varies only according to the earth's surface temperature, emissivity, and atmospheric conditions (Fig. 10.4). Since back radiation neither increases sufficiently at low latitudes nor decreases sufficiently at high latitudes to compensate for the unequal solar heating of the earth, there must be another mechanism that is acting to redistribute heat from an area of surplus heat to an area of deficit.

The transfer and redistribution of heat with latitude are accomplished by atmospheric motion and the currents of the oceans. The north-south motions in the atmosphere and in the surface waters of the oceans redistribute the excess radiative heat gained at the equator, modifying surface temperatures

Table 10.1 Heat budget of the earth and heat transported from lower to higher latitudes

Latitude, degrees, Northern Hemisphere	Annual average heat received, g-cal cm^{-2} min^{-1}	Annual average heat lost, g-cal cm^{-2} min^{-1}	Average northward transport of heat across latitude, g-cal $min^{-1} \times 10^{-16}$
0	0.40	0.35	0
10	0.42	0.36	1.98
20	0.41	0.37	3.60
30	0.37	0.35	4.56
40	0.30	0.30	4.80
50	0.25	0.29	4.32
60	0.20	0.27	3.18
70	0.14	0.26	1.80
80	0.11	0.24	.36
90	0.10	0.23	0

Calculated from data in J. London, *A Study of the Atmospheric Heat Balance*, Final Report, New York University Department of Meterology and Oceanography, New York, 1957.

over the entire earth. Table 10.1 shows the heat budget of the earth averaged over land, water, and time, as well as the heat transported across parallels of latitude in the redistribution process. Heat is carried by the atmosphere, the most important mechanism for heat transport, as both potential energy and sensible and latent heat. Only at very low latitudes is the sea as important or more important than the atmosphere in redistributing heat over the earth.

THE HEAT BUDGET OF THE OCEANS

If we consider the heat budget of the oceans on a worldwide scale, so that latitudinal differences are averaged out, it can be expressed as

$$Q_s - Q_b - Q_h - Q_e = 0, \qquad (10.1)$$

where Q_s is the shortwave radiation absorbed by the oceans from the sun and sky (an incoming heat flux and therefore considered positive); Q_b is the effective net longwave back radiation flux from the ocean's surface to the atmosphere or space (a loss of oceanic heat and considered negative); Q_h is the heat flux from the oceans which raises the temperature of the atmospheric, sensible heat; and Q_e is the heat given up by the oceans during the process of evaporation and is related to the latent heat of vaporization.

If an isolated region of the oceans is considered rather than the total oceans of the world, two additional terms, Q_v and Q_θ, appear in the heat budget equation. The first, Q_v, represents *advective heat*, the heat carried into a region by currents from another part of the ocean; it is considered positive. The second term, Q_θ, represents *sensible heat*, the heat which changes the temperature of the water; it is considered negative if it increases the water temperature. The Q_v and Q_θ terms are not included in the total ocean budget because Q_v represents a redistribution of heat by ocean currents and Q_θ represents a local increase or decrease in temperature which must be accompanied respectively by a decrease or increase elsewhere. The net value of each of the two terms is therefore zero when averaged over the oceans.

There are other, minor ways of supplying heat to the oceans which we shall neglect. They include condensation of water vapor on the sea surface, conduction of heat through the sea floor from the interior of the earth, transformation of kinetic energy into heat in viscous dissipation of energy, and heating due to chemical and biological processes in the sea. We have already mentioned the conduction of heat through the sea floor, which is about 1.3×10^{-6} g-cal cm^{-2} sec^{-1}, or 7.8×10^{-5} g-cal cm^{-2} min^{-1}. The kinetic energy imparted to the sea by the wind acting on the sea surface when converted to heat is about 10^{-4} times the heat supplied by solar radiation. The dissipation of tidal energy into heat can be as great as 2×10^{-3} g-cal cm^{-2} min^{-1} in the shallow border regions and in tortuous channels of the sea, but its values are very low in the open ocean. Heating due to exothermic chemical processes is similarly insignificant on the oceanic scale.

It is important to know the apportionment of energy flux in the oceans among Q_s, Q_b, Q_h, and Q_e to understand the relative impact of these sources of heat transfer on the sea. Averaged with time over the world's oceans, Q_s, the incoming solar radiation taken up by the sea, is about 0.221 g-cal cm^{-2} min^{-1}. Since the incident solar radiation at any particular location on the earth's surface is affected by many factors (such as the altitude of the sun, turbidity of the atmosphere, cloud cover, and reflectivity of the earth's surface), it is important that we consider Q_s as only that part of the solar radiation actually absorbed in the water.

Averaged with time over the oceans, Q_b, the net heat lost by longwave radiation, is about 0.090 g-cal cm^{-2} min^{-1}. This radiation is absorbed in the atmosphere and in part lost to space. In general, a good absorber of heat is also a good radiator. Water, for example, radiates efficiently at a rate proportional to the fourth power of its absolute temperature whether it is in the sea or the atmosphere. Moisture in the atmosphere absorbs radiation and then reradiates it as longwave radiation, some of which strikes the earth. Therefore, Q_b represents the difference between incoming and outgoing longwave radiation and is the net *effective back radiation.* On the world average, more longwave radiation is lost from the sea than is returned from the atmosphere. Therefore, Q_b averaged over the oceans is a negative flux of heat. Effective back radiation decreases as the temperature of the sea surface decreases and atmospheric humidity and cloud cover increase.

Averaged over time and the oceans' area, Q_h, the conduction of heat to the atmosphere, is estimated at a flux rate of 0.013 g-cal cm^{-2} min^{-1}. This heat transfer occurs when cool air moves over warm water; it is typical of the manner in which the lower atmosphere is heated from the air-earth surface. When cool air is over warm water it is warmed by heat transfer at the boundary, producing an unstable air column which promotes the transfer of heat into the atmosphere by free convection. Convection requires that air aloft be supplied to the surface and in turn modified. This vertical circulation continuously resupplies air, usually of lower moisture content and lower temperature, to the sea surface, where it picks up moisture and heat to be carried upward. The free convection increases the transfer of both heat and water (evaporation). The transfer of water to the atmosphere during evaporation carries latent heat of vaporization, which is supplied by the sea and released in the air when the water vapor condenses. The heat lost by the oceans in the evaporation process is Q_e. It is estimated that, averaged over the earth, 90% of the sea's heat is transferred to the atmosphere as latent heat and 10% by direct conduction. Thus Q_e, on the average, should be about ten times greater than Q_h. The average value for Q_e is about 0.118 g-cal cm^{-2} min^{-1}.

The sum of the average values of Q_e, Q_h and Q_b should equal Q_s and yield a zero balance in the heat budget of the world's oceans. If the sum did not equal Q_s to balance the budget, the oceans would have to show a trend of gradual warming or cooling, depending on whether Q_s were larger or smaller respectively than the sum of the negative heat-loss terms. The processes that govern the heat losses from the oceans are not equivalent, since evaporation extracts more heat than does any other process. This heat is stored in the atmosphere to be released when condensation occurs and raises the air temperature; it may be transferred from an area of evaporation to an area of precipitation by moving air masses.

The oceans absorb incident radiation easily because they have a small *albedo*, which is a reflective index equal to the ratio of the reflected radiation divided by the incident radiation. When the altitude angle of the sun lies between 40° and 90°, the albedo of water is about 0.05, indicating that about 5% of the incident radiation is lost due to reflection from the sea surface. Table 10.2 gives typical albedos for other earth surface materials.

It has been estimated that the annual supply of incoming shortwave radiation exceeds the effective back radiation for all latitudes of the earth. At low latitudes (0° to 10° N) the excess of incoming radiation is about 0.170

Table 10.2 Albedos of earth surface materials

Material	*Albedo*
Flat grass-covered ground	0.25–0.33
Flat rock	0.12–0.15
Sand	0.18
Dry black loam soil	0.14
Ice or snow	0.18–0.70
Earth and atmosphere combined	0.35

From O. G. Sutton, *Micrometeorology*, McGraw-Hill, New York, 1953.

g-cal cm^{-2} min^{-1}, while at 60° to 70° N it is about 0.040 g-cal cm^{-2} min^{-1}. The latitudinal distribution of this surplus shows the effect of higher incident radiation at low latitudes, where heat must be given up to the atmosphere and transported poleward. The atmosphere is very important in the redistribution of the surplus, since it obtains most of its heat from the oceans by evaporation. Thus water transport to the atmosphere acts to control not only the surface temperature of the oceans but also the surface salinity.

THE SURFACE THERMAL STRUCTURE OF THE OCEANS

Because of its uniquely high thermal capacity (specific heat), water can absorb or give off large quantities of heat with little change in temperature. Therefore, the oceans act as a climatic buffer, exercising thermostatic control on land temperatures. Figure 10.5 shows the average oceanic surface temperatures during summer and winter. Note that the highest surface temperatures exist at low latitudes, while lower surface temperatures are found at high latitudes. Equally important is the annual range of temperature, which is very small at low latitudes and reaches a maximum at midlatitudes (Fig. 10.6). Thus the ocean is most effective as a seasonal temperature buffer at midlatitudes. The maritime climate of the Pacific Northwest, with its prevailing winds from the Pacific Ocean, is quite different from the subarctic continental climate of eastern Canada at the same latitude. The Pacific Ocean has an annual surface temperature range of about 8° C at this latitude,

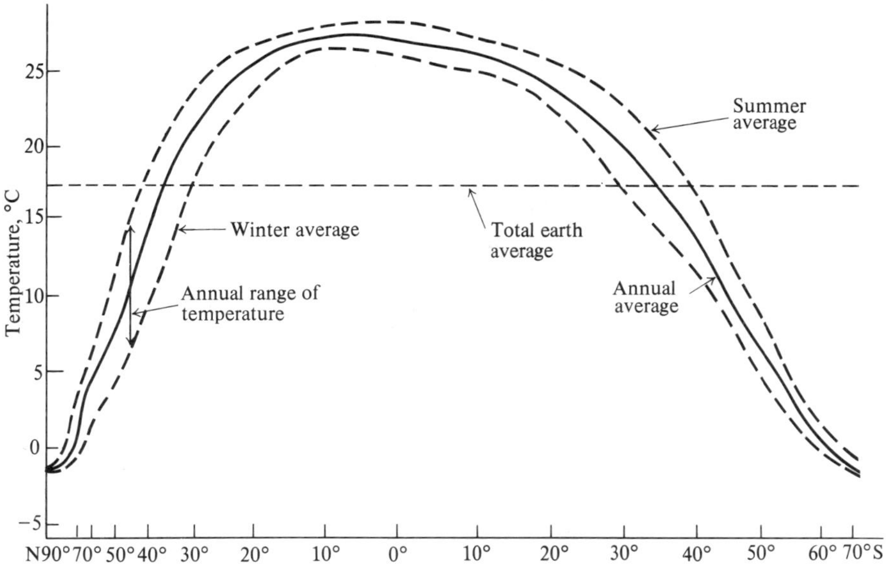

Fig. 10.5 The average oceanic surface temperatures, summer, winter, and annual, and their variation with latitude. (After J. W. Hedgpeth, ed., *Treatise on Marine Ecology and Paleoecology*, Ecology, vol. 1. The Geological Society of America, Boulder, Colorado. 1957.)

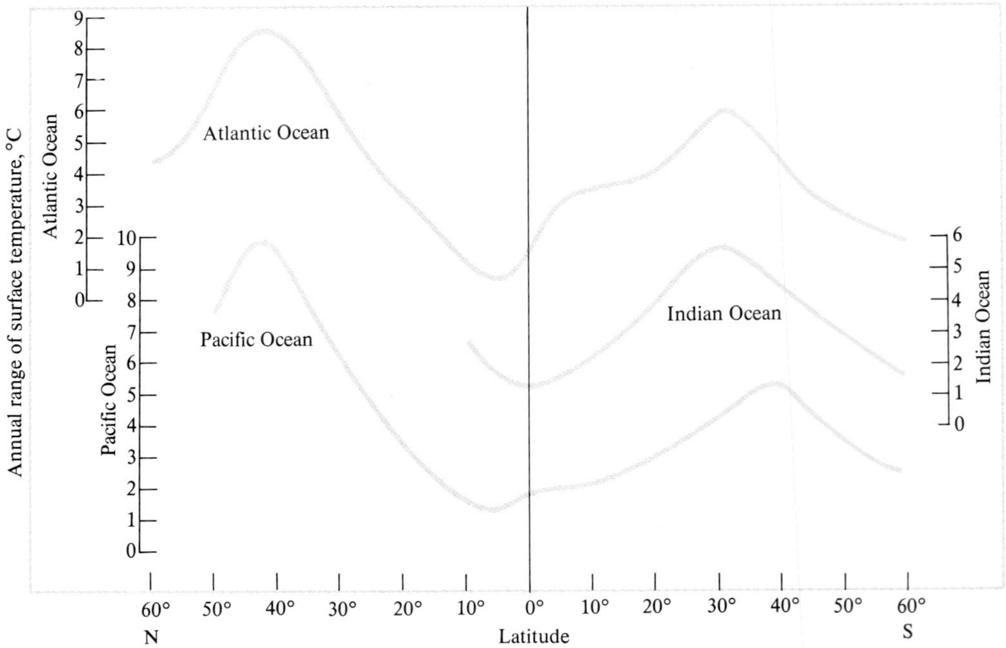

Fig. 10.6 The maximum annual range of sea-surface temperature occurs at mid-latitudes and is larger in the Northern Hemisphere than in the Southern Hemisphere due to a greater abundance of land masses in the Northern Hemisphere. (From H. U. Sverdrup, M. W. Johnson, and R. H. Fleming, *The Oceans*, © 1942. By permission of Prentice-Hall, Inc., Englewood Cliffs, N.J.)

but the temperature range on the continental land mass can be greater than 50° C.

We have noted that land occupies more of the earth's area in the Northern Hemisphere than in the Southern Hemisphere and that is has lower heat-storage and heat capacities than water. Thus heating and cooling at midlatitudes produce larger temperature changes in the Northern Hemisphere than in the Southern Hemisphere. The greatest range in sea-surface temperature occurs at midlatitudes on the lee side of large land masses, where prevailing winds cause the air to move away from the continental influence of the land and out over the sea. As the air moves farther out across the oceans, it is modified by the oceanic environment and becomes maritime in nature with high moisture content and a temperature dictated by the sea-surface temperature.

Klaus Wyrtki has compiled data on the distribution of North Pacific sea-surface temperatures and their annual range (Figs. 10.7 and 10.8). Note the uniformity of the average annual temperature structure, east to west across the ocean.

The north-south change in temperature is controlled by the distribution of incident solar radiation, as noted in reference to Fig. 10.3. The annual range in sea-surface temperature (Fig. 10.5) is greater in isolated regions than

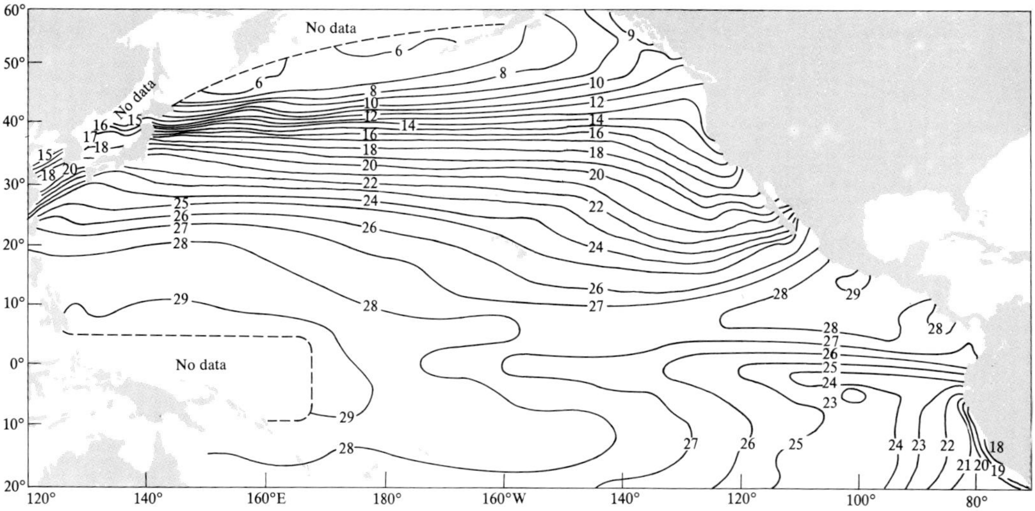

Fig. 10.7 The average annual sea-surface temperature in the Pacific Ocean in °C, in the years 1947–1960. [From K. Wyrtki, "The Annual and Semiannual Variation of Sea Surface Temperature in the North Pacific Ocean," *Limnology and Oceanography*, 10 (3), 307–313 (July 1965).

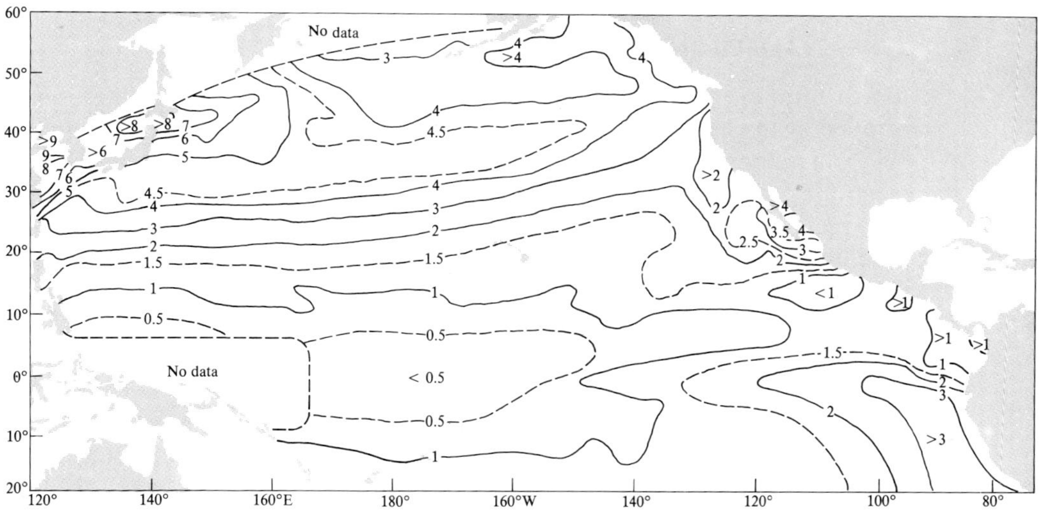

Fig. 10.8 The amplitude of the average annual variation of sea-surface temperature in °C. The average annual range in temperature is about twice the values shown. [From K. Wyrtki, "The Annual and Semiannual Variation of Sea Surface Temperature in the North Pacific Ocean," *Limnology and Oceanography*, **10** (3), 307–313 (July 1965).

in midocean. For example, the region near Japan having a large annual temperature fluctuation demonstrates the influence of continental modified air on the sea as the air moves eastward in the prevailing westerly wind belt.

According to Wyrtki's analysis, the amplitude of the annual temperature variation for the North Pacific Ocean at 43°N Lat is not as large as that stipulated by Fig. 10.6. However, one must recall that the amplitude is one half of the range.

At high latitudes, freezing of sea ice also affects the heat budget. Sea ice (fresh) has a higher albedo than water and, when snow-covered, reflects a large percentage of incoming radiation. Once ice is formed there is a sudden decrease in absorbed radiation but not in the effective longwave back radiation; therefore, the surface ice temperature drops rapidly due to the heat lost by unbalanced back radiation and the thickness of the ice increases until its insulation properties curtail its rate of production. This temperature depression would continue until the back radiation is reduced to the point that it nearly balances the absorbed incoming radiation. The cooling of the air above the region of ice formation also aids in increasing the lateral extent of the ice as the cool air spreads outward from the initial point of freezing.

The reverse process occurs during melting. When the air or land at the boundaries of the sea ice is warmed, a small amount of ice-free surface forms. The albedo is then immediately lowered over the exposed water and incoming radiation is again more readily absorbed, warming the water and melting the ice at its edges.

THE EVALUATION OF HEAT-BUDGET TERMS

Each term of the heat-budget equation represents a flux of heat that is controlled by the heat-transfer processes—conduction, convection, and radiation. It is therefore possible either to measure the flux directly or to evaluate the transfer process and calculate the resulting flux. Instruments can measure Q_s, the incident shortwave solar radiation absorbed by the sea. A hemispherical radiometer can sense the available shortwave radiation at the sea surface and then can be turned over, so to speak, to measure the shortwave radiation reflected from the sea surface and scattered upward in the water column. The difference in the incoming and outgoing fluxes of shortwave solar radiation is Q_s, the absorbed solar energy flux.

We can use a radiometer sensitive to longwave radiation to evaluate longwave radiation emitted by the sea surface and that received by the sea from above. The difference is Q_b. In some instances, total radiation (both longwave and shortwave) may be measured by the same sensor. In such cases, the instrument usually records the difference or net total radiation flux—the sum of Q_s and Q_b. Net total radiation is usually positive during the day, when Q_s is larger than Q_b, and negative at night, when Q_s is less than Q_b. We can also estimate Q_b by applying the blackbody radiation law if we know the temperature and emissivity of the sea surface.

The values of Q_s and Q_b both decrease as cloud cover and atmospheric moisture increase; Q_b changes little in response to the day-night cycle of the earth, whereas Q_s changes from zero at night when the sun goes down and reaches its maximum at local apparent noon if atmospheric conditions are constant. In the temperate latitudes, surface temperatures drop rapidly at night during the winter if the sky is clear and the humidity is low. Under these conditions, Q_b is high and the temperature drops rapidly in response to this outward flux of radiant energy. If there is a cloud cover, the outward radiation is absorbed by the clouds and reradiated back toward earth, decreasing the effective back radiation and preventing so rapid a temperature drop. This reradiation is an example of the aforementioned greenhouse effect.

The conductive and convective transfer of heat to the atmosphere, Q_h, is proportional to the specific heat of air times the coefficient of eddy conductivity times the temperature gradient of the air immediately above the sea. The eddy conductivity coefficient is difficult to determine by direct measurement. Turbulent convective and conductive heat transfer is suppressed if the water is colder than the air and no wind is blowing, and greatly enhanced if the water is warmer than the air, producing an unstable air column, and the wind is blowing. At sea, we can measure the wind speed and thermal gradient to estimate the coefficient of thermal conductivity and calculate Q_h.

When the movement of air transfers heat upward from a warmer sea surface, it also transfers water upward into the atmosphere. Thus Q_e, since it is determined by the rate of evaporation, is part of this transfer process. Even though the transfer of the latent heat of vaporization requires a gradient of vapor pressure, the thermal instability produced in the air column when cold air lies over warm water speeds up the transfer of water vapor. Like Q_h, Q_e can be computed empirically as a function of the latent heat of vaporization times an eddy conductivity coefficient, a constant over the atmospheric pressure and the vapor pressure gradient. We have the same problem in directly evaluating the eddy coefficient for Q_e as for Q_h.

In practice, we can often overcome the difficulty of determining Q_h and Q_e by finding the *Bowen ratio*, Q_h/Q_e. We used this ratio earlier when we stated that, on the world average, 90% of the oceans' heat is transferred from the sea by evaporation and 10% by convection and conduction as sensible heat, a Bowen ratio of 0.1. The Bowen ratio is usually small with little seasonal change at low latitudes; at midlatitudes, however, it has a value of about 0.5 in the winter and very small values in the summer. Negative values of the Bowen ratio are found in certain areas in the summer, indicating that heat is conducted from the atmosphere to the sea. Since there would have to be a greater rate of water vapor condensation than evaporation from the sea surface, it is physically unrealistic for a negative Q_e to produce a negative Bowen ratio, while a negative Q_h is realistic.

The Bowen ratio, R, can be determined from P, the atmospheric pressure, in millimeters of mercury; T_a, the air temperature, and T_w, the water temperature, in degrees Celsius; e_w, the saturation vapor pressure of the water at T_w in millimeters of mercury; and e_a, the vapor pressure of the air at T_a. The Bowen ratio is related to these measurement by

$$R = \frac{Q_h}{Q_e} = 0.49\,\frac{\mathrm{P}}{760} \times \frac{T_w - T_a}{e_w - e_a}. \qquad (10.2)$$

We can evaluate Q_e on land by directly determining the evaporation rate (E), but this technique is extremely difficult at sea since it is almost impossible to eliminate the ship's effect and measure evaporation from pans floating on the sea surface. For a particular location at sea, the heat-budget equation can be used to indirectly determine the local evaporation rate. The heat budget for a particular region at sea is

$$Q_s - Q_b - Q_e - Q_h + Q_v + Q_\theta = 0,$$

or (10.3)

$$Q_e + Q_h = Q_s - Q_b + Q_v + Q_\theta.$$

We can rewrite this equation as

$$Q_e\left(1 + \frac{Q_h}{Q_e}\right) = Q_s - Q_b + Q_v + Q_\theta. \quad (10.4)$$

If we now use the relationships $R = Q_h/Q_e$ and $Q_e = E \times L \times \rho$, where E is the evaporation rate in centimeters per unit time, L is the latent heat of vaporization in gram-calories per unit mass of water, and ρ is the density of the water, then

$$Q_e\left(1 + \frac{Q_h}{Q_e}\right) = EL\rho\,(1 + R) \quad (10.5)$$

and

$$E(\text{cm per time}) = \frac{Q_s - Q_b + Q_v + Q_\theta}{L\rho(1 + R)}. \quad (10.6)$$

The time base used in the heat fluxes in the numerator of the latter equation determines the time base of E. In most cases, the density ρ does not appear and we assume that the latent heat of vaporization is given in gram-calories per unit volume of water or that $\rho = 1$. It is retained here to make the units of the equation consistent.

This equation enables us to calculate E from the measurable variables. Once E is known, Q_e is easily calculated, leaving only one unknown flux term (Q_h) in the heat budget. However, we can now evaluate Q_h either by assuming that it must equal the flux rate necessary to balance the heat-budget equation or from $Q_h = RQ_e$.

EVAPORATION, CONDENSATION, AND FOG

Throughout this chapter, we have related heat flux to processes governing heat transfer in and over the oceans. The distributions of sea-surface temperatures, sea-surface salinities, and the Bowen ratio over the world's oceans are all interrelated and must be reflected in climate and weather patterns. *Fog* is among the weather phenomena related to these transfer processes. It is produced whenever the air temperature drops below its dew point and water vapor condenses in the layer of air just above the earth's surface. The production of fog liberates the latent heat of vaporization and warms the air.

Two basic types of fog form at sea, the most prevalent and important of which is *advective fog*. Advective fog forms when warm, moisture-laden air passes over cold water and is cooled from below to its dew point. Areas of the oceans that form boundaries between warm and cold water masses so that warm, moist air moves over cold water are famous for these fogs, e.g., the Grand Banks region, where the cold waters of the Labrador current meet the warm waters of the Gulf Stream system. Regions in which there is strong upwelling of deep ocean water, which is cold relative to the surrounding surface water, are also noted for advective fog (e.g., the northern coast of California in July). The warm, moist air passing over the cold water forms a stable air column next to the sea surface and convective turbulence is retarded. This type of fog gives heat to the sea from the atmosphere by conduction and by condensation of the water vapor on the sea surface. (The former process is more important.)

The second type of fog, *sea smoke*, is not widespread; it usually lasts only for short periods in nearshore regions. It is caused by cold, dry air moving over water that is considerably warmer than the air. The air column in this case is made unstable by the temperature structure. The air next to the water warms, gains moisture from the sea, and rises rapidly into the colder air above the water, where its temperature drops and condensation occurs. The observer sees streamers of fog rising from the water's surface. Heat is rapidly removed from the sea by both the high evaporation rate and convection of sensible heat. Although both advective fog and sea smoke are fogs, they are formed under different physical conditions and produce fluxes of heat that move in opposite dirrections and are of considerable different magnitudes.

One usually thinks that evaporation occurs when the water is warm. In the oceans, however, evaporation rates are high during the winter, when the surface water is coldest. During this time of year the air, which moves seaward from the land masses, is on the average colder than the sea over which it moves. Warming from below enables the air to hold more moisture and sets up free convection. Both factors promote evaporation. In the summer, the surface air moving off the land is usually warmer than the sea surface. The air becomes stably stratified as it moves out over the sea, where it is cooled; thus its capacity to hold water vapor decreases. Both stable stratification and reduction of the saturation value of the air tend to retard evaporation. Annual changes in the Bowen ratio result from the variations in the transfer of heat by evaporation or conduction. The magnitude of R and whether it is positive or negative is determined by the relative importance of Q_h and Q_e and the direction in which the heat is transferred (sea to air and air to sea). These values in turn may be used to qualitatively describe the general climatic conditions.

LOCALIZED THERMAL POLLUTION

Industries and power plants located along the ocean shores or on rivers that empty into the sea often use large quantities of water as a coolant. In the process of cooling, the heat is transferred to the coolant water, raising its temperature. This water is then returned to the coolant source at a point located the same distance from the intake, raising the temperature of the water at that location. The amount of heat added to the world's oceans by this process is insignificant and does not noticeably increase the average temperature of the oceans; however, the local elevation of temperature in a restricted body of water can be great enough to harm biological life. Thus, in its effect on the biota, this rejected heat is as much of a pollutant as any chemical material.

Nuclear power plants that use water as a coolant can cause extreme thermal pollution. It is estimated that the Vermont Yankee nuclear power plant in southeastern Vermont will require about 60% of the maximun flow of the Connecticut River water at the plant site as coolant and discharge heat at a rate sufficient to raise the water temperature 15° to 20°C.* Not all manmade devices thermally pollute water so dramatically, however; in fact, thermal pollution can be created without the direct discharge of heat into the system by the damming of rivers for power, water storage, or flood control. Dams retain water in reservoirs that have considerably larger surface areas exposed to the atmosphere than does the original river. This increased surface area allows more heat exchange between the atmosphere and the water. Since Q_s, Q_e, Q_b, and Q_h are all rates of heat transfer per unit surface area per unit time, the ratio of exposed surface area to volume of water is quite important in determining how these heat fluxes control the thermal pattern of the dammed-up water. The reduction of water turbulence in these reservoirs also allows a temperature-controlled density stratification not found in moving streams or rivers.

STUDY QUESTIONS

1 The heat budget equation $Q_s - Q_b - Q_h - Q_e = 0$ describes the balance of the inflow and outflow of heat of the total oceans. Determine the effective back radiation of the total oceans when $Q_s = 0.25$ g-cal cm^{-2} min^{-1}, $R = 0.1$, and $Q_h = 0.013$ g-cal cm^{-2} min^{-1}.

2 Describe the climatic environment indicated by a Bowen ratio R of (1) small positive value, (2) small negative value, (3) large positive value, and (4) large negative value.

3 Calculate the evaporation rate in centimeters per year given the following average heat budget terms in gram-calories per centimeter^{-2} per year: $R = 0.07$, $Q_s = 0.27$, $Q_b = 0.071$, $Q_v = +.020$, $Q_\theta = -.015$.

---

* B. Nelson, "Thermal Pollution: Senator Muskie Tells the AEC to Cool It," *Science*, **158** (1), 755–756 (Nov. 10, 1967).

4 Discuss the role of heat transfer during conditions that generate advective fog and sea smoke. What values would you expect R to have in these two cases?

5 What factors combine to decrease the effective back radiation from the earth?

6 Explain clearly why incident solar radiation reduces in intensity when it impinges on a rotating spherical surface and why the available incident radiation further decreases with increasing latitude.

7 Name several geographic regions that are famous for advective fog. Describe the circumstances that cause the fog to form in each case.

8 Why can we neglect Q_v and Q_θ when considering the total oceans? Give a separate reason for each term.

RECOMMENDED READING

Hill, M. N., ed., *The Sea*, vol. 1, Interscience Publishers, New York, 1965.

Neumann, G., and W. J. Pierson, Jr., *Principles of Physical Oceanography*, Prentice-Hall, Inc., Englewood Cliffs, N.J., 1966.

Pickard, G. L., *Descriptive Physical Oceanography*, Pergamon Press, New York, 1963.

Sverdrup, H. U., M. W. Johnson, and R. H. Fleming, *The Oceans*, Prentice-Hall, Inc., Englewood Cliffs, N.J., 1942.

CHAPTER 11

THE EARTH'S ATMOSPHERE AND ITS CIRCULATION

INTRODUCTION

The "ocean" of atmosphere confined to the earth by the gravitational field is much like an ocean of water. Both are mobile fluids and therefore can be set in motion by processes that alter their density distributions. However, unlike the sea, which is inefficiently heated from its upper surface, the atmosphere is heated from below. Surface heating suppresses thermally driven vertical circulation in the ocean, while cooling enhances it; in the atmosphere, heating at the earth's surface enhances thermally driven vertical convective circulation, while cooling suppresses it.

Driven by the unequal heating of the earth's surface, rising at the equator where the earth's surface temperature is high, and sinking at the poles where the temperature is low, air acts as a major mechanism for redistributing heat. Air cannot circulate uniformly in a single convection cell, even though it is held captive by the earth's gravity and driven by the gross temperature gradient between pole and equator. The earth's rotation and the resulting Coriolis effect prevent the atmosphere from moving in a direct north-south path from low to high latitudes, and force the air to circulate in several discrete closed cells; in addition, the distribution of land and water produces irregularities in the surface temperature distribution within zones of constant latitudes.

The circulation of the atmosphere and its seasonal changes significantly affect the earth and man. In the oceanic environment, moving air produces waves and drives the major surface currents. Once the surface waters are set in motion, they create distributions of sea-surface temperatures which in turn alter the winds that drive the currents. Although atmospheric motions are complex, if mean motion is considered, the atmosphere appears to behave in a regular fashion. The mean motions will concern us here.

THE ATMOSPHERE

The gaseous atmospheric envelope surrounding the earth extends upward from the earth's surface through the *ionosphere*, where it gradually loses its identity as a gas and becomes indistinguishable from the void called space. Although the earth's surface is usually considered the lower boundary of the atmosphere, atmospheric gases are dissolved in the oceans and even penetrate to some small depth the porous covering of the earth.

Like sea water, air is a mobile fluid. Forces arising from changes in the atmospheric density distribution cause the air to move from place to place, mixing it thoroughly and giving it a nearly uniform composition throughout the world. Thus air is also similar to seawater in that its composition is constant;

Table 11.1 Composition of dry air

Gas	Percent of mixture by molecules (volume)
Nitrogen	78.08
Oxygen	20.95
Argon	0.93
Carbon dioxide	0.03
Neon	1.8×10^{-3}
Helium	5×10^{-4}
Krypton	1×10^{-4}
Hydrogen	$< 1 \times 10^{-5}$
Xenon	1×10^{-5}
Other trace gases	
	~100%

From O. G. Sutton, *Micrometeorology*, McGraw-Hill, New York, 1953.

it is not a single gas but a mixture of gases. Unless affected by local additions or extractions of gases, its composition is approximately as listed in Table 11.1.

Note that we have listed the components of *dry* air. Water vapor, an atmospheric gas found in variable amounts, is not listed. Although it makes up only about 3% by volume of the atmosphere (even in very humid conditions), it is extremely important in determining the behavior of the lower atmosphere and climate. The transport of water vapor by the moving air completes the hydrologic cycle described earlier. Without this gas, there would be no cycle and the land masses would receive no precipitation. As water vapor condenses into drops and again returns to its gaseous state, it acts to add or subtract the latent heat of vaporization to or from the atmosphere. It also absorbs radiation and reradiates it back to earth and space to maintain the heat budget. Water vapor and the other gases listed make up 100% of the natural atmospheric gas mixture. Thus the gases listed in Table 11.1 constitute slightly less than 100% of moist air.

The density of air, which is determined by the number of gas molecules per unit volume of air, decreases with increasing temperature, water vapor content, and altitude above the earth. The latter effect is due to the decrease with altitude in pressure exerted on the air by the thinning overlying gaseous envelope. The atmospheric pressure of static air at the earth's surface is described by the hydrostatic equation, $P = \rho_a gz$, where ρ_a is the air density, a function of z; g is the acceleration of gravity, also a function of z; and z is the height of the overlying air column. Thus pressure decreases with elevation as ρ_a, g, and z, the remaining height of the air column above the measurement point, decrease.

Approximately 6 km above the earth, the air pressure decreases to one-half its sea-level value, while at about 11 km the pressure is one-fourth the

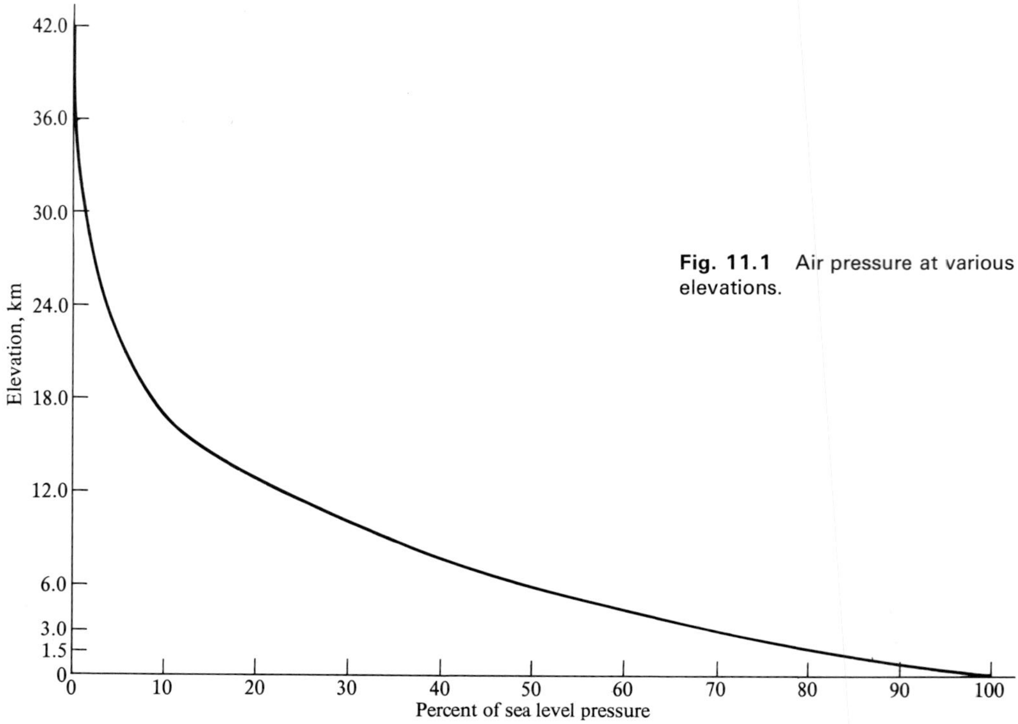

Fig. 11.1 Air pressure at various elevations.

sea-level value (Fig. 11.1). At sea level, normal or standard atmospheric pressure is 1015 millibars (mb), where 1 bar of pressure equals 1×10^6 dyn/cm². Since this pressure is equivalent to that produced by a column of mercury 760 mm high, it is often referred to as "760 mm of mercury." In the English system, 14.7 psi is the standard atmospheric pressure.

Knowing the temperature distribution with elevation, we can relate the pressure and density of the air to elevation more precisely if we assume that air behaves like a perfect gas. Simple barometric altimeters can be constructed with a correction for average temperature distribution so that they read elevation as a function of the pressure drop. These altimeters are used by mountain climbers and surveyors, as well as in airplanes; they are designed to allow correction of their zero point (local sea-level pressure set equal to zero elevation).

The atmosphere is divided into discrete layers. The layer closest to the earth is the *troposphere*. It is the densest layer of the atmosphere, containing approximately three-fourths of the atmospheric mass. Above the troposphere are the *stratosphere*, the *mesosphere*, the *ionosphere* (which embraces part of the mesosphere), and the *thermosphere*. The boundaries of the layers are not arbitrary, but are determined by temperature maxima and minima (Fig. 11.2). The relationship between altitude, pressure and boundaries is tabulated in Table 11.2.

The *tropopause*, the upper boundary of the troposphere, is found at an average elevation of about 13 km above the earth. (The height varies from

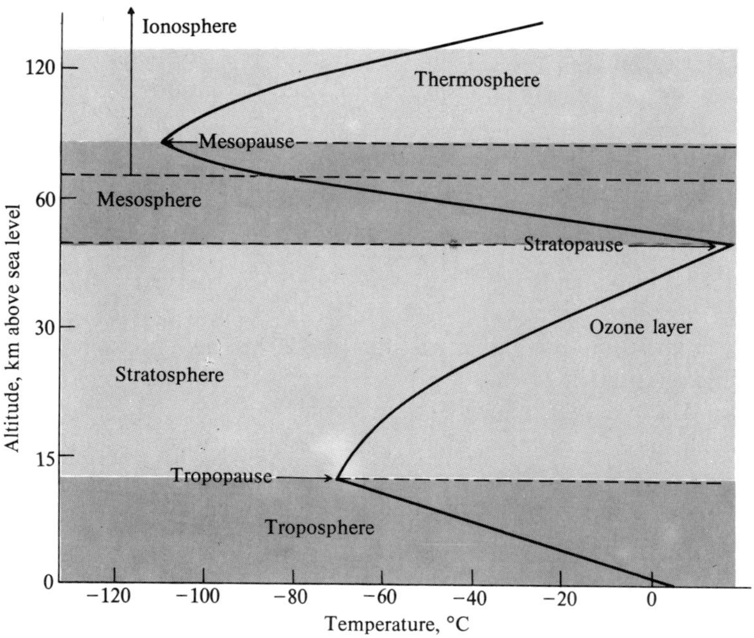

Fig. 11.2 The layers of the atmosphere. (After J. Spar, *Earth, Sea, and Air*, 2d ed., Addison-Wesley, Reading, Mass., 1965.)

Table 11.2 Average pressure, in percent of sea-level pressure, versus altitude above sea level

Altitude, *km*	*mi*		*Pressure, %* *(approximate)*
0	(0)		100
1.5	(1)		85
3	(2)		70
6	(3.5)		50
10	(6)		30
12	(8)	Average tropopause	20
16	(10)	Tropical tropopause	10
30	(20)		1
50	(30)	Stratopause	0.1
80	(50)	Mesopause	0.001
160	(100)		0.0000005

From J. Spar, *Earth, Sea, and Air*, Addison-Wesley, Reading, Mass., 1962.

about 7 km at the poles to 15 km at the equator.) The atmosphere, like the oceans and the earth, is acted upon by many forces. Atmospheric "tides" are superimposed on the geoid shape of the tropopause; they are produced by the gravitational attraction of the celestial tide-raising bodies and by the daily and annual cycles of heating and cooling of the atmospheric envelope. The annual back-and-forth migration of the earth-sun line of centers from one hemisphere to the other causes an annual shift in the elevation of the tropopause due to density changes in the air column and the gravitational attraction between the atmosphere and the sun. This shift in turn produces a seasonal transfer of mass from low to higher latitudes and vice versa.

In Chapter 2, we noted that there is a gradual decrease in the earth's rotation rate which is obvious from the gradual lengthening of the sidereal day. There are also small annual fluctuations in the rate of decrease which are caused by the seasonal shifting of the atmospheric mass. When an atmospheric mass is transferred from a low to a higher latitude, it carries its low-latitude angular momentum to a higher latitude where the angular momentum is smaller. At lower latitudes, air has a greater eastward velocity (associated with the rotation of the earth) than it does at higher latitudes. Thus the displacement of an air mass from low to higher latitudes also causes a transfer of eastward acting momentum. This momentum is imparted to the earth by friction in the air. Since the displaced air mass at higher latitudes rotates eastward a little faster than it does at low latitudes because it is closer to the earth's axis of rotation, the earth speeds up. One can demonstrate this phenomenon by gradually drawing his arms in from an outstretched position to close to his body while spinning on a piano stool. When the air mass is transferred back to a low latitude, it travels from a region of low momentum to one of higher momentum; therefore its friction with the earth causes the earth's rotation rate to slow slightly.

The troposphere is the most dynamic region of the atmosphere because it interacts directly with the earth's surface to receive and give up moisture, gases, heat, and momentum. The oceans and the atmosphere are mirror images of each other with respect to the input of, and dynamic response to, energy; i.e., the troposphere is the counterpart of oceanic surface layers above the main thermocline (the zone in which temperature changes rapidly with depth), while the less dynamic stratosphere is the counterpart of the deeper regions of the sea.

THE ATMOSPHERE IN MOTION

The temperature of the earth's surface and the transfer of heat and water vapor to the atmosphere are important factors controlling the density of the lower atmosphere and thus its vertical convective motions. In Chapter 3, we noted that the atmosphere contains latitudinal zones of predominantly ascending or descending air driven by Archimedean forces. We characterized these zones as regions of prevailing low or high barometric pressure and high or low precipitation respectively. However, air cannot continuously ascend in one area and descend in another unless there are horizontal flows to maintain continuity. These flows are the winds of the earth at its surface and aloft; combined with the vertical flow of air, they constitute the basic circulation pattern

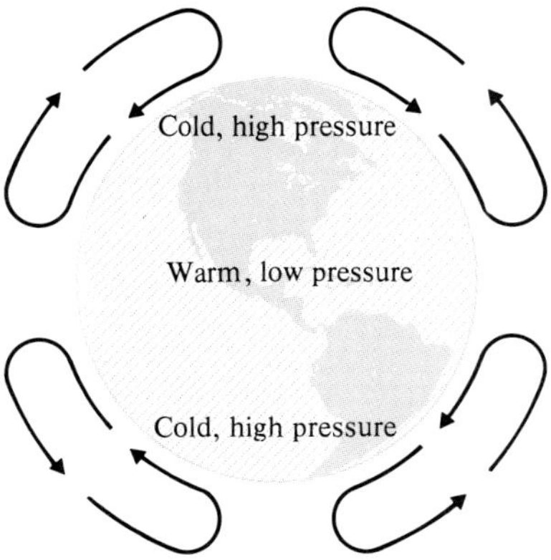

Fig. 11.3 The imbalance between incoming and outgoing radiation at each latitude means that the earth is warmer at low latitudes than at high latitudes. On a nonrotating earth, this temperature structure would cause the air to circulate in a single closed cell in each hemisphere. (After Earth Science Curriculum Project, *Investigating the Earth*, Houghton Mifflin, Boston, 1967.)

of the earth's lower atmosphere. To understand atmospheric circulation and the forces that drive it, we must first consider the processes that govern the heating of the earth and supply the energy to maintain air circulation. In Chapter 10, we pointed out that the incoming and outgoing radiation creates a surplus and deficit of heat at low and high latitudes respectively (Fig. 10.4). We also stated that the surplus and deficit do not have a marked effect, continuously increasing the earth's temperature at low latitudes and further cooling it at high latitudes, because another mechanism acts to redistribute the heat.

This mechanism is the mobile atmosphere, which transports and redistributes heat while using some of the heat energy to make the air circulate. On a nonrotating, water-covered model earth, air warmed from below at lower latitudes rises due to its lessened density and flows aloft toward the poles, where its increased density causes it to sink and flow back along the earth's surface toward the equator (Fig. 11.3). This simple circulation presumably redistributes heat from regions of surplus heat to regions of deficit. However, such a unidirectional flow of air—i.e., from pole to equator along the earth's surface and from equator to pole aloft, thus creating a single circulation cell—cannot occur on the real rotating earth.

In Chapter 2, we showed that fluids moving over the earth with low friction do not maintain a straight course, but are deflected by the Coriolis force. Thus if we rotate our water-covered model earth, we find that air moving away from the North Pole, where the Coriolis effect is great for horizontal motion, is deflected to become an easterly flow. As the deflected air moves toward lower latitudes, it warms slightly and its moisture content increases, thus decreasing the air's density. The air then rises and short-circuits the simple pole-to-equator flow of the single-cell nonrotating system.

The rotation creates a similar short-circuiting at the equatorial end of the single-cell system. The ascending warm moist air cools as it rises and is deflected as it moves aloft toward the poles; it then sinks and flows toward the

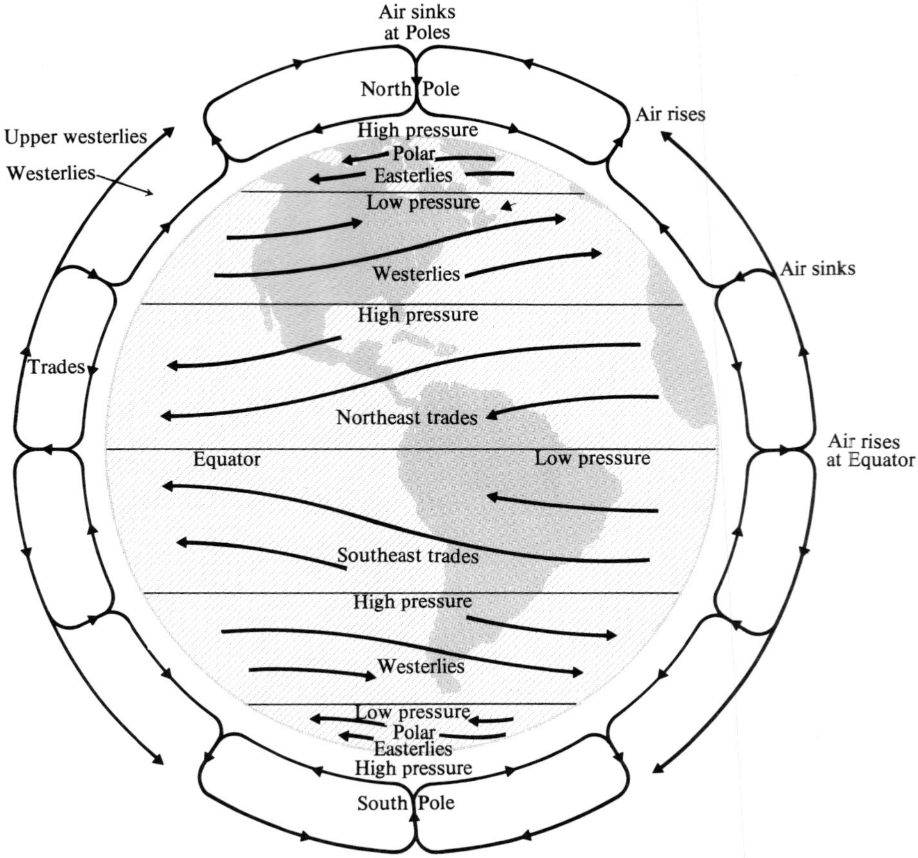

Fig. 11.4 A schematic representation of the cellular circulation of the atmosphere on a rotating earth. The driving force for this motion is the thermal imbalance between pole and equator. The division of the circulation pattern into three cells between pole and equator rather than a single cell is caused by the earth's rotation. (After Earth Science Curriculum Project, *Investigating the Earth*, Houghton Mifflin, Boston, 1967.)

equator, forming a second small cell within the large single cell. The Coriolis force causes a short-circuiting of the flow at each end of the single cell, producing a set of three motion cells with horizontal air flows that are oblique to the north-south direction (Fig. 11.4).

The uniformity of size of the three cells between the equator and poles requires that (1) the specific heat of the earth's surface materials be constant, (2) the albedo of the surface be constant, and (3) there be no changes in elevation. These conditions are not satisfied on the real earth; however, if the earth were water-covered it would have a three-cell circulation as depicted.

On the real earth, the moving air follows paths that resemble right- and left-handed coil springs stretched around the latitude belts. It is not constrained to stay in a particular closed cell, but may move from cell to cell along their common boundaries. On the model earth, air ascends at the equator and at 60°N or S and descends at the poles and at 30°N or S. The vertical motions are produced when dense air aloft sinks through less dense air below and less dense air below rises through dense air above. The convergence of horizontal air flows (surface winds blowing toward each other) occurs where air ascends, while their divergence (surface winds blowing away from each other) occurs where it descends.

Because air in regions of surface divergence is dense, dry, and cool, they have a higher than average barometric pressure, while regions of ascending warm, moist air have low barometric pressure. The thermal structure of the troposphere—decreasing temperature with elevation—also has a role in altering the vertically moving air. Because ascending air is cooled due to the adiabatic temperature drop with decreasing pressure and mixing with cooler air aloft, it is less capable of holding moisture. Water vapor condenses and water is often lost as precipitation under these conditions. Regions of ascending air are noted for cloudiness and high precipitation as well as for low pressure and unsteady surface winds. Similarly, because descending air is warmed and therefore more capable of holding moisture, condensed water is readily converted to water vapor, resulting in low precipitation and few clouds. The three-cell model of atmospheric circulation shows belts of low atmospheric pressure, clouds, and precipitation surrounding the earth at 0° and 60°N and S; belts of high atmospheric pressure, reasonably clear skies, and low precipitation are found at 90° and 30°N and S.

The deflection of the north-south flow of air along the earth's surface in the three cells of the model produces the familiar prevailing wind systems. Figure 11.4 shows the trade winds on either side of the equator, the prevailing westerlies in the next higher latitude system, and the polar easterlies. Where these horizontal flows converge or diverge, surface winds are light and variable; in these regions, the doldrums of the equator and the horse latitudes at approximately 30°N, sailing ships are likely to become becalmed. Hence throughout the days of sailing ships, these areas were feared by mariners.

The atmosphere, driven by heating from below, is sensitive to seasonal changes in the earth's surface temperature. One would expect the equatorial zone of ascending air to follow the sun in its migration between the Tropics of Cancer and Capricorn, aligning itself with the region of maximum surface heating and forming an "equator" of atmospheric circulation. This *meteorological equator* (the doldrum belt) moves northward from its mean position during northern summer and southward during northern winter. Its mean position would be coincident with the equator of the water-covered earth model. In reality, however, it is approximately 5°N and its migration limits in response to the sun are about 0° and 11°N. The northward displacement of the meteorological equator on the real earth is associated with hemispherical differences in average surface temperatures, radiation budgets, and the ratio of land to water.

THE LAND-WATER EFFECT ON ATMOSPHERIC CIRCULATION

The hypothetical water-covered earth used to illustrate the gross features of atmospheric circulation is not sensitive to changes in surface temperature caused by equal quantities of incident radiation striking earth materials of different specific heats or heat-storage capacities. We noted earlier that, unlike water, land undergoes large and rapid changes in temperature, and that its greatest area lies in the Northern Hemisphere. The distribution of land and water significantly affects the mean circulation of the atmosphere and alters the circulation pattern from that of the water-covered model.

Since for equal amounts of solar radiation land warms more readily and reaches a higher temperature than does the sea, the meteorological equator (the low-pressure doldrum belt) should be displaced north or south of its mean position, staying above regions with higher annual mean temperatures. Large land masses of low elevation north of 0° deflect the meteorological equator northward, while land masses south of 0° deflect it southward. The overall excess of land in the low to midlatitudes of the Northern Hemisphere results in permanent northward displacement of the meteorological equator.

The uniform high-pressure dry belt at 30°N and S on the water-covered model are also different from those on the real earth. Since at this latitude average oceanic surface temperatures are lower than average land temperatures, high-pressure zones extend toward lower latitudes over the sea and low-pressure zones extend toward higher latitudes over the land. This displacement tends to create isolated oceanic high-pressure cells rather than maintain a continuous high-pressure belt such as that shown on the model. At the higher latitudes, where emitted radiation exceeds incident radiation, the oceans have higher average surface temperatures than does the land at the same latitudes. Thus the oceanic areas are predominately low-pressure zones while high pressure predominates over the land masses.

The low-pressure zone around the equator is quasi-permanent because of the high annual influx of incident radiation and small annual range of surface temperatures. Similarly, the lows over the oceans and highs over the land at high latitudes are nearly permanent due to the constant low level of, and small annual variation in, radiation and surface temperatures respectively. At midlatitudes, there are large annual variations in incident radiation; thus the highs and lows of pressure are not permanent. Under summertime conditions, there tend to be lows over land and highs over water (Fig. 11.5a), but in the winter this situation may be reversed (Fig. 11.5b). The seasonal shifting of pressure systems at midlatitudes is considerable; thus we can compare only the annual mean conditions of the existing earth- and water-covered model.

The seasonal temperature distribution over the earth's surface, more than any other factor, controls atmospheric circulation and pressure distribution. In general, the following rule holds true: If an area of the earth is, on the average, warmer than other regions at the same latitude, it will have a low

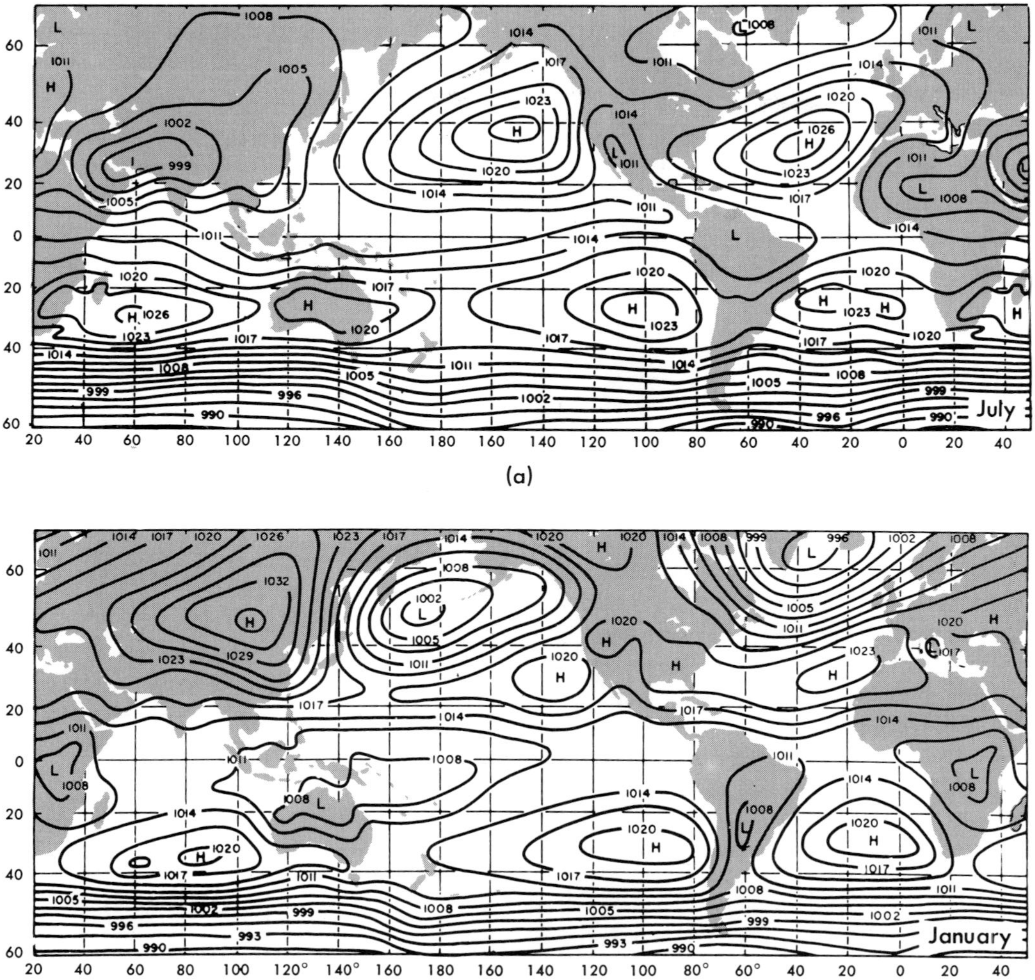

Fig. 11.5 Mean sea-level pressures expressed in millibars for (a) July and (b) January. (From W. S. von Arx, *Introduction to Physical Oceanography*, Addison-Wesley, Reading, Mass., 1962.)

prevailing barometric pressure; if it is cooler than other regions, it will have a high barometric pressure.

Considering this generalization and the atmospheric circulation pattern of the water-covered model, we can quite easily discern similarities between the annual mean wind and pressure distribution on the hypothetical and real earths. Although there is north-south warping of the atmospheric pressure belts in response to changes in surface temperatures of the land and sea surface along a constant latitude, the model's basic pattern of atmospheric pressure

Fig. 11.6 The surface wind pattern associated with the annual mean sea-level pressure distribution.

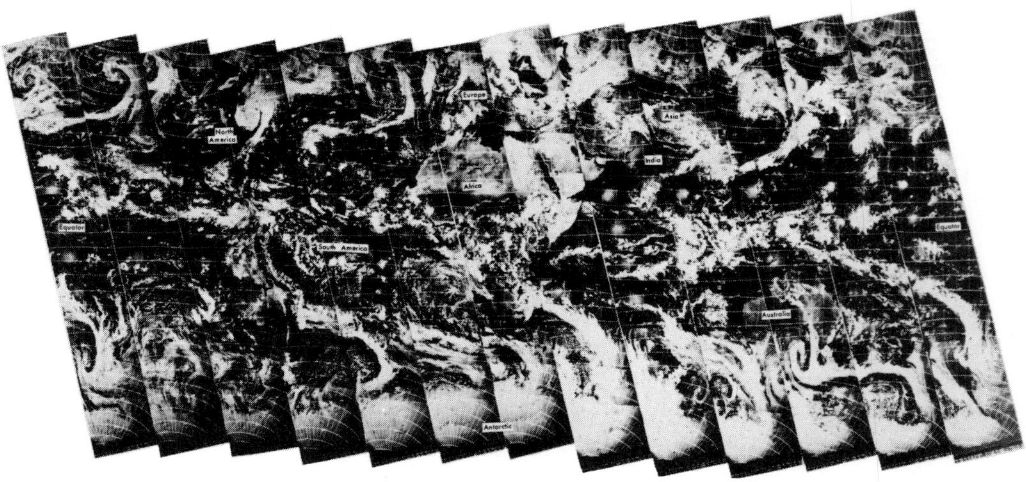

Fig. 11.7 Tiros composite photo. Courtesy of National Oceanic and Atmospheric Agency, U.S. Dept. of Commerce.

holds in general for the actual earth (Fig. 11.6). In composite pictures of the earth's cloud cover taken from orbiting satellites, the intense high-pressure areas stand out clearly as cloud-free, while the low-pressure bands are cloudy (Fig. 11.7). The pictures, taken daily, aid meteorologists in interpreting the position and migration of important storm centers. Of particular interest are cloud patterns, which form individual vortices or swirls rather than solid continuous bands of clouds, as we might expect from the water-covered model. The swirls are indicative of the next smaller scale of atmospheric motion, the circulation of air about closed cells of high and low pressure, and the location and types of fronts formed by converging air masses.

CYCLONIC AND ANTICYCLONIC FLOW

Although the average annual sea-level pressure distribution of the earth appears as alternating latitudinal bands of high and low pressure, over a short period of time the winds circulate about discrete closed pressure cells (Fig. 11.8). The cells may or may not be approximately in the "correct" latitude zone according to the annual mean distribution. The instantaneous distribution of wind and pressure is to the mean distribution as daily weather is to climate; i.e., although daily weather often seems unusual or unseasonal, the mean climate for an area remains essentially the same.

In a closed high-pressure cell system, air descends and moves outward toward the low-pressure area. As it moves it is deflected by the Coriolis force, to the right in the Northern Hemisphere and to the left in the Southern Hemisphere. Air flowing down the pressure gradient from high to low creates an *anticyclonic flow* of air about the high-pressure cell. Regardless of the direction of their rotation (clockwise in the Northern Hemisphere and counter-clockwise in the Southern Hemisphere), high-pressure cells are called *anti-*

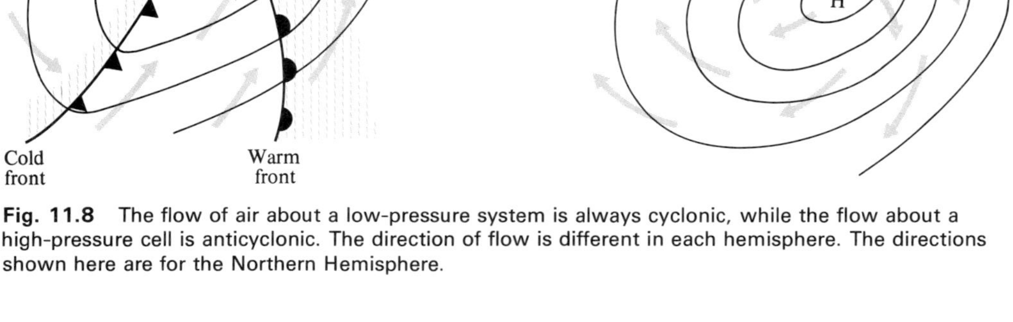

Fig. 11.8 The flow of air about a low-pressure system is always cyclonic, while the flow about a high-pressure cell is anticyclonic. The direction of flow is different in each hemisphere. The directions shown here are for the Northern Hemisphere.

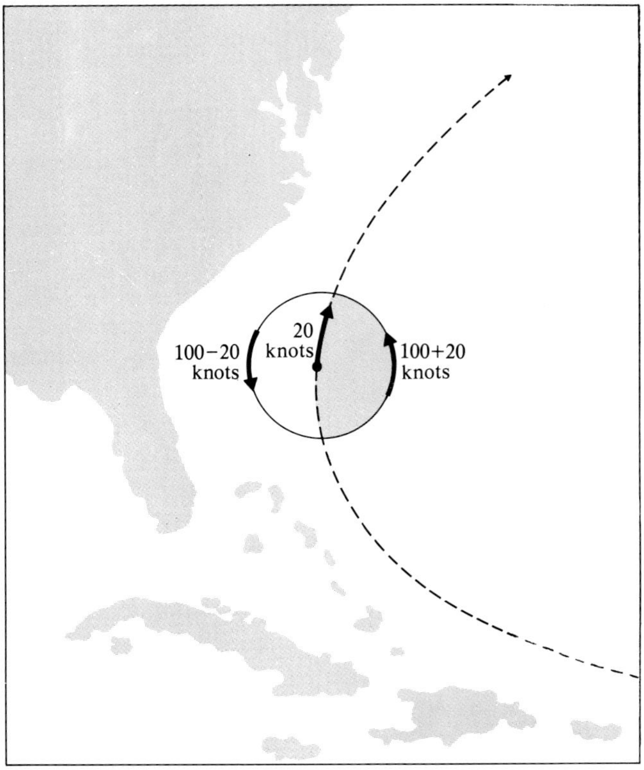

Fig. 11.9 The surface winds of a moving, intense cyclonic disturbance are not symmetrical about the storm center.

cyclones. As air flows into the low-pressure system and ascends, it causes a *cyclonic circulation* about the cell. Low-pressure cells are therefore called *cyclones* regardless of their direction of rotation (counterclockwise in the Northern Hemisphere, clockwise in the Southern Hemisphere).

The speed at which air moves from a high-pressure region to one of low pressure depends on the pressure gradient, the driving force. As we have learned, the deflection of the air to the right or left by the Coriolis force depends on latitude and wind speed, as well as on the amount of friction between the moving air and the earth's surface. If friction is great, the Coriolis deflection is reduced and the air moves more directly from a high-pressure area to a low-pressure area, cutting sharply across the *isobars* (lines of equal pressure) and flowing nearly directly down the pressure gradient. If friction is small, the air is deflected more and moves nearly parallel to the isobars and at right angles to the pressure gradient.

Since the surface winds are more likely to be acted upon by friction than are the winds aloft, the former are deflected less by the Coriolis force. Therefore, the wind direction in the Northern Hemisphere shifts slightly clockwise with increasing altitude. The meteorologist can compute the speed of the upper winds (*gradient winds*) from the *pressure gradient* (change in pressure over distance). He converts the gradient winds to surface winds by considering the effect of friction in reducing wind speed and in changing wind direction at the lower altitudes. He then corrects for the speed at which pressure-cell systems that create the winds are moving over the earth's surface.

In the case of severe cyclonic storms, such as the hurricanes that move up across the Gulf of Mexico toward the Gulf states, the sector to the right of the storm track usually carries more severe winds than does the sector to the left. If 100-knot winds are circulating about the low cell and the storm is moving at 20 knots, there are 120-knot winds in the right-hand sector and 80-knot winds in the left-hand sector (Fig. 11.9). The most dangerous area in the path of an intense cyclonic storm is the side of the storm path to which the storm would normally be deflected by the Coriolis force.

THE MONSOON EFFECT

The strength, distribution, and deflecting influence of pressure systems help control the earth's wind system. The processes that establish the pressure systems do not suffer large annual changes except at midlatitudes, where the pressure distribution is continually changing and has a definite seasonal trend. There are low-pressure areas above large land masses at midlatitude during the summer and high-pressure areas above them during the winter, since the land warms and cools relative to adjacent water bodies. This seasonal alteration of pressure and the resulting reversal of wind systems create the *monsoon effect*.

Under these conditions, a steady supply of moist maritime air can be carried landward, creating heavy precipitation in the summer when the low is over the land; in the winter, dry continental air flows seaward. Thus there is a *wet monsoon* in summer and a *dry monsoon* in winter. This reversal of winds and pressure systems is evident if we examine the monthly mean air pressure and winds for summer and winter periods (Fig. 11.10). Asia, Australia, and eastern Africa are particularly affected by seasonal monsoons, which have a great influence on local agriculture and travel conditions.

The monsoon phenomenon may also operate over a much shorter time period and a smaller area. Daytime heating and nighttime cooling of land adjacent to a body of water produce a flow of air from sea to land during the day and from land to sea during the night. This land-sea breeze effect changes the direction of the local wind, and the Coriolis effect controls the direction of its rotation. When there are no large atmospheric systems to mask this diurnal wind shift, the late-evening breeze may carry land smells 20 miles or more out to sea.

The presence of even a small speck of land may create small-scale atmospheric effects in an immense body of water. Midoceanic islands with some mountains have considerably more rainfall on their windward sides, where moist air rises because of both the topography and higher land-surface temperatures. Heating of localized land masses in the tropics produces sufficient vertical motion in the air to ensure cumulus clouds and frequent after-

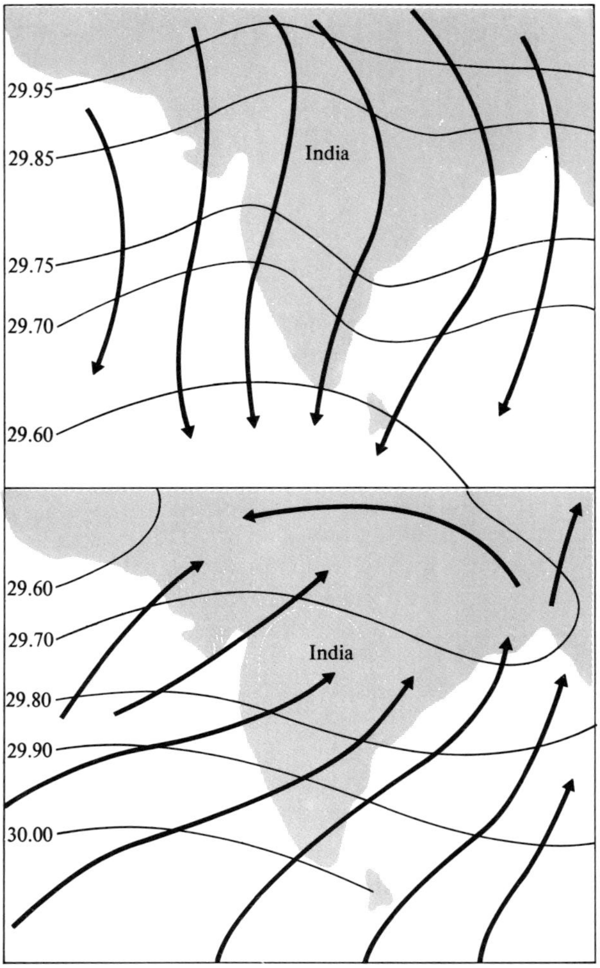

Fig. 11.10 The seasonal reversal in wind patterns associated with the Asiatic winter (or dry) monsoon, above, and the summer (or wet) monsoon, below. Isobars are in inches of Hg.

noon thundershowers. Even coral atolls, which are small and only slightly above sea level, can be discerned while still below the horizon because of their cloud covers.

At this point we may conclude that the large oceanic areas at lower mid-latitudes are nearly always overlain by a large anticyclonic system, the subtropical high-pressure cells. Air circulates about these cells clockwise in the Northern Hemisphere and counterclockwise in the Southern Hemisphere, affecting the behavior of the underlying water. The friction between the moving air and water creates a stress and transfers momentum from the air to the water, setting it in motion. The stress causes the surface waters under the subtropical highs to rotate in nearly the same direction as the moving air, producing large closed cells of moving surface water called *gyres*. Gyres are the most conspicuous features of the ocean's surface circulation. At 40 to 50°S, where the wind exerts its influence on the water in a more unidirectional manner, large westerly drifts of water circumnavigate the earth scarcely impeded by land masses. These currents responding to the winds form the backbone of the oceans' surface circulation, carrying water from the influence of one climatic region to mix with water from another climatic area. By this means water rising from the oceanic depths is transported along the surface to be modified, become dense, and sink again. Both horizontal and vertical flows are influenced by the surface winds acting on the sea surface. These currents in turn affect the distribution of sea-surface temperatures, which aid in keeping the air in motion. The dynamic interaction of winds producing currents which redistribute surface water which in turn alters the driving winds, is indicative of the interactions that occur in complex nature systems.

STUDY QUESTIONS

1 How are the boundaries between the layers of the atmosphere established?

2 What atmospheric properties are determined by water vapor?

3 If a barometer reads 760 mm of Hg at sea level and 665 mm of Hg at the top of a mountain, how high is the mountain?

4 Would you expect a barometer to register the same change in pressure at an altitude of 2 km above sea level as at a depth of 2 km in a mine?

5 What process compensates for the imbalance between incoming and outgoing radiation with latitude?

6 What climatic factors are associated with predominately low atmospheric pressure belts and with predominately high atmospheric pressure belts?

7 Why is the north-south circulation of the atmosphere broken into three discrete cells per hemisphere rather than a single cell per hemisphere?

8 If the general horizontal flows of air over the earth were swifter, would you expect more or fewer circulation cells per hemisphere? Why?

9 Describe the pressure distribution you would find in the summer if you went from ocean to land to ocean to land along a constant midlatitude.

10 If you stood with your back to the wind in the Southern Hemisphere, would you expect to find the low-pressure system to your left or right?

RECOMMENDED READING

Hill, M. N., ed., *The Sea*, vol. 1, Interscience Publishers, New York, 1965.

Neumann, G., and W. J. Pierson, Jr., *Principles of Physical Oceanography*, Prentice-Hall, Inc., Englewood Cliffs, N.J., 1966.

Starr, V. P., "The General Circulation of the Atmosphere," *Scientific American*, **195**, 6 (December 1956).

Strahler, A. N., *The Earth Sciences*, Harper and Row, New York, 1963.

Von Arx, W. S., *Introduction to Physical Oceanography*, Addison-Wesley, Reading, Mass., 1962.

CHAPTER 12

THE CIRCULATION OF THE OCEANS

INTRODUCTION

The title of this chapter implies a discussion of all the factors that make the waters of the sea move from one place to another, that is, the currents that transport water. However, to understand the circulation of the sea we must be aware not only of the forces that create the currents but also of the forces that modify them once they are generated.

We have stated previously that the moving atmosphere, driven by the variations in heating over the earth's surface, exerts a drag or stress on the sea surface. This causes surface water to flow from one climatic regime to another, where its density may be modified by surface heating or cooling and the addition or removal of water by evaporation, precipitation, and other processes. Dense water produced at the surface by climatic factors in turn introduces a vertical circulation that is density-driven. The wind may also act to produce a vertical circulation that is not density-driven, since an area in which there is a wind-created divergence of surface water (waters moving apart) must be supplied by seawater from depth, while in an area of wind-created convergence (waters flowing together) the accumulated surface water sinks.

Some oceanographers assert that all ocean currents except those of tidal origin are essentially wind-driven, either directly or indirectly. We will not explore the pros and cons of this theory here. It can be said, however, that the thermohaline circulation* is related in part to wind-driven circulation; and further, that once a circulation is started it acts to modify the temperature structure of the sea surface, which in turn modifies the winds, which move the water. The process of cross-coupling or feedback often occurs in nature, creating second-order modifications that are difficult to evaluate. There is feedback in the oceans because winds cause currents, currents redistribute surface thermal patterns, thermal patterns alter wind fields, and wind fields alter currents.

Once currents are set in motion, forces such as the Coriolis effect modify them by deflecting the moving water and sending it off in a slightly new direction. Thus the currents are bent and may be acted upon by horizontal centrifugal forces. The continental land blocks form obstacles, blocking or

* *Thermohaline* circulation is induced by density changes due to changes in the temperature and salinity of the ocean waters.

deflecting current flow. The ocean's bathymetry may also deflect currents into curved paths; the Coriolis force and the centrifugal force may act in the same or in opposite directions.

There are all kinds of currents, driven and formed by all kinds of conditions. We shall discuss only the major currents and their effect on the ocean.

THE MAJOR CLOSED-CELL SURFACE-CURRENT SYSTEMS

Large, closed-current gyres are centered about a latitude of 30°N in both the Atlantic and Pacific Oceans. These anticyclonic gyres rotate in a clockwise direction, driven by prevailing westerly winds to the north and trade winds to the south. The closed-cell circulation is maintained in part by the Coriolis force, which continuously deflects the moving water toward the cell's center, and by a condition of continuity, i.e., water carried away from one region by the surface wind stress must be resupplied from another.

We pointed out earlier that the Coriolis force acts at right angles to the moving water; in the case of anticyclonic gyres, it forces the moving surface water toward the gyres' center. This movement creates an accumulation of the less dense surface water, which builds up in the center until it elevates the sea surface to the point at which downslope gravity forces acting outward from the cell's center counterbalance the inward-acting Coriolis force. The local sea-surface slope of the dome of water is dependent on the local current about the dome, decreasing as the current decreases and increasing as the current increases.

In the North Atlantic, the sea-surface slope on the Gulf-Stream side of the gyre may produce a rise in sea level of about 1 m over the width of the Gulf Stream; the sea-surface slope on the eastern side of the gyre, where the current is less, is about 1/100 as great. The accumulation of surface water in anticyclonic gyres and the elevation of the sea surface also produce an inversely related density structure at depth. At a moderate depth in the sea, we find that the pressure head of the overlying seawater is nearly constant, even though a column within the gyre is higher than the column outside the gyre. Pressure, however, is the product $\rho g z$; therefore, the seawater density distribution must be different in the two columns if z (the elevation above a given depth) is to be different and pressure the same.

Since density differences in seawater are small compared with the density difference between air and water, constant-density surface slopes in the sea must be many times greater than the slope of the surface. For a pressure balance to exist at depth, the density surfaces have to lie at an inverse angle to the surface slope (Fig. 12.1). The internal density distribution makes a sensitive indicator for detecting the surface slope, which cannot be measured directly in the open sea, and for determining the average current that must be present to produce a surface slope.

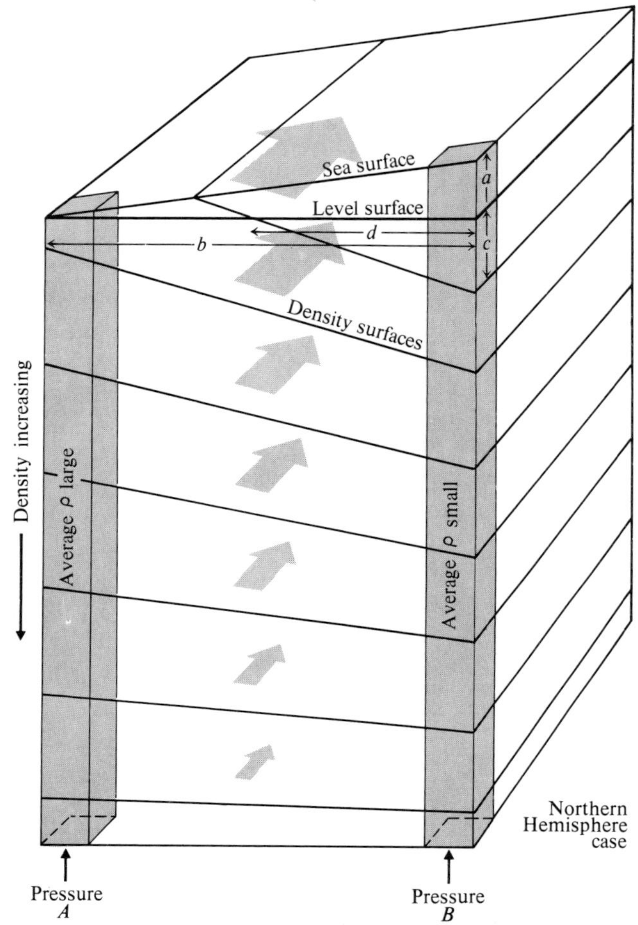

Fig. 12.1 In geostrophic flow, moving water is acted upon by the Coriolis force and deflected until an equal and opposing pressure gradient force is formed. When this balance is struck, the water moves at a right angle to the pressure gradient. The slope of the sea surface required to form this pressure gradient is a/b. The density surfaces take on a reverse slope c/d which is much greater than the sea-surface slope and which is readily measured. The adjustment in the mass field allows the hydrostatic pressure at some depth to be constant; Pressure A equals Pressure B even though the columns of water are of unequal lengths.

THE CLIMATIC MEAN OCEAN CURRENTS

If we consider the fundamental factors producing steady water motion, we can show that currents are caused by both external and internal forces. Internal forces arise from the mass field (density distribution) within the sea; they act whenever constant density surfaces are inclined to level surfaces. A component of gravity drives water horizontally and vertically along the inclined density surfaces, seeking to level all density surfaces and create a distribution such that density increases with depth to the same degree everywhere. External forces that produce currents are the winds, tides, and the removal or addition of water locally by evaporation or precipitation. There must be a flow of water to compensate for the last effect, but it is small and will be neglected. However, the resulting alteration of density at the sea surface is not insignificant and is considered as part of the internal force field. We shall consider tidal force only briefly here; it will be treated as a special problem in Chapter 17. The

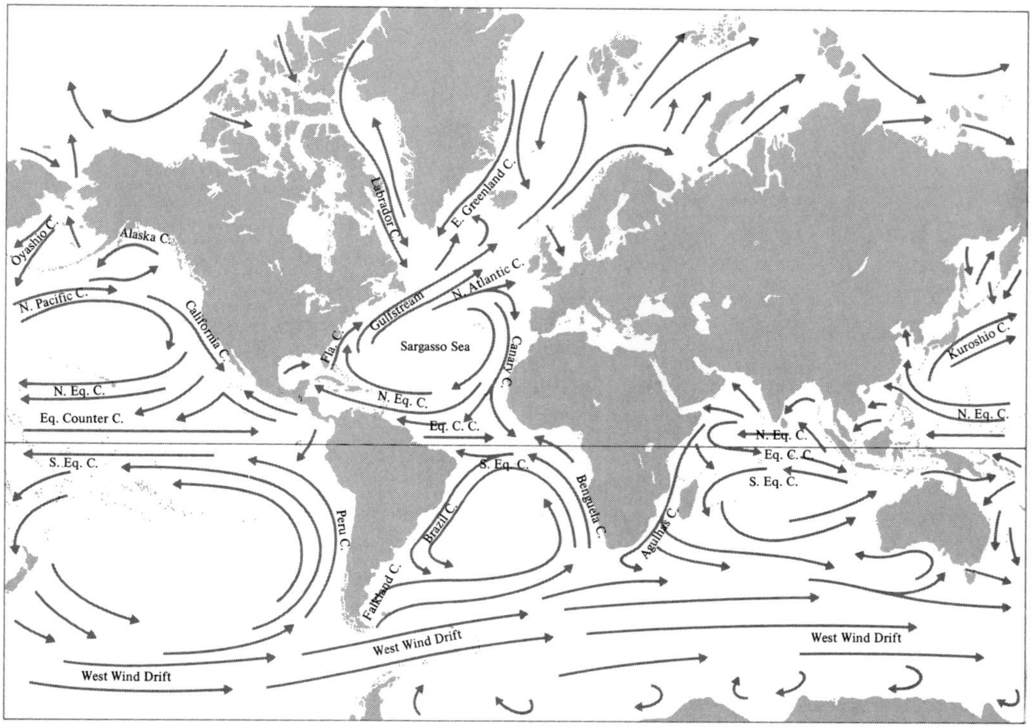

Fig. 12.2 The major surface currents of the oceans. The arrows indicate the direction of principal flow but not the speed.

most important external force, the wind acting on the sea, will be dealt with here in more detail (see Fig. 12.2).

The internal mass force field and the external wind force field produce currents in entirely different ways; they cannot be considered independently, however, since they affect each other. The mean circulation of the oceans is a result of these two force fields, both of which are dependent on the distribution of solar energy over the earth. The interaction of forces and currents and their variation with time are among the fundamental problems yet to be fully solved in oceanography. Currents modify the distribution of temperature and salinity, distribute dissolved gases, and in part control the distribution of marine life. Although oceanographers have yet to describe fully and definitively the dynamics of currents, they have gained much insight into the mechanisms of the mean currents and current systems.

Fig. 12.3 The surface currents tend to make the surface waters of the sea both diverge and converge Major zones of divergence and convergence are shown here. They are AAD (Antarctic divergence), AAC (Antarctic convergence), SC (subtropical convergence), EQ (equatorial convergence and divergence), and AC (Arctic convergence).

The major surface current systems of the oceans can be described as average or mean motion systems based on many direct observations of the speed and direction of moving water over a long period of time. The U.S. Naval Oceanographic Office has determined average values of surface currents from data on the surface drift of vessels collected over many years. Sufficient data are available to determine not only an annual mean current system but also monthly mean current systems, which describe annual changes throughout the oceans. Figure 12.2 shows the major mean surface currents, the names of which should be learned. The relationship of these currents to the predominant air circulation patterns seems evident, but one need not know it to produce current charts based on direct measurements.

The mean steady ocean surface currents show surface waters moving from low latitudes towards high latitudes (warm currents) and from high latitudes toward low latitudes (cold currents). Areas of horizontal convergence and horizontal divergence are also evident (Fig. 12.3). We can also see that currents are deflected by landmasses and that they exist in regions in which horizontal flows of air are small, e.g., the doldrums and horse latitudes.

WINDS AS A DRIVING FORCE

Since the development of the relationship between surface winds and the currents requires considerable mathematical and physical knowledge, it is beyond the scope of this book. However, we shall introduce this concept to show that the association between wind field and currents is more than just coincidence.

We stated earlier that since water and air are fluids, they move over the earth with little frictional effect; however, friction is necessary for moving air to set water in motion. Moving air exerts a stress that is a force per unit area applied tangentially to the water's surface. If there were no friction between the air and water, there would be no stress. The stress has been determined in several ways, including measurement of the wind's ability to blow water against a landmass and maintain a water slope with a gravity force acting downslope against the wind's force.

Wind stress is proportional to $\rho' W^2$ the air density times the square of the surface wind speed. The constant of proportionality is called the drag coefficient, C_D; it is related to the roughness of the water surface. Thus we have

$$\text{wind stress} = C_D \rho' W^2 \, . \tag{12.1}$$

Measurements of wind stress show that the drag coefficient is slightly dependent on the wind speed; it is also believed to be affected by the stability of the air column above the water surface. As is obvious even to a casual observer, water develops one roughness configuration when acted upon by light winds (less than 15 knots) and another roughness configuration when acted upon by stronger winds (above 15 knots). Light winds produce waves with small ratios of height to wavelength, while at about 15 knots the waves steepen and develop into "whitecaps," creating a rougher sea surface for the wind to grip. The drag coefficient, therefore, increases slightly with wind speed and has two values; one for light winds and one for stronger winds.

In the open ocean with moderate winds, a value of 2.6×10^{-3} is realistic for C_D if the wind speed (W_{10}) is measured in m sec^{-1} at 10 m above the sea surface. The stress on the sea surface is then

$$\text{wind stress} = 2.6 \times 10^{-3} \rho' W_{10}^2 \text{ g cm}^{-1} \text{ sec}^{-2}. \tag{12.2}$$

Wind acting on the water sets it in motion through this stress coupling, but the water moves much more slowly than the air. The water speed is determined by an empirical relationship:

$$V = \frac{0.0127}{\sqrt{\sin \phi}} W_{10}, \tag{12.3}$$

where V is the surface water speed expressed in the same units as W_{10}, the wind speed, and ϕ is the latitude.

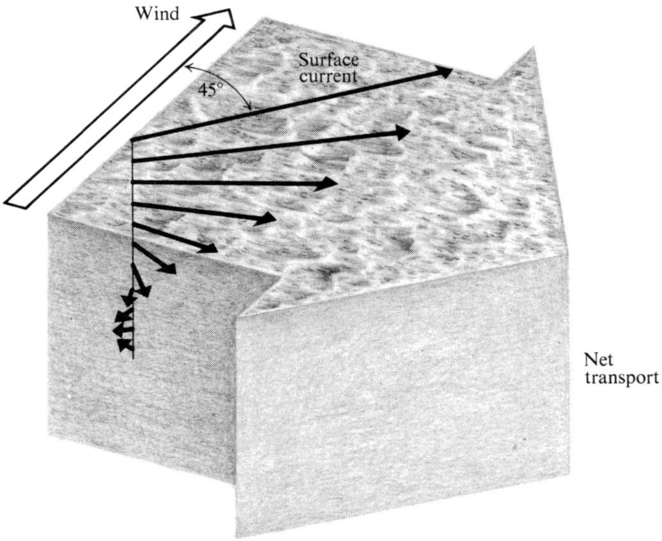

Fig. 12.4 Water is set in motion by the wind. The direction of flow and speed of flow change with depth to form the Ekman spiral. This change with depth is a result of the earth's rotation and the inability of water to transmit a driving force downward with 100% efficiency due to low friction. The net transport over the wind-driven column is 90° to the right of the wind (Northern Hemisphere).

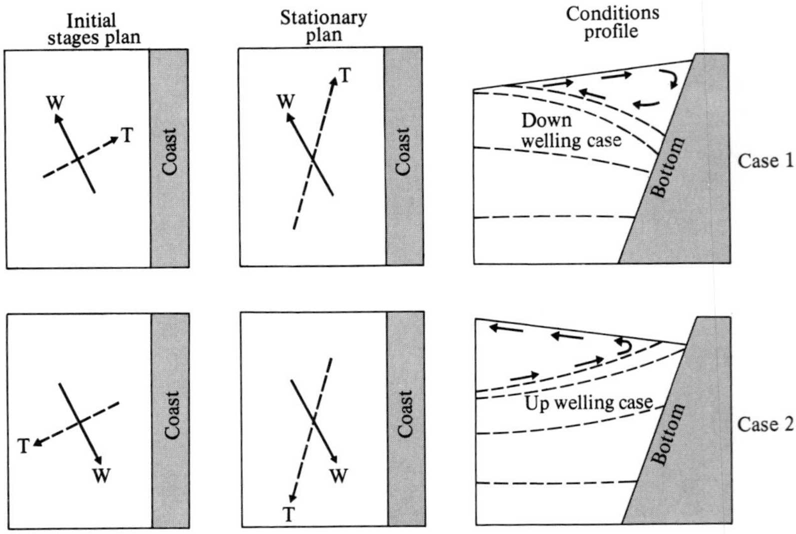

Fig. 12.5 The influence of a coast alters the Ekman wind-driven transport. When transport of water is onshore (above) it is forced to flow slightly alongshore as it forms a wedge of surface water along the beach. When transport is offshore (below) a depression is formed at the beach. Examples shown are for the Northern Hemisphere. (From H. U. Sverdrup, M. W. Johnson, and R. H. Fleming, *The Oceans*, © 1942. By permission of Prentice-Hall, Inc., Englewood Cliffs, N.J.)

We must also consider the Coriolis effect. In Chapter 2, we found that the Coriolis force increases with the velocity of the moving fluid and with increasing latitude:

$$\text{Coriolis force/unit mass} = 2\omega\,(\sin\phi)\,V.$$

We also showed that when material moves more slowly, the deflection suffered by a moving mass is greater; i.e., the ratio of the distance the object is displaced at right angles to the original distance moved increases. In the problem considered above, since the surface water moves more slowly than the wind, it will be deflected from the path of the driving wind.

In a classic analysis published in 1902, V. Walfrid Ekman developed a mathematical model to show how water moves in response to wind stress. Ekman assumed a homogeneous ocean of infinite depth with no density structure and no boundaries. Conditions in midocean come closest to meeting these requirements and pure Ekman wind drift of the water is approximated. According to Ekman's analysis, the wind sets the surface water in motion, the latter undergoing a Coriolis deflection that is greater than that of the wind. The surface water, moving obliquely to the wind, exerts a drag on the underlying water through internal friction coupling and sets it in motion at a still slower rate so that it suffers an even greater deflection. Layer by layer, the water is set in motion; each layer moves more slowly than the one above and therefore suffers a greater deflection. The total effect can be shown by a device known as the Ekman spiral (Fig. 12.4) from which the reduction of water speed with depth and the increased turning caused by the Coriolis deflection are quite evident. From the direction of the deflection, we can see that the spiral is applicable to the Northern Hemisphere; its sense of rotation would be reversed in the Southern Hemisphere. A total reversal of the direction of the surface current occurs at D, the *depth of frictional resistance*, which is

$$D = \frac{3.67\sqrt{W_{10}^{3}}}{\sqrt{\sin\phi}}. \tag{12.4}$$

In the Northern Hemisphere, the Ekman surface current flows in a direction 45 degrees to the right of the wind. However, if we consider the total transport of surface and near-surface water set in motion to depth D, the total Ekman transport of wind-driven water is 90 degress to the right of the wind. In the Southern Hemisphere, these angles would be the same, but to the left of the wind.

We shall now consider what happens to the Ekman spiral, the direction of surface current, and the total transport of wind-driven water when Ekman's conditions are not met in nature. Consider the proximity of a coast as a lateral boundary in the Northern Hemisphere (Fig. 12.5).

In case 1, the initial total transport is 90 degrees to the right of the wind and toward the coast. As water is transported onshore and held there, a slope develops in the sea surface. Gradually, the slope will force the total transport to veer and flow farther up along the coast at an angle of less than 90 degrees to the wind. Under these conditions, a longshore current up the coast will develop and downwelling of surface water will occur at the surface convergence formed against the beach.

In case 2, the initial transport is 90 degrees to the right of the wind and away from the shore, creating a void of surface water along the coast (surface divergence) which must be replaced by dense upwelling water. This configuration causes a sea-surface tilt with a high water level offshore and low level at the coast, forcing the total transport to move closer to the wind direction.

Wind not only transports the near-surface water horizontally, but also (in the presence of land barriers) produces vertical motions in the water; downwelling, as in case 1, and upwelling, as in case 2. The latter, of course, supplies the chemical nutrients to the sunlit surface zone and promotes biological growth. We will show later that other conditions can also produce divergence and upwelling that promote biological productivity.

The northern coast of California in the summer is an excellent example of case 2. The predominantly northerly winds during summer move the surface water offshore, producing coastal upwelling. Since the upwelled water is cold in relation to the offshore surface water and the moist overlying air, fog is common along the coast and biological production is high. The cooling effect of the upwelled water justifies San Francisco's boast that it is an "air-conditioned" city; its July and August temperatures average about 15° and 17°C (59° and 63°F) respectively.

We should also consider the effect of depth on the Ekman spiral. Since water velocity is zero at the sea floor, the turning in the spiral with depth decreases and the surface current and total transport are at angles less than 45° and 90° respectively to the wind when depths are small. In very shallow water, the wind-driven current and transport are nearly identical in direction to the wind stress. The increasing density of seawater with depth in the open sea has little effect on wind-drift current; thus it can usually be ignored.

Since we can estimate the speed of the wind-driven surface current and its orientation relative to the wind for either hemisphere, we may now compare the mean oceanic surface currents with the mean wind field of the earth (Fig. 12.2). The relationship between orientation of wind and wind-driven current is obvious. Even if the monsoon effect is considered, the reversing currents in the Indian Ocean seem to be direct consequences of the reversing seasonal winds.

ANOMALOUS FEATURES OF THE SURFACE CURRENTS

We can readily see the influence of land boundaries in deflecting currents, e.g., the initiation of the Humboldt Current by the tip of South America and the oblique cant of the equatorial currents of the Atlantic Ocean. However, there is one phenomenon that puzzled oceanographers for many years: the presence of the swiftest flowing surface currents, such as the Gulf Stream and the Kuro-Shio, in regions where the wind stress on the water is much less than it is in the trade wind zones or the prevailing westerlies, e.g., the horse latitudes.

If water in the central North Atlantic circulates clockwise about the Sargasso Sea in a closed system, then the water transported (in cubic meters per second) under the driving force of the prevailing westerly winds must equal the total amount of water transported southward in the eastern half of the ocean, westward by the North Equatorial System, and northward by the Gulf Stream. This constant transport requirement around the gyre must be met for continuity of mass to hold for the closed system, i.e., water removed from one area must be replaced by an equal counterflow.

The surface currents indicate that the circulating water gyre is not an ideally closed system. Water from southern latitudes flows toward the Caribbean and into the Gulf Stream; part of it then exits from the Gulf Stream extension or the North Atlantic Westwind Drift and flows toward the Scandinavian countries. Convergences at zones in which the Labrador current meets the Gulf Stream require surface water to sink, while divergences require deeper water to rise and mix with the surface water. The transport of surface water from southern to northern latitudes requires a return flow which obviously is not present at the sea surface and therefore must exist at depth. Although the gyre is not an ideally closed system, a transport budget can be worked out to account for water entering and leaving it.

The transport of surface water at any section across the current is the product of the average speed of the water times the cross-sectional areas through which it moves. Therefore, a section with a small cross-sectional area and a swift current can transport as much water per unit time as one with a large cross-sectional area and a lethargic flow. For example, the narrow Gulf Stream transports the same amount of water per unit time as the wide southerly flow across the eastern half of the central Atlantic Ocean. These two extremes in current balance each other; they are required to maintain the continuity of flow around the gyre, which is wind-driven primarily on the northern and southern sides. However, the continuity requirement does not explain the presence of a swift current of small cross section on one side of the gyre and a slow current of large cross section on the other. We need an additional factor in order to understand this characteristic asymmetry of flow.

WESTERN INTENSIFICATION OF SURFACE CURRENTS

The westerlies and the northeast trade winds exert a torque, or turning moment, on the surface water, causing it to rotate in the subtropical gyre about the Sargasso Sea. A vertical column of water moving around this gyre tends to maintain its angular momentum, which in this case is the product of the column's rotation rate and its mass of water. The former has two components:

planetary vorticity, $2\omega \sin \phi$ (commonly referred to as f) and *relative vorticity*, ζ. Vorticity refers to the horizontal circulation of a particle of fluid about a vertical axis. Consider a solid cylinder floating upright in water and moving with a current. If it rotates as it moves along, the flow is said to have vorticity; the cylinder is assumed to approximate the rotation of the column of displaced water. The cylinder rotates because water speeds at its boundaries are different from the speed of its axis as it moves with the flow. In other words, spatial variations in speed occur in the horizontal directions and current shear is present.

The relative vorticity is thus expressed as

$$\zeta = \frac{\Delta Vy}{\Delta X} - \frac{\Delta Vx}{\Delta Y}; \qquad (12.5)$$

that is, the change in the y-component of the velocity in the x-direction minus the change in the x-component of the velocity in the y-direction equals the relative vorticity. The interaction of the velocity shears, $\Delta Vy/\Delta x$ and $\Delta Vx/\Delta y$, produces a torque or twisting moment on a fluid parcel (or cylinder, in this case) to give it rotation relative to the earth, or relative vorticity.

Fluids on the earth that have no relative vorticity are nonetheless subject to rotation due to planetary vorticity. In Chapter 2, we pointed out that the earth rotates around any arbitrary assumed vertical at a rate of $2\omega \sin \phi$, as measured by the Foucault pendulum. Therefore, any particle of water at rest with respect to the earth rotates about a vertical axis at the rate $2\omega \sin \phi$ in reference to a fixed space or inertial reference system. Thus the absolute vorticity of a column of water referred to fixed space has two parts, the planetary part f and the relative part ζ. The planetary vorticity can be zero only at the equator, while the relative part may be zero if horizontal current shears are zero or compensating $\Delta Vy/\Delta X = \Delta Vx/\Delta y$.

The absolute vorticity $(f + \zeta)$ expresses the rotation rate of a column of water (mass of water) relative to fixed space. If the column foreshortens so that its radius must increase to contain the same volume (same mass if density remains constant), then the rotation rate $(f + \zeta)$ will decrease. If the column of water elongates, then its radius must decrease and $(f + \zeta)$ will increase. This relationship, a result of conservation of angular momentum, can be stated as

$$\frac{f + \zeta}{D} = \text{constant}, \qquad (12.6)$$

where D is the length of the vertical column of given volume or mass.

At a fixed latitude, with f held constant, any change in D will cause a change in ζ. If D is held constant and the latitude is varied, a change in f will occur which must be balanced by an equal and opposite change in ζ. With this information, we can now explain why current gyres intensify at their western boundaries.

Consider the oceans to be of uniform depth (D is a constant); then $(f + \zeta)$ is also a constant. On the western boundary of the gyre, f increases as a water column moves from lower to higher latitudes and thus it must acquire clockwise relative vorticity (anticyclonic relative vorticity); on the eastern side, f decreases as a water column moves from higher to lower latitudes and it will acquire counterclockwise relative vorticity (cyclonic relative vorticity).

Considering this relationship in conjunction with the vorticity imposed on the water by the wind shear, prevailing westerlies in the northern half of the gyre, and southeast trades in the southern half, we find that changing latitude has an additive effect on the relative vorticity on the western side. Both wind shear and latitude displacement produce anticyclonic relative vorticity on the western side; on the eastern side they are subtractive, however, since wind shear produces anticyclonic relative vorticity and latitude displacement produces cyclonic vorticity.

Because of the conservation of angular momentum, absolute vorticity remains constant in an ocean of uniform depth D, if friction does not decrease the water column rotation or if an external force, e.g., wind stress, does not enhance or decrease rotation. For example, on the average, the gyre surrounding the Sargasso Sea is in a steady state. For a steady state to exist, the vorticity on the western side of the gyre cannot rise indefinitely when wind stress and relative vorticity act together and in the same direction. Hence there must be considerable friction at the western boundary that produces friction-generated cyclonic corticity to keep the absolute vorticity at a normal steady-state level. More friction is present at the western boundary of the

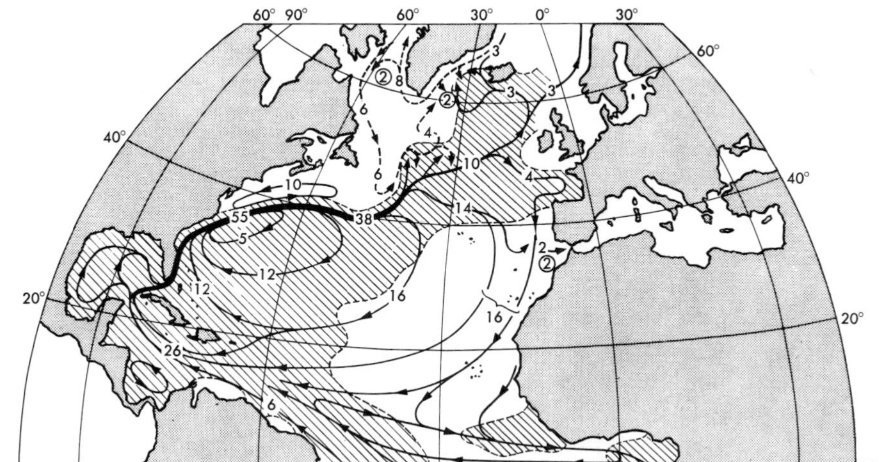

Fig. 12.6 Continuity of water transport is maintained in the surface currents. Where currents are swift, transport is large. Where currents are slow, transport is small. However, broad, slow current can transport the equivalent of a narrow, swift current. Transport values are given in millions of cubic meters of seawater per second. Regions of the sea where surface temperatures are generally higher than the average sea-surface temperature for a given latitude are shaded. (From W. S. von Arx, *Introduction to Physical Oceanography,* Addison-Wesley, Reading, Mass., 1962.)

gyre, where the current shear is large, than at the eastern, where wind-stress-induced vorticity and relative vorticity oppose each. The alternating increases and decreases in relative vorticity and friction combine to produce a large asymmetry in the currents about the gyre.

Figure 12.6 shows the asymmetry in flow about the subtropical gyres and the water budget that must be maintained. The transport of wind-driven water near the surface is given in 10^6 m^3/sec. If the gyre in the figure is transected by a line, the transport across the line in one direction approximately equals the transport in the other. The Gulf Stream transport of 55×10^6 m^3/sec to the northeast therefore equals the transport to the southwest at the opposite side of the gyre.

VERTICAL CIRCULATION

Thus far we have considered aspects of the surface circulation of the oceans that are primarily wind-driven. This circulation displaces water with a given set of characteristics to another location, where it is further modified by climatic conditions and mixed with other water types. The modifications alter the water's density, setting up vertical circulations which are density-driven due to thermal and salinity changes. In Fig. 12.2, a substantial transport of surface water is shown heading northward into high-latitude regions. Obviously, the water cannot keep accumulating at these high latitudes, but must return by some route other than surface flow. Convective vertical circulation that is set up by dense water sinking and gradually flowing southward at depth accounts for the return transport.

If one considers only the effect of the sun on density-driven vertical circulation, it would seem reasonable that solar heat would cause water to rise near the equator and flow toward higher latitudes, where it would cool and then sink to return to the equator—thus creating a mirror image of the single-cell atmospheric system discussed in Chapter 11. However, solar heating also affects both the evaporation of seawater and the alteration of surface water density by changing its salt concentration. High solar energy levels at either side of the equator combined with high rates of evaporation tend to increase the salinity and impose a circulation opposite to that caused by thermal conditions alone.

Thermohaline convective motions in the sea are actually combinations of the two extremes of pure haline and pure thermal circulation; they are further modified by the influence of the atmosphere. Convective vertical flows, some of which are periodic and associated with seasonal changes at high latitudes, have low speeds compared to those of surface currents. Despite the slowness with which the flows supply water to depth (where it gradually moves laterally to rise again in an area of surface divergence), their cross-sectional area is so great that they can transport large volumes of water.

Since seawater from one area is carried along at the surface while water below it is carried from another area, it is obvious that the ocean must have a

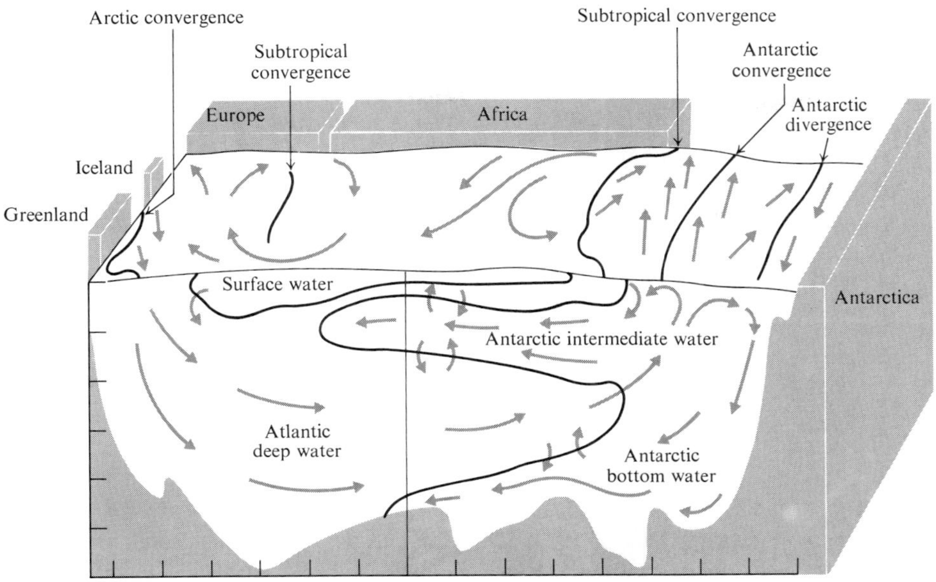

Fig. 12.7 The anatomy of the Atlantic Ocean in schematic form. The role of the surface currents is evident where they bring waters together at the major convergences to mix and sink to the appropriate density level. This continuous supply of water to convergences adds water of a specific quality to each layer, making it extend horizontally at depth until mixing processes erode it at a rate equal to the supply. The density of each layer increases with depth. The densest water is the Antarctic bottom water. (After A. Defant, *Physical Oceanography*, vol. 1, Pergamon Press, New York, 1961.)

layered structure. Each layer moves at the rate and in the direction required to maintain flow continuity. However, rate and direction of flow are not the only factors that permit us to detect the water layers. Since each flow has a set of physical characteristics (e.g., salinity, temperature, and density) determined by local surface climatic conditions and is altered only slowly by mixing with adjacent water, layering may also be determined by the vertical distribution of these measurable variables.

Of all the major oceanic areas, the Atlantic Ocean has the most dramatic and most studied structure. Figure 12.7 shows the generalized major features of the Atlantic Ocean in north-south vertical profile. Its densest water is the Antarctic bottom water, which forms in the shallow Weddell Sea to flow downslope to the sea floor. Next in density is the North Atlantic deep water, which originates at the convergence of the warm Gulf Stream water and the cold currents moving southward from the Greenland region. Moving along the density scale from the bottom toward the surface, we encounter the Antarctic Intermediate water, which forms at the Antarctic convergence at about 40°S (Fig. 12.3). Above the Intermediate water are several superficial layers, which we shall classify as surface water.

If one were to sample the Atlantic Ocean to determine the temperature and salinity distribution at discrete depths at various latitudes, he would penetrate a number of water bodies, each with its own set of identifiable characteristics and direction of flow. These layers, which also have different thicknesses and are centered about different depths, create a vertical structure in the oceans that is nearly constant over time. During the International Geophysical Year, 1957–1958, measurements of salinity and temperature with depth were made through the layered Atlantic at the same positions sampled during the German Meteor Expedition in 1925–1927. The structures of the Southern Atlantic determined from the two investigations were nearly identical.

Despite the dynamic processes in the ocean, its vertical circulation is a slow and relentless process driven by the climatic conditions at the sea surface. The water that sinks and moves horizontally to rise elsewhere in one season is followed in the next by water of similar properties, thus maintaining an apparently steady state. The horizontal net drift in these deep thick layers is very small, about 1 to 3 cm/sec, yet extended over a large cross-sectional area it accounts for considerable water transport. The age of the deeper water, time-elapsed since it was last in contact with the surface, is judged to be about 800 years. One should not assume that the slowness of this circulatory flow is indicative of current speeds in the deep sea, since currents of approximately 1 knot or more have been recorded. However, they are usually oscillatory and do not produce a net circulation of water from the surface to depth and back to the surface at another location.

CURRENTS AND THE MASS FIELD STRUCTURE OF THE OCEANS

As we have stated, horizontal currents cause a tilt of density surfaces in the sea, while the vertical circulation aids in maintaining the layered structure; both are detectable by determining salinity, temperature, and density with depth. The distributions of salinity and temperature with depth define the density structure and thus the mass field of the oceans. If the currents have a known relationship to the mass field, it is more practical to measure its distribution than to try to determine the currents by measuring them directly. We must thoroughly understand the mass field, however, before we can use this reverse technique.

The permanency of the oceans' gross vertical structure, maintained by vertical circulation, allows oceanographers to identify regions of the sea from their vertical salinity and temperature distributions. If we measure salinity (S) and temperature (T) with depth at a given location and plot their values on a T-S diagram, we obtain a characteristic signature curve (Fig. 12.8). Each sampled depth is a point on the T-S curve. Near the surface, the curved line segments separating the plotted points of S and T are long, indicating a considerable change in salinity and temperature over short

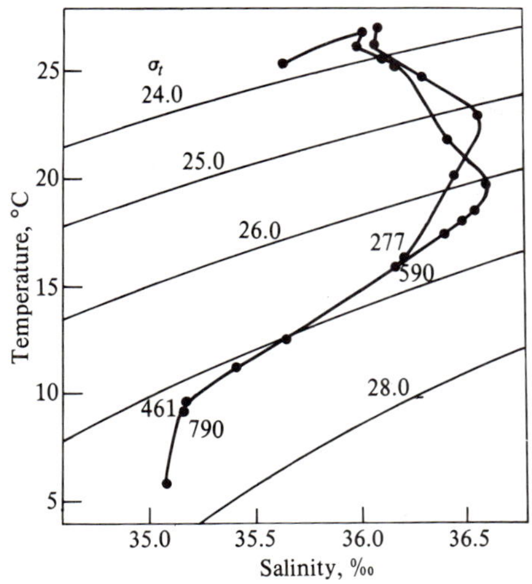

Fig. 12.8 *T-S* curves plotted from observed salinities and temperatures at two geographic locations. Water having the same values of *S* and *T* was found at 277m depth at one location and 590m at the other. (From H. U. Sverdrup, M. W. Johnson, and R. H. Fleming, *The Oceans*, © 1942. By permission of Prentice-Hall, Inc., Englewood Cliffs, N.J.)

vertical distances. At greater depths, the segments become very short, indicating that changes in *S* and *T* are small over large vertical distances and the water is more homogeneous. If the water were truly homogeneous, with the same *S* and *T* over a large vertical distance, it would be represented by a single point on a *T-S* diagram and could properly be called a *water type*.

The oceans have been divided into water mass areas and transition areas between masses (Fig. 12.9). A water mass area is a region in which the vertical distribution of *S* and *T* will produce either the same *T-S* curve or a family of curves which closely approximate each other on a *T-S* diagram. The envelopes embracing the *T-S* curves for each of the water mass areas are shown in Fig. 12.10.

Some of the water masses have cool, low-salinity water near the surface, while others have warm, high-salinity water. Each area's *T-S* curve is affected by surface climatic control and the distribution of water layers with depth. In the hands of an oceanographer, *T-S* diagrams are powerful diagnostic tools. They can be used to detect erroneous data obtained from faulty measurements since a point badly out of line on a *T-S* curve or showing a layer of less dense water below denser water to form a density inversion is a clear warning that something is wrong. They can also indicate the proportions of two adjacent water masses that are present in a sample taken from a transition zone.

We could discuss many fine points concerning *T-S* curves and their shape, use, and change with horizontal space. It will suffice, however, to stress that they characterize the mass field structure of oceanic subareas. They are also used to detect the presence of water of a given set of *S* and *T* values in a given location. Near-surface water which appears on a *T-S* curve as related to one water mass may be found in a location normally occupied by a different water mass; this displacement may be caused by surface currents formed

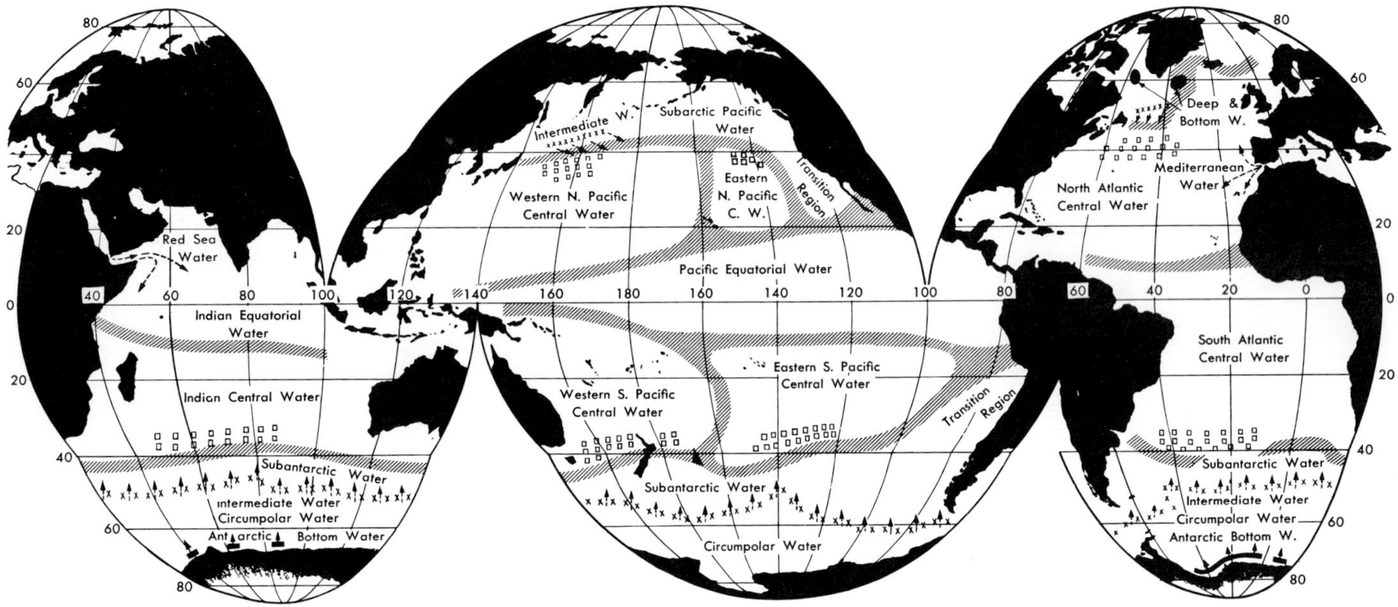

Fig. 12.9 The geographic distribution of water masses that have common vertical salinity and temperature distributions, as denoted by their T–S curves. Squares mark the regions in which the central water masses are formed; crosses indicate the lines along which the Antarctic and Arctic intermediate waters sink. (From W. S. von Arx, *Introduction to Physical Oceanography,* Addison-Wesley, Reading, Mass., 1962.)

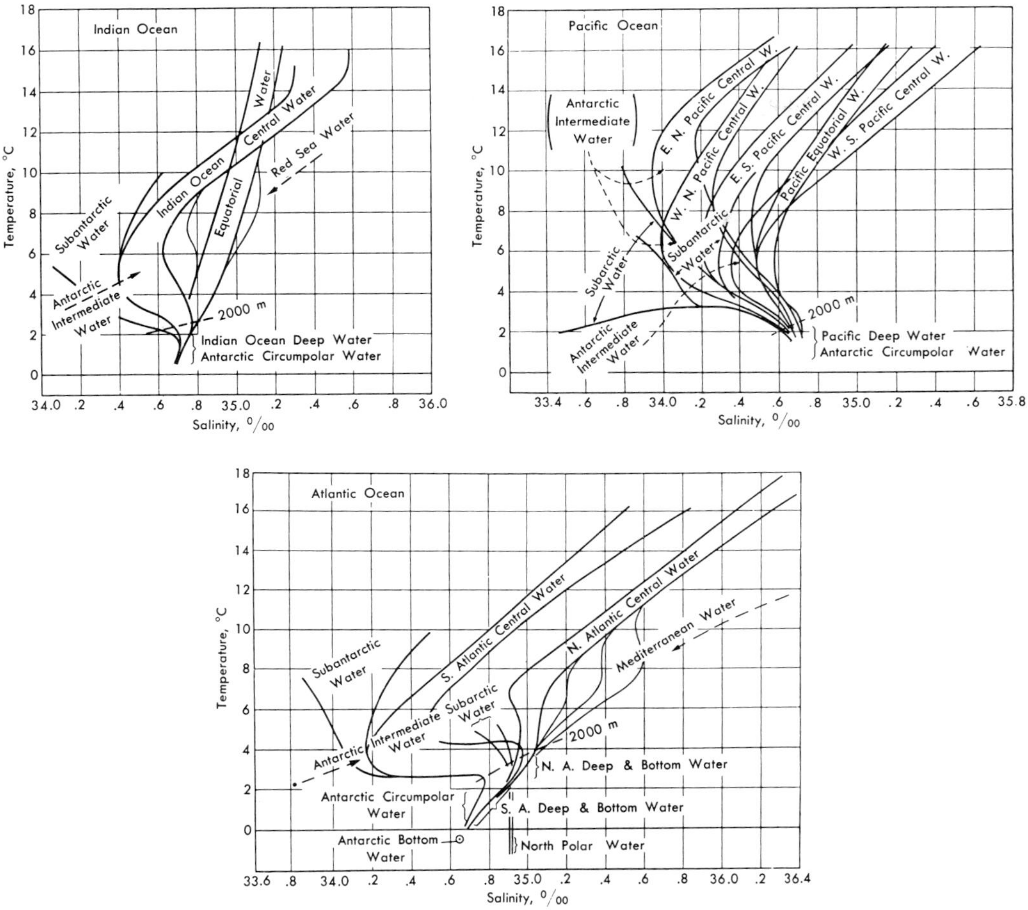

Fig. 12.10 The *T–S* curves that describe the vertical structure of the recognized oceanic water masses. From W. S. von Arx, *Introduction to Physical Oceanogrphy*, Addison-Wesley, Reading, Mass., 1962.)

periodically or aperiodically as a result of changing surface wind stresses. For example, Subarctic Pacific water is often found south and east of Baja California.

One can also easily detect whether or not a given water exists at depth. In Fig. 12.10, the dense Antarctic bottom water appears as a water type (a point on a *T-S* diagram) at the deepest depths in the southern Atlantic and Indian Oceans, but not in the Pacific Ocean. The shallow source areas of this dense water, e.g., the Weddell Sea, are located so that submarine ridges confine the sinking water to the Southern Atlantic and Indian Oceans. Even in the Southern Atlantic, some deep basins are so effectively isolated by their

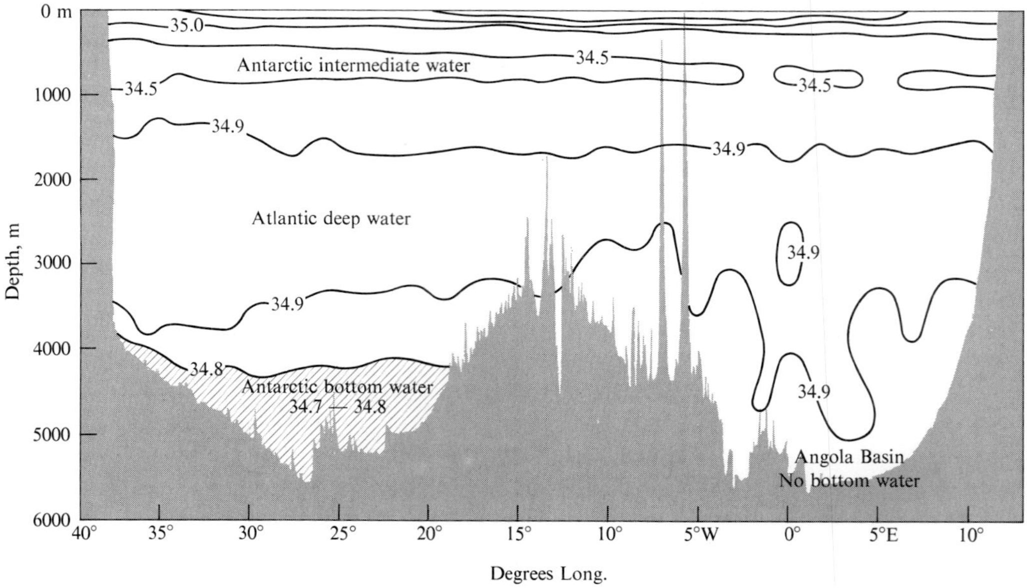

Fig. 12.11 A vertical profile of salinity along 16°S Lat. shows that the submarine ridge system excludes Antarctic bottom water from the Angola Basin. Adapted from the *Atlantic Ocean Atlas*, Woods Hole Oceanographic Institute.

bounding ridges that the Antarctic bottom water never reaches sufficient height to spill over the ridge and fill the basins. The Angola Basin, protected by the Walvis and Mid-Atlantic Ridges, has no bottom water. Oceanographic data west of the Mid-Atlantic Ridge and east of it at the latitude of the Angola Basin show the effect of exclusion of the deep Antarctic bottom water (Fig. 12.11).

At the beginning of this chapter, we stated that moving water causes a tilt in the sea surface and a greater, reverse tilt in the underlying density surface. Thus we may use measurements of salinity and temperature with depth to determine the inclination of the density surfaces and the current required to cause it. The relationship between the currents and the mass fields, which allows the current to be calculated from the density distribution, incorporates basic assumptions: the current is assumed steady (nonaccelerating) and the Coriolis force is assumed constant at a given location and balanced by the downslope gravitational forces. When these assumptions hold, the current is said to be *geostrophic.*

The faster the current, the greater the Coriolis force; thus the steeper the surface slope required for balance and the steeper the tilt of density surfaces within the sea. Figure 12.12 shows a profile of temperature across the Gulf Stream and into the center of the gyre. The isotherms, which approximately

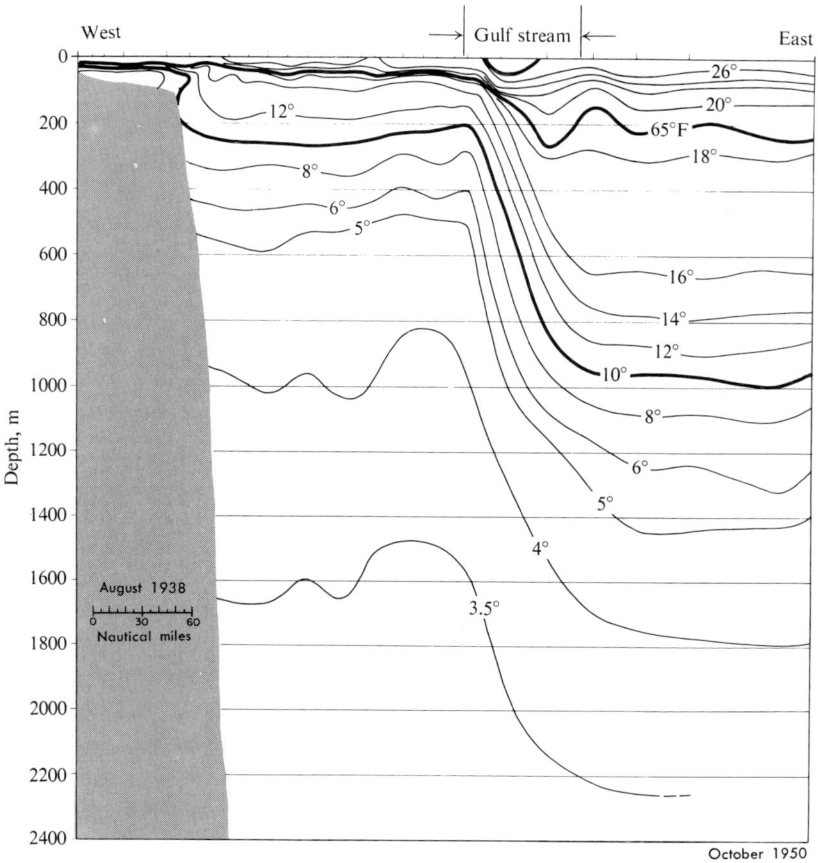

Fig. 12.12 The tilting of the density surfaces as approximated by the isotherms under the narrow, swift-flowing Gulf Stream is evident, whereas a tilting of the sea surface across the current would not show. (From W. S. von Arx, *Introduction to Physical Oceanography*, Addison-Wesley, Reading, Mass., 1962.)

represent the density surfaces, are steeply tilted in response to the surface-slope which is caused by less dense water, moving to the right of the current. Density or σ_t surfaces (isopycnals) would show the same pattern. At sea it is impossible to observe the surface slope directly; however, the vertical density structure is readily obtained from oceanographic data.

If we gather vertical data at two stations across the surface flow to determine the density structure we may calculate the surface current relative to the current existing at a deeper depth, assuming the pressure at a given depth at one sample location equals the pressure at the same depth at the other location. Thus the average density times gravitational acceleration times the length of the water column must be the same for both stations. Figure 12.12 shows that across the Gulf Stream the average density of the vertical water column cannot be the same at the two edges of the stream. The sum of the ρgz products over small increments of depth is taken over the two columns until the pressure is the same in both cases, requiring that the two columns are of different lengths; the difference in length represents the elevation of the sea surface at one column position over that at the other. We can then calculate a tilt or slope of the sea surface from the change in elevation divided by the distance between the two columns. From this we can determine the current speed at the surface relative to that at the depth at which the pressures were equal, as well as the slope of the sea surface relative to the slope of the density surfaces at the same depth.

The situation would be ideal if there were no slope in the density surface at depth. The calculated free surface slope would than approximate the actual free surface slope and the surface current calculated relative to the current at depth (which would be nil) would approximate the true surface current. Relative surface currents calculated from the mass properties of the water columns are called *geostrophic* currents. Their speed is given by

$$V_0 - V_z = \frac{g(\Delta H/L)}{2\omega \sin \phi}, \qquad (12.7)$$

where g is gravity, ΔH is the difference in column height, L is the horizontal distance between sample locations, $2\omega \sin \phi$ is the Coriolis acceleration, and V_0 and V_z are the current at the top and bottom of the sampled water column respectively. The relative geostrophic current calculated in this manner is the average surface current at right angles to the line between the two sample stations.

There are refinements of the method described above. It is even possible to determine the speed of current of one depth relative to another over an entire water column, producing a velocity profile with depth. A relative velocity profile over a water column is likely to show reversals in current, indicating that water at one depth is flowing in one direction across the line between station pairs and in the opposite direction at another depth. Consider Fig. 12.8, which shows the layered Atlantic Ocean with its contra-flows. Reversals in flow direction require that zero velocities occur at specific depths. It is difficult to establish the precise depth of no motion and to shift the relative velocity curve back and forth so that it has zero speed at the proper depths. When this is done, however, the relative velocities can be converted to absolute values. Direct measurements of currents and calculation of water budgets in opposing flows are used as aids in this adjustment. However, despite the assumptions, geostrophic flow calculated from mass distribution in the ocean is still a more successful method of determining mean ocean currents and water transport over large oceanic areas than is the direct measurement of flow using current meters.

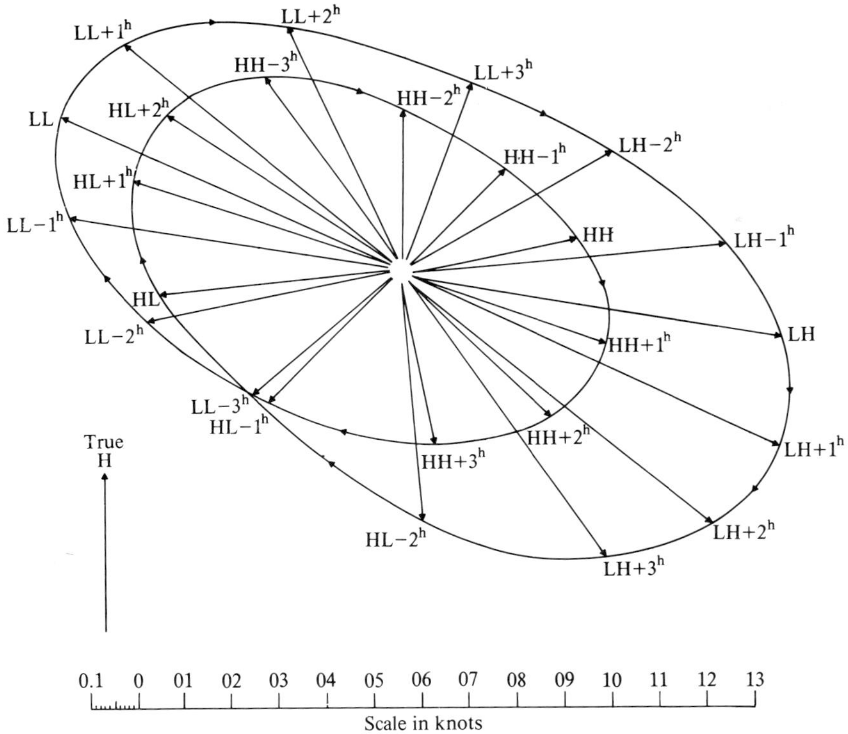

Fig. 12.13 The tidal current curve for Swiftsure Bank, off the entrance to the Straits of Juan de Fuca, showing its rotary nature. The direction and magnitude of the current were observed hours after or before recognized mixed tide levels. LL is lower low water, HH is higher high water, LH is lower high water, and HL is higher low water. (From N. Bowditch, *American Practical Navigator*. H.O. Publication No. 9, U.S. Naval Hydrographic Office, Washington, D.C., 1962.)

ROTARY CURRENTS IN THE SEA

In addition to the mean motion current systems of the sea, there are oscillating currents which are related to tides and inertial motion. These rotary currents may be considerable; however, they do not create an effective net displacement of water along a path as do the mean current systems, and the excursion distance of water undergoing rotary motion is small.

Tidal currents are present at all depths in the open sea and may have speeds of 1 knot or more. Their rotary nature causes them to change direction continually (Fig. 12.13). If we take current measurements over a considerable time span, we may note a small displacement of the center of the rotary current vector diagram, resulting from the superimposition of mean drift on the rotary motion. A water particle, therefore, may move considerably about its mean spatial position due to rotary currents, although its mean position changes slowly. The displacement per time of the mean particle position approximates the 1 to 3 cm/sec drift of the deep ocean layers. The same phenomenon is present in the near-surface waters, where horizontal flows are much greater. The presence of a mean drift gives the current, when portrayed on a hodograph,* the appearance of a stretched spring (Fig. 12.14). If we subtract the mean flow vectorially from the current, the pure rotary current becomes more evident.

Tidal rotary currents have periods of rotation that match the local tidal periods; however, other rotary currents also have periods associated with the pendulum day. The latter are *inertial currents* and form when water set in motion is deflected by the Coriolis force and sent into a closed circle of motion with an outward-acting (centrifugal) force that counterbalances the inward-acting Coriolis force:

$$\frac{V^2}{r} = 2\omega (\sin \phi) V. \tag{12.8}$$

At a given latitude ($\sin \phi$ is constant), the radius r of the inertia circle is controlled by the velocity V. The period for the current to travel a full circle in inertial motion is

$$T = \frac{2\pi r}{V} = \frac{\pi}{\omega \sin \phi}, \tag{12.9}$$

which is one-half of the local pendulum day. The hodograph in Fig. 12.14 is for an inertial current having a period of about 14 hours and stretched by the presence of a mean current.

Whether a current is tidal or inertial, its rotation is caused by the Coriolis force. However, the period of tidal current rotation is controlled by the period of the tides, whereas the period of inertial rotation is controlled by the earth's rotation about the local vertical. Unlike tidal currents and wind-driven currents, inertial currents are not continually forced; rather, they are free oscillatory currents that form when the water is set in motion by an impulsive force and is then left to seek its own type of motion.

Currents distribute sea waters of differing properties far and wide; their influence on the oceans and on the climate is unmistakable and allows their detection through cause-and-effect relationships. The effect of currents on the anatomy of the seas was recognized long after sailing ships detected their existence and usefulness. Benjamin Franklin, who studied the Gulf Stream using a thermometer rather than a current meter, recognized that it had boundaries. He commissioned a Nantucket sea captain, Timothy Folger, to draw a map of the Gulf Stream so that merchant ships could stay within its boundaries and make better time from the Colonies to Europe.

No description of these mighty flows is more dramatic than that of Matthew Fontain Maury. In *The Physical Geography of the Sea*,† he states:

* The plot of a path described by the extremity of a vector drawn from a fixed origin and representing the linear velocity of a moving point.

† 1855, Harper and Brothers.

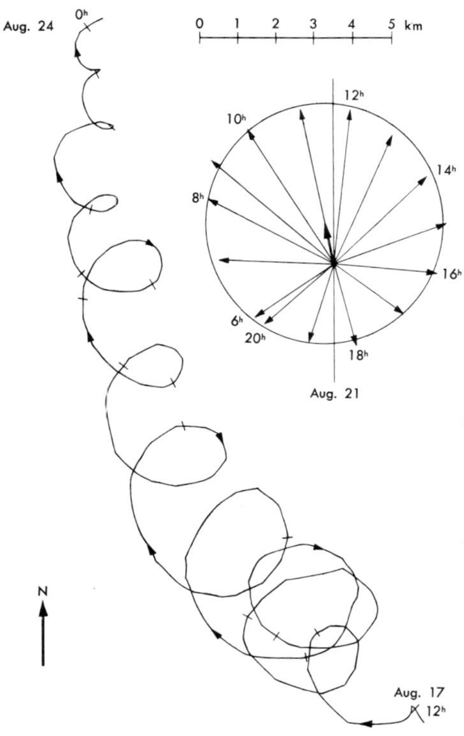

Fig. 12.14 A rotary inertial current having a rotary period of one half-pendulum day. A net drift is superimposed on the rotary current as shown by the progressive vector diagram. When this drift is removed, the rotary nature of the current appears more regular. Observations were made in the Baltic Sea in 1933. (From H. U. Sverdrup, M. W. Johnson, and R. H. Fleming, *The Oceans*, © 1942. By permission of Prentice-Hall, Inc., Englewood Cliffs, N.J.)

"There is a river in the ocean. In the severest droughts it never fails, and in the mightiest floods it never overflows. Its banks and its bottom are of cold water, while its current is of warm. The Gulf of Mexico is its fountain, and its mouth is the Arctic Seas. It is the Gulf Stream."

STUDY QUESTIONS

1 Wind can produce a stress acting on the sea surface and build up a slope of water against a lee shore until the pressure acting seaward and downslope balances the stress. What sea-level slope can a 50-knot onshore wind support in shallow water?

2 The speed of an ocean surface current is related to the wind speed. What will be the speed and the direction of set of a current in the Southern Hemisphere if the driving wind blows from the south at 20 knots?

3 Geostrophic flow implies a balance between the Coriolis force, produced by a steady current, and a pressure gradient force acting against the Coriolis force. If the current accelerates or decelerates, what happens to this balance and the direction of flow of the current relative to the isolines of sea-surface slope?

4 Why does the current continually change direction and decrease with depth in the Ekman spiral?

5 Why can the wind, through stress coupling with the water, produce wind tides of larger amplitude in a region having extensive shallow coastal zones than in a region having a deep shoreline?

6 Since the absolute vorticity of a column of water tends to remain constant (due to conservation of angular momentum of the rotating water column), on which side of the oceans in the Southern Hemisphere would you expect the currents to intensify?

7 If the currents do not intensify in the Southern Hemisphere oceans, what additional factors have entered into their development that are not accounted for?

8 Study the *T-S* curves for the major water masses and state what type of climatic regime is present above them. Base your conclusions on the salinities and temperatures shown as near-surface values in Fig. 12.8.

9 Why is Antarctic bottom water found at the deeper depths of the south Atlantic along 30°S Lat on the western side but not on the eastern side?

10 What would you say about *S* and *T* data that produce a *T-S* curve on which σ_t first increases with depth, only to suddenly reverse and decrease with depth?

11 The surface slopes of the sea produced by geostrophic currents are not directly detectable. The counter-tending density surface slopes at depth are many times larger, however, and may be easily detected. Discuss their significance in determining the elevation or depression of deeper water into or out of the photic zone. How do they promote production of organic material in the sea?

RECOMMENDED READING

Defant, A., *Physical Oceanography*, vol. 1, Pergamon Press, New York, 1961.

Knauss, J. A., "The Cromwell Current," *Scientific American*, **204**, 4 (April 1961).

Munk, W., "The Circulation of the Oceans," *Scientific American*, **193**, 3 (September 1955).

Neumann, G., and W. J. Pierson, Jr., *Principles of Physical Oceanography*, Prentice-Hall, Inc., Englewood Cliffs, N.J., 1966.

Robinson, A. R., ed., *Wind Driven Ocean Circulation*, Blaisdell Publishing Co., New York, 1963.

Stommel, H., "The Anatomy of the Atlantic," *Scientific American*, **192**, 1 (January 1955).

———, *The Gulf Stream*, University of California Press, Berkeley, Calif., 1965.

Sverdrup, H. U., M. W. Johnson, and R. H. Fleming, *The Oceans*, Prentice-Hall, Inc., Englewood Cliffs, N.J., 1942.

Von Arx, W. S., *Introduction to Physical Oceanography*, Addison-Wesley, Reading, Mass., 1962.

CHAPTER 13

MARGINAL SEMI-ENCLOSED SEAS AND BASINS

INTRODUCTION

The continental, inland, or semi-enclosed seas bordering the oceans are of particular interest to oceanographers and man in general, since our greatest contact with the marine environment is through these seas. These semi-isolated areas are like miniature oceans and therefore can be monitored and used as study models. In some respects, however, they are different from the open ocean; they are influenced to a greater extent by the climate of the surrounding land masses, for example. Via the seas, freshwater drainage from the land masses enters the ocean, where it is mixed with seawater by the turbulence caused by tidal forces, wind, and hydraulic flow from rivers.

The bordering seas range in size from small river estuary systems to the larger mediterranean types. They may be narrow, deep, and elongate, as are the glaciated fjords; shallow regions of the continental shelf isolated from the ocean by offshore islands and bars; or major basin systems of near-oceanic proportions, embraced by the continental masses and isolated from the ocean by islands and shallow sills.

The Mediterranean Sea is a classic example of an isolated major sea having very restricted contact with the ocean (only through the Straits of Gibraltar and the Suez Canal). The term *mediterranean* ("in the middle of the land") is descriptive of this isolation. The Gulf of Mexico, another example of a major isolated body of seawater, is often referred to as the American Mediterranean Sea in European literature. The Baltic Sea, Caribbean Sea, Sea of Japan, North Sea, Bering Sea, and many others are also large, semi-enclosed mediterranean-type seas.

The water in the bordering seas is similar to oceanic water and supports comparable marine populations. However, the coastal populations may differ in species composition due to the larger fluctuations in temperature and salinity imposed by continental climatic conditions and land drainage. Furthermore, the combined effect of coastal upwelling, the supply of nutrients from land, mixing, and shallow depth increases the productivity of coastal areas over that of the open ocean. In general, though, the similarities outweigh the differences sufficiently that the smaller adjacent seas may be used as convenient models to study the changes in the biopopulation — indeed, changes in any parameter, such as the chemical, salt, or heat content — in the open ocean.

ISOLATION: AN AID IN STUDYING THE ENVIRONMENTAL BUDGETS

In the open sea, a volume of water, say 1 km^3, can be delineated by geographic grid lines. If this volume has the air-sea interface, through which fluxes are minimal, as one of its sides, we must monitor the remaining five sides of the rectangular volume to determine the activities that govern changes within it. Over time, currents and diffusion through the sides of our volume will lead to an increase or decrease in the parameter in question. If the parameter is conservative, the sum of all additions to it and removals from it yields the parameter's rate of change averaged over volume. If the parameter is nonconservative, we must also consider additions to and removal from its dissolved state within the water volume to ascertain the rate of change averaged over the volume. Needless to say, it is very difficult to accomplish this task with any accuracy under open-ocean conditions.

The partially isolated basins, especially those with simple channels of communication to the open ocean, are better suited to budget studies. Their water is bounded by the impervious land sides, i.e., the bottom and head of the embayment. The flux of material in and out of the system occurs primarily across the seaward face unless material is contributed by river inflow, which is also relatively easy to monitor. Thus it is much easier to make the required measurements in isolated basins.

Measurements of inflow and outflow due to diffusion and currents at the mouth of the embayment are sufficient to describe the rate of change in the embayment if the material is conservative. If the material is nonconservative, we must also measure the extraction or liberation of the material *in situ* to ascertain the change with time. In practice, the following relationships are considered:

1 Conservative case:

$$\text{local change} = \text{diffusion} - \text{advection}. \tag{13.1}$$

2 Nonconservative case:

$$\text{local change} = \text{diffusion} - \text{advection} + \text{biological effect}. \tag{13.2}$$

The *local change* is the change in material (concentration) with time averaged over the volume of water, which is centered about a fixed point in space; *diffusion* is the sum of the material added or subtracted from the volume due to diffusive transport processes; *advection* is the sum of the material carried in and out of the volume by currents; and the *biological effect* is the change over the volume due to *in situ* biological processes which remove or liberate materials from or to the soluble state. The change in material per time is usually measured as a change in concentration with time.

For example, we may study a body of water to determine how its nutrient content varies in time (local change). If the volume is sampled repeatedly and the total nutrient content is observed to decrease in time, the local change is negative. If diffusion and advection at the mouth are equal and do not produce a net change in the nutrient level of the embayment, the observed local change must be due to *in situ* biological removal of the nutrient from its soluble state. Since nutrients and carbon are in specific ratios to each other in organic material (see Chapter 6), we can calculate the rate of organic production from the rate of nutrient chemical depletion in the water. Thus we can estimate organic productivity without direct measurement of the biological processes.

The conservative and nonconservative budget equations are in effect continuity equations, since they refer to the conservation of matter or mass in a system. The continuity condition of an embayment can be expressed in another form with the assumption of steady state (local change is zero). Consider an embayment with which you are familiar. If it has a considerable tidal range, its seawater volume is continually changing over tidal periods; over long periods, however, it is nearly constant, as indicated by its constant mean sea level. Similarly, the average salinity of the embayment, when determined over a long time span, is relatively constant even though it may change either over tidal periods or seasonally, if there is fluctuation in the freshwater contribution or evaporation. We can say, therefore, that the embayment is in steady state over long time periods; its volume or mass of water and its total salt content are constant with time. Since steady state exists, the local change of water and salt is zero over a long time period. Diffusion and advection may now be combined as simple transport and divided into processes that contribute salt and water and remove salt and water. Since they must be equal to maintain the steady state, we have

$$T_o + E = P + R + T_i, \tag{13.3}$$

and

$$T_o \times \rho_o \times S_o = T_i \times \rho_i \times S_i. \tag{13.4}$$

Equation (13.3) states that T_o, the average volume of water transported from the embayment to the open sea, plus E, the average volume of water removed by evaporation, equals P, the average volume of water added by precipitation, plus R, the average volume of river runoff, plus T_i, the average volume of water added to the embayment from the sea. In Eq. (13.4), T_o times the density and average salinity of the outflowing water equals T_i times the density and average salinity of the inflowing water. The difference between the two average densities, ρ_o and ρ_i, is usually considered insignificant in view of the assumptions on which the equations are based. Thus, dividing out the densities, we have

$$T_o \times S_o = T_i \times S_i. \tag{13.5}$$

The two equations for continuity of water and salt can then be combined to give

$$T_o\left(1 - \frac{S_o}{S_i}\right) = (P + R) - E, \tag{13.6}$$

$$T_i\left(1 - \frac{S_i}{S_o}\right) = E - (P + R). \tag{13.7}$$

We can further modify the above equations by assuming that R is either small relative to E and P and thus negligible or that it can be combined with P to yield P', the total average contribution of freshwater to the system. We then find

$$T_o = \frac{S_i(E - P')}{S_o - S_i} \tag{13.8}$$

and

$$T_i = \frac{S_o(E - P')}{S_o - S_i} \tag{13.9}$$

The values used in these equations are time averages of volume exchange, salinity, evaporation, and precipitation.

Equations (13.8) and (13.9) give us some interesting facts about the isolated embayments bordering the oceans. In an area in which P' is greater than E, $E - P'$ is negative and the accumulated freshwater must be removed by T_o. Since T_o is positive, $S_o - S_i$ must also be negative, requiring S_i to be greater than S_o. Thus dilute water of volume T_o and salinity S_o must move seaward, while incoming water of greater salinity must move inward at volume T_i and salinity S_i. If salinity alone determines density, then S_o water is less dense than S_i water and the flow at the entrance to the sea must be seaward at the surface and inward at depth when $P' > E$. Similarly, T_o must be greater than T_i since

$$T_o = T_i \times (S_i/S_o). \tag{13.10}$$

This circulation pattern of inflow at depth and outflow at the surface is typical of a system in which $P + R$ is greater than E. The inflowing water at depth supplies the salt necessary to maintain the salt balance, replacing the salt which is mixed into the less dense surface water and transported seaward.

When E is greater than $P + R$, the opposite occurs, that is, T_i is greater than T_o and S_o is greater than S_i. Since the S_i water is less dense than S_o water, inflow occurs at the surface and outflow at depth. Thus the average circulation pattern is the inverse of that given by Eq. (13.10). Figure 13.1 shows the two circulations in schematic form.

Examples of areas in which $P' > E$ are the Puget Sound region of the Pacific Northwest and the Chesapeake Bay system. The Mediterranean Sea and the Red Sea are systems with the reverse flow, i.e., $E > P'$. Interestingly

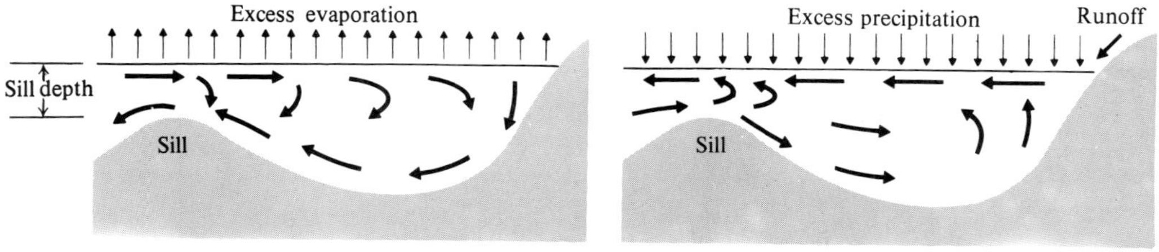

Fig. 13.1 A schematic representation of the gross circulation in basins having entrance sills when evaporation exceeds precipitation (left) and when precipitation and river runoff exceed evaporation (right). (After M. G. Gross, *Oceanography*, Charles E. Merrill, Columbus, Ohio, 1967.)

enough, the Black Sea, which joins the Mediterranean, does not have an inverse circulation; sufficient river contribution from the local land region make P' greater than E.

THE CLASSIFICATION OF SEMI-ISOLATED SEAS AND COASTAL EMBAYMENTS

Other factors also enter into the control of circulation in the seas. Among them are the degree of isolation, sill depth, ratio of width to depth, amount of fresh water runoff, tidal current strength, and in some instances wind. These factors, together with the relative values of E and P', determine the classification of the seas.

One family of seas includes deep basins that are well isolated from the open sea both horizontally and vertically, being connected to the ocean by narrow entrances which are partially blocked by sills. If $E > P'$, the basin circulation is such that dense saline water is produced locally at the surface and sinks to depth to be carried seaward by T_o. This circulation continually flushes the basin, preventing stagnation at depth, even though the water density below the sill depth is greater than that outside the sill (Fig. 13.1a). This overturn supplies oxygen to depth, maintaining the water quality needed to support biota throughout the water column. The Mediterranean Sea is an example of this type of sea. The ability of this circulation to remove nutrients that accumulate at depth is evident.

If a deep basin with a shallow entrance sill is in a climatic area in which $P' > E$, another problem arises. The water in the basin is stably stratified and there are no surface processes creating a continual density-driven overturn. If the dilution of the surface water is greater ($P' \gg E$), S_o is very low and T_o is large. The swift outward surface flow must be confined to depths less than the sill depth and above the layer of incoming water at the sill (Fig. 13.1b). The changing velocity structure in this shallow upper zone creates considerable turbulence which blends the incoming water into the outgoing water at shallow depths, allowing very little of the former to mix with or displace the deeper basin water. Basins such as the Black Sea which have this set of conditions often

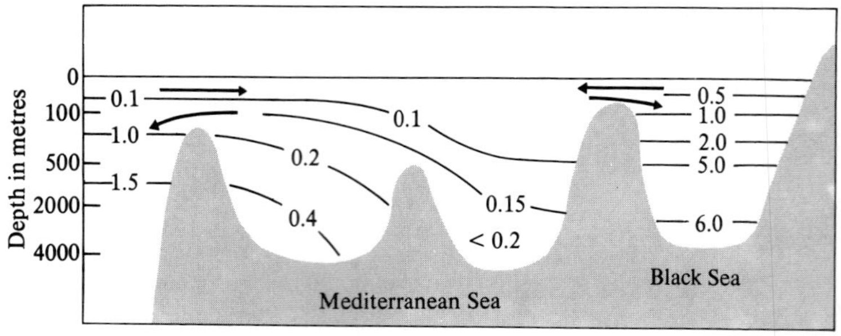

Fig. 13.2 The gross circulation in the Mediterranean and Black Seas is opposite. The Mediterranean Sea is well flushed at depth and does not accumulate a large amount of soluble nutrients at depth. The Black Sea is poorly flushed at depth, with its deeper regions acting as a nutrient trap. The contours show diagrammatically the concentrations of inorganic phosphate phosphorus found dissolved in the water columns. (After M. N. Hill, ed., *The Sea*, vol. 2, Interscience, New York, 1963.)

stagnate and may even become anoxic at depth, where organic material and nutrients are accumulated (Fig. 13.2). Flushing is affected when water on the oceanic side of the sill becomes more dense than the deep basin water and rises above the sill depth to spill over and displace the deep basin water.

When $P' > E$, the transports in and out, T_i and T_o, are small compared to the transports when $E > P'$. In the former situation, the inflow is usually a small fraction of the outflow, which is nearly equal to $(P + R) - E$. The Black Sea, the fjords of Scandinavia and Alaska, and the Baltic Sea are examples of such systems. Dense water from the Mediterranean Sea seldom displaces deep water in the Black Sea, and then only partially; thus the isolated water below sill depth in the Black Sea becomes anoxic, loaded with nutrients, and nearly devoid of marine life. Marine animals requiring oxygen are restricted to the upper layers of the water above the sill depth.

Additional factors besides the E-P' relationship determine the characteristics of the smaller shallow seas that are partially isolated by island chains and extend inland into rivers. The tidal and freshwater flow, the slope of the bottom, and the width, length, and depth of the system all control water distribution. There are four basic types of bordering embayments, or *estuaries*.

The first is the bar-built estuary, a region of dilution along the continental shelf that marks the transition zone between dilute coastal waters and offshore oceanic waters. The coastal bar-built estuaries do not require major river embayments, but can occur along rather uniform shorelines if there is freshwater drainage from continental lowlands and isolating offshore bars. The southeastern and eastern central coasts as well as the Gulf Coast region of the United States have bar-built systems. Offshore islands and bars act as partial barriers to contain the estuaries, as their name indicates.

Another type of estuary forms when the ratio of freshwater runoff flow to tidal volume is large and the ratio of width to depth is small. These embayments,

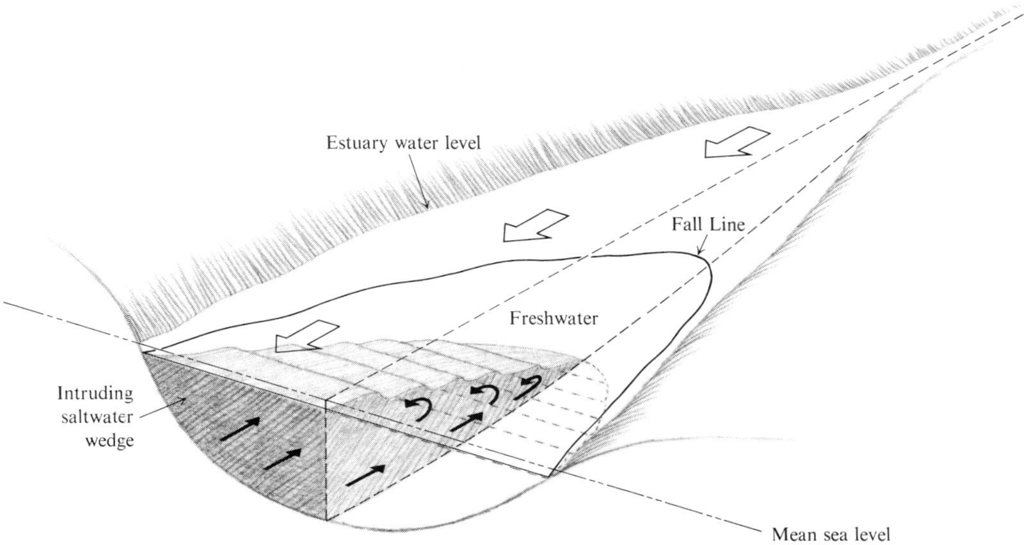

Fig. 13.3 A type I estuary. The discharge of freshwater is sufficient to maintain a steep face on the intruding salt wedge. Friction between the outflowing water and the salt wedge may produce waves that break and deliver salt upward into the discharging water, a process known as salt entrainment. This friction also prevents the salt wedge from intruding upstream to its fall line. Tidal flow is small compared to freshwater discharge.

which are usually the mouths of large river systems that are influenced by tides, are known as type I salt-wedge river estuaries. The tide forces an intruding wedge of saltwater in along the channel bottom; the river flow then works out over the top of the wedge. The current shear located near the boundary of fresh- and saltwater produces turbulence which helps mix the saltwater with the surface runoff and carries it seaward. There is considerable friction between the wedge of intruding saltwater and the swift surface outflow. It decreases the distance that the salt wedge can penetrate upstream, steepens its front, and maintains a relatively sharp boundary between the two water types. The mixing of the underlying saltwater and the dilute surface water is governed primarily by the velocity shear created by the swift outflow. Tidal mixing is slight since tidal volume is small and the resulting tidal currents are much less than those produced by the river runoff (Fig. 13.3).

This situation may be altered if the tidal volume increases so that the salt wedge intrudes farther into the estuary, migrating farther upstream because of the larger tidal excursion. Under these conditions, tidal flow is considerably larger than the freshwater flow, and embayment is called a type II salt-wedge estuary. Its average surface outflow is comparable to the average inflow at depth, and both are usually considerably larger than flow created by the river contribution. The predominate flow is tidal and supplies the energy and turbulence to mix the inflowing and outflowing waters. The boundary between

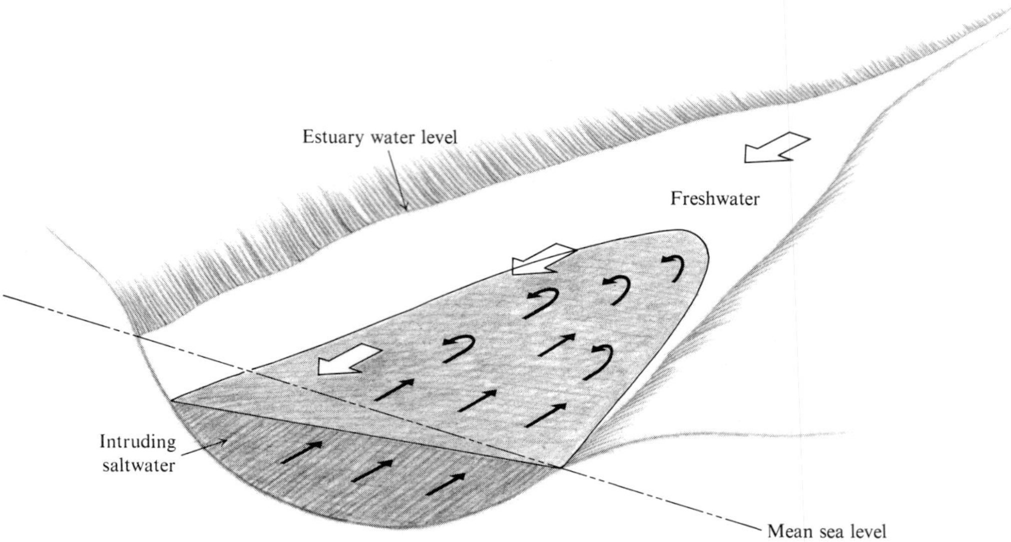

Fig. 13.4 A type II estuary. The surface between the intruding salt wedge and outflowing freshwater is tilted (Northern Hemisphere case). Freshwater discharge is less and tidal flow greater than in type I. Transfer of salt from depth to the surface water is by tidal mixing. The salt concentration of the wedge decreases upstream as the wedge seeks to reach its fall line.

the two types is diffuse, since turbulence is supplied at all depths by the tidal current rather than just at the boundary and in the upper dilute layer, as is the case in type I estuaries. The large average tidal flows at depth and the surface are acted upon by the Coriolis force, which lends an average cross-channel tilt to the density surfaces. In the Northern Hemisphere, the incoming saltwater is inclined to favor the right side of the channel (as one looks upstream); the outflowing surface water favors the left side of the channel (Fig. 13.4). There is not always a tilted interface, however, since curvature of the channel may create centrifugal forces which reverse, cancel out, or reinforce the cross-channel density slopes.

If tidal velocities are so great the river outflow rate is insignificant, the cross-channel tilt may intersect the surface and the channel bed, thus creating a type III estuary. In such an embayment the change in salinity is small with depth but considerable laterally across the channel (Fig. 13.5). The near-vertical density surfaces make the estuary's right side saline and the left side dilute in the Northern Hemisphere. An estuary with nominal river runoff and large tidal volume may be of type III structure near its mouth and type II structure farther upstream.

Thus far we have implied a distinction between deep basins that are protected by shallow sills and well isolated from the oceans and shallow estuaries in

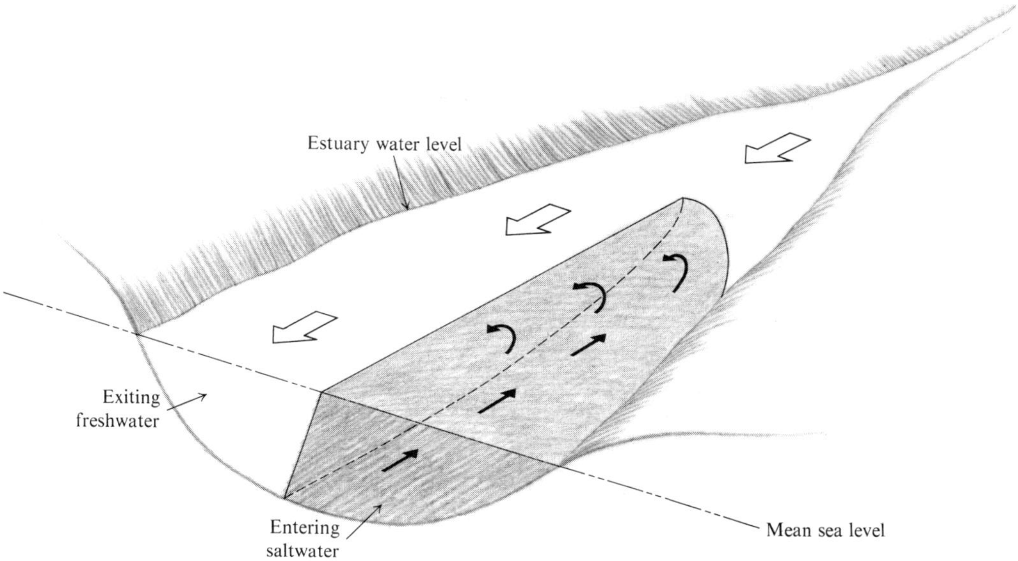

Fig. 13.5 A type III estuary. The surface between the saltwater and the freshwater is nearly vertical (Northern Hemisphere case). Tidal flow and mixing in this case are further increased over discharge to destroy vertical gradients of salinity. Only cross-channel variation in salt concentration is found. Saltwater may intrude nearly to its fall line. If the channel is particularly narrow, then cross-channel tidal mixing may destroy this distribution to produce a horizontally and vertically homogeneous estuary.

which $P' > E$ that are isolated less severely by lateral boundaries only. However, the two are not always clearly distinguishable. The source of the problem is the generic terms, basins and estuaries, which are not clearly defined. In 1964, a scientific conference was held which was devoted to the subject of estuaries and resulted in the publication of a symposium volume entitled *Estuaries.**

In the first article in the volume, *What is an Estuary?*, Donald W. Pritchard, an authority on estuaries, tackles the generic name problem. He points out that dictionary definitions supply the first clue as to what constitutes an estuary. The Concise Oxford Dictionary states, "Tidal mouth of a large river." Webster's New 20th Century Dictionary offers, "An arm of the sea; a frith or firth; a narrow passage, or the mouth of a river or lake, where the tide meets the current." Webster's New International Dictionary lists, "1(a) A passage, as the mouth of a river or lake where the tide meets the river current; more commonly, an arm of the sea at the lower end of a river; a firth. (b) In physical geography, a drowned river mouth, caused by the sinking of the land near the coast." The French dictionary Larousse elaborates by adding, "a coastline bight which is covered only at high tide. A gulf formed by the river mouth."

* George H. Lauff, ed., *Estuaries*, American Association for the Advancement of Science, New York, 1967.

There are similarities in these definitions. Most mention rivers; the French usage, however, implies that shallow coastal embayments not having rivers may also be considered estuaries. None of the definitions specify the degree of isolation, size, or the conditions that P' be greater than E or that the estuary cannot be a deep basin behind a sill.

Pritchard, however, states: "From a physical standpoint, the definition of an estuary should recognize certain basic similarities in the distribution of salinity and density, as well as the circulation pattern and the mixing processes; it should point out also the importance of the boundaries which control the distribution of properties and the movement and mixing of waters." He then offers a definition of an estuary: "An estuary is a semi-enclosed coastal body of water which has a free connection with the open sea and within which sea water is measurably diluted with fresh water derived from land drainage."

This definition, which has found acceptance among oceanographers, allows considerable freedom of interpretation. For example, how enclosed is semi-enclosed? How free and open is the connection with the open sea? Is an estuary still an estuary when evaporation exceeds precipitation and river runoff provided dilution by P and R is measurable?

Most of these questions invite arbitrary answers. If an embayment appears as a distinct indentation of the coastline, it is likely to be considered semi-enclosed. If the connection with the open sea is such that the exchange of water occurs continuously rather than only during extremely high tides, there is free exchange. Pritchard's dilution condition is considered to mean net dilution. On the average, therefore, P' must be greater than E, and regions in which $E > P'$ are not considered estuaries. Not long ago, however, embayments were classified as positive estuaries, neutral estuaries, and negative or inverse estuaries based on $P' > E$, $P' = E$, and $P' < E$, respectively. Now, the first condition must hold for an embayment to be considered an estuary. This classification excludes the Mediterranean Sea; however, it and other regions in which $E > P'$ are not excluded from the mass budget analyses applicable to estuaries. Fjords, which are isolated deep basins and thus similar to the Mediterranean Sea, are considered estuaries since $P' > E$.

Now that we have established a basic definition of an estuary, we shall reconsider the four basic types presented earlier. The coastal bar-built estuaries need not be embayments in terms of a land indentation; they need only be partially isolated from the ocean by offshore islands and bars and diluted by coastal drainage. Bar-built estuaries are usually shallow, elongate, narrow reaches of water running parallel to the coasts. Pamlico Sound and Albemarle Sound along the North Carolina coast are examples. The salt-wedge estuaries, subtypes I, II, and III, are found in coastal indentations associated with river systems. As we noted, the subtypes arise from dynamical processes; however, we stated nothing concerning their geomorphology.

Estuaries may also be divided into structural types according to their manner of formation, or geomorphology. In this classification, coastal bar-built estuaries again appear as a specific type. Drowned river valleys typified by the dendritic Chesapeake Bay and Delaware Bay systems on the eastern coast form another morphological subgroup. Fjords and fjordlike regions carved by glaciers and estuaries formed by tectonic processes are also distinguishable types. One additional group of estuaries is associated with rivers that are clean, stream-cut channels rather than drowned river valleys. The Columbia River estuary and the Mississippi River channel are examples of this group.

We shall now present specific examples that combine the geomorphological and dynamic methods of classifying estuaries. The Delaware Bay and Chesapeake Bay are typical drowned river valleys; the former is a type III salt-wedge estuary in its wider portions near the mouth, while the latter and some of its tributaries are type II salt-wedge estuaries. Fjord and fjordlike estuaries are likely to be type II salt-wedge estuaries, as is the Puget Sound in Washington State, or type II in their upper levels if there is a layer of dense stagnant water lying at depth. Excellent examples of estuaries formed by tectonic processes are San Francisco Bay and the northern portion of the Gulf of California, classified primarily as a type II system. The estuaries of the Columbia and Mississippi Rivers form type I salt-wedge systems since their wedges penetrate only short distances upstream and maintain sharp boundaries.

Man requires an extensive knowledge of estuaries and their behavior, since it is through them that he has his greatest effect on the sea. Nearly all land-derived wastes carried by surface waters enter the ocean through estuaries; they also form major seaports and are centers of industry. In the past, before excessive pollution by man, the estuaries were highly productive in shellfish and fish. Although it is a byproduct of civilization that cannot be entirely avoided, pollution must be controlled before man suffers irreversible losses.

THE FLUSHING OF SEMI-ISOLATED EMBAYMENTS

The control of pollution in a specific area is indeed difficult, since we must consider both present and future conditions. To determine the ability of an estuary to absorb reasonable levels of pollutants, we must first know the rate at which it flushes and exchanges its water with the open sea. An estuary's flushing ability is measured in terms of *flushing time*. If the flushing time is short, water exchange with the sea is good and contaminants can be dispersed rapidly; thus an estuary with a short flushing time can handle more pollutants than one with a long flushing time.

We may express flushing time as the ratio of F, the total freshwater volume of the estuary at a given time, to R, the rate at which freshwater is being added:

$$\text{Flushing time} = F/R. \tag{13.11}$$

Thus flushing time is the time required to exchange all the freshwater in the estuary with the riverwater. A pollutant must enter the system at the same position as does freshwater for the flushing times of both to be the same; if the pollutant enters at a different position, its flushing time may be different.

From the above equation, we can see that flushing time varies if river discharge varies, i.e., if R is a variable. One should also realize that at a time of high river discharge, F is likely to increase, while at low discharge, F will probably decrease. Thus F and R tend to offset each other. However, the total volume of the estuary is the upper limit for F. During excessive flood stages, this limit may be approached in physically small estuaries and the flushing time reduced considerably. Ideally, pollutants should be introduced into the estuary system when the flushing time is the shortest.

In estuaries with large tidal volumes and moderate freshwater additions, T_i, the inflow at depth, is nearly equal to T_o, the outflow at the surface. Pollutants should be dispersed in the surface layers, where they will be well mixed with the surface water by tidal action and carried seaward. We can calculate a flushing time appropriate to these conditions from

$$T_o = \frac{S_i(E - P')}{S_o - S_i}, \tag{13.12}$$

and

$$T_i = \frac{S_o(E - P')}{S_o - S_i} \tag{13.13}$$

The flushing time is the time required for V, the total volume of the estuary, to be removed by T_o:

$$\text{flushing time} = \frac{V}{T_o} = \frac{V(S_o - S_i)}{S_i(E - P')} \; . \tag{13.14}$$

Since estuaries are restricted bodies of water and, unlike open ocean areas, highly affected by man, they have a difficult time maintaining and cleansing themselves under modern conditions. They are often used for recreational as well as commercial purposes. One might reflect carefully on the multipurpose use of these areas; he might also be aware of the activities in one estuary affecting those of another. For example, several river estuaries enter the Long Island Sound region, itself an estuary. Currents carry pollutants from one river system to another, creating a situation analogous to that of neighbors who dump garbage over each other's fences and feel satisfied if they gain only as much as they get rid of. It would seem more reasonable to abstain from such practices in the first place.

STUDY QUESTIONS

1 What properties must an arm of the sea have to be defined as an estuary?

2 Design an experiment to determine the local change of a parameter in the sea.

3 What condition must exist for a system to be in steady state?

4 Given steady-state conditions in a region of the sea, what must we know to ascertain the biological effect on the *in situ* concentrations of a nonconservative material?

5 Show that when $S_o \ll S_i$, the transports in and out of an embayment are approximated by

$$T_o \simeq (P - E)\left(1 + \frac{\overline{S_o}}{\overline{S_i}}\right), \qquad T_i \simeq (P - E)\,\frac{\overline{S_0}}{\overline{S_i}}.$$

6 Why are T_i and T_o larger when $E - P = 50$ than when $E - P = -50$? Discuss how this fact affects the capability of an embayment to flush itself with open seawater.

7 What can you say about an estuary with a large freshwater runoff that is only slightly diluted and whose average salinity is only a little below that of the outside seawater?

8 A salt wedge can intrude into a river estuary only a short distance compared to the tide wave, especially if the tidal range is large and riverflow is fast. How does the tidal rise and fall of the river affect its flow? How does it affect the rise in water temperature due to the discharge of waste heat into the lower reaches of the river by a power plant?

9 Why are the deeper waters of a typical fjord-type estuary subject to stagnation and possible anoxia?

10 Would you expect the two methods presented in this chapter to yield the same flushing times for a given estuary? Why or why not?

RECOMMENDED READING

Hill, M. N., ed., *The Sea*, vol. 2, Interscience Publishers, New York, 1963.

Ingle, R. M., "The Life of an Estuary," *Scientific American*, **190**, 5 (May 1954).

Lauff, G. H., ed., *Estuaries*, Publication No. 83, American Academy for the Advancement of Science, New York, 1967.

Pickard, G. L., *Descriptive Physical Oceanography*, Pergamon Press, New York, 1963.

Sverdrup, H. U., M. W. Johnson, and R. H. Fleming, *The Oceans*, Prentice-Hall, Inc., Englewood Cliffs, N.J., 1942.

Williams, J., *Oceanography*, Little, Brown, Boston, 1962.

Photo courtesy of
G. Dallas Hanna

CHAPTER 14

THE MARINE BIOLOGICAL ENVIRONMENT

INTRODUCTION

We have already discussed some of the chemical, physical, climatological, and geological aspects of the oceanic environment. In studying its biological aspects, we must consider the effects of the biopopulations on the physical-chemical environment and vice versa. We have also discussed motion and circulation, the dynamic aspects of the oceans, which help to maintain the environment by making it less responsive to local changes caused by the biopopulations. The environment of the oceans changes seasonally in the upper near-surface regions, in which there is considerable climatological control. It may also change markedly in regions in which the sea is partially isolated and greatly affected by the bordering land masses. However, the greater extent of the oceans, which lies below the surface waters, is a stable environment where physicochemical processes balance biological processes to produce a steady state.

In this chapter we shall discuss the factors that produce environments in the ocean, as well as why some environments are more conducive than others to the maintenance of biopopulations. As we shall see, the oceans may be divided into environmental subsections which reflect the biopopulations in the water column or on the sea floor.

CLASSIFICATION OF ENVIRONMENTS IN THE SEA

We can observe little or no time-dependent environmental change below the permanent thermocline of the oceans, except that caused by catastrophic events. Above this level, there are seasonal changes in temperature, light energy, turbulent mixing due to winds, gas content, nutrient supply, and the sea surface state. Thus depth divides the ocean into a time-dependent surface region and a steady-state deeper region. However, since latitude also affects the shallow environment, subdivision by depth alone is insufficient.

The latitude effect is associated with the amount of solar radiation and the resulting north-south temperature structure of the sea surface. We learned earlier that seasonal changes in radiation and sea-surface temperatures are small in the tropical and polar latitudes and large in the temperate latitudes. We can establish latitude zones related to sea-surface temperatures and radiation which are symmetrical about the equator and extend poleward from it. The zones are usually referred to as *tropical*, *subtropical*, *temperate*, *subpolar*,

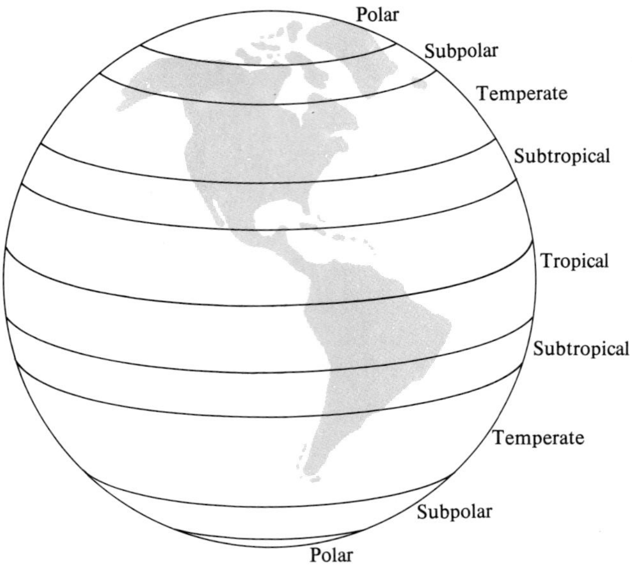

Fig. 14.1 The sea-level climatological zones of the earth.

and *polar* (Fig. 14.1). Their boundaries are not distinct, since currents carrying warm water to higher latitudes or cold water to lower latitudes can cause the latitude displacement of a zone. The temperature zone, therefore, may be displaced poleward on the western sides of the major ocean basins and equatorward on the eastern sides.

As we have noted, biological oceanographers, who are keenly interested in the interactions between marine organisms and the environment, have developed a descriptive nomenclature. The marine environment is divided into the water region (*pelagic*) and sea-floor region (*benthic*). The former is subdivided into the *neritic* zone, which lies above the continental shelves (low tide to 200 m depth); and the *oceanic* zone, which lies seaward of the shelves. Thus the ocean is divided into one area that is highly influenced by land masses and one that is not. The oceanic zone is further divided into depth regions: *epipelagic*, from the sea surface to about 200 m depth; *mesopelagic*, 200 to 1000 m; *bathypelagic*, 1000 to 4000 m; and *abyssopelagic*, the deepest parts of the ocean (Fig. 14.2).

The epipelagic and neritic regions correspond roughly to the euphotic or photic zone, in which there is sufficient solar radiation for active photosynthesis; the deeper regions are aphotic. The photic zone may extend to about 180 m in clear oceanic water at low latitudes or may be as shallow as a few meters in the more turbid coastal regions of temperate latitudes. Its depth is

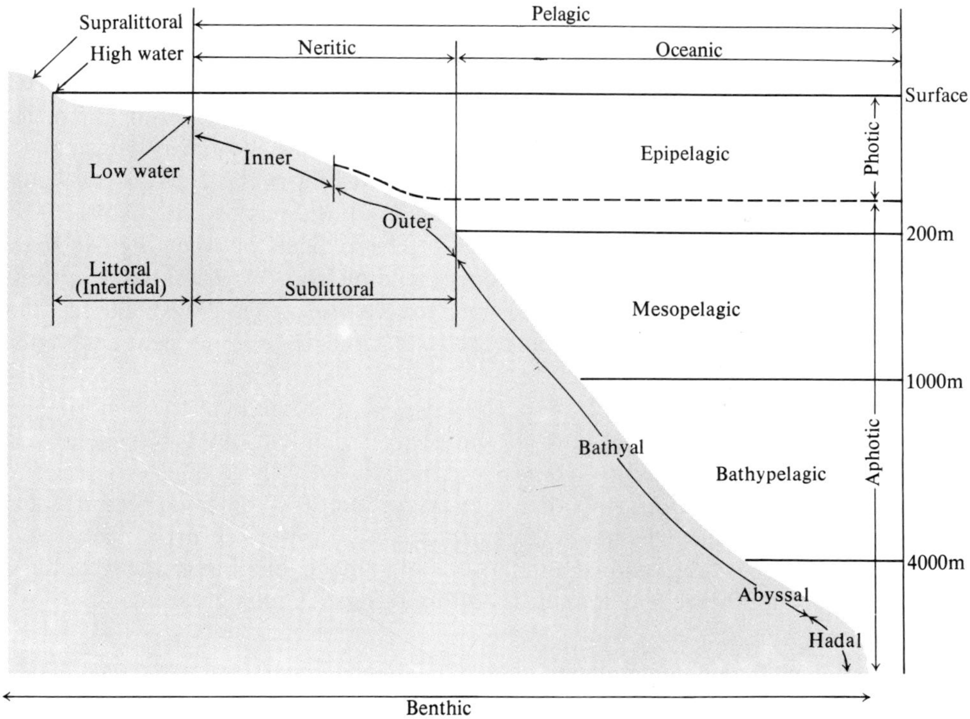

Fig. 14.2 The classification of the benthic and pelagic marine environment. (After J. W. Hedgpeth, ed., *Treatise on Marine Ecology and Paleoecology*, Ecology, vol. 1. The Geological Society of America, Boulder, Colorado, 1957.)

determined by the minimum energy level required for photosynthesis, and thus by the attenuation rate of solar energy and the available radiation level at the sea surface. The photic zone is deepest, therefore, in tropical and subtropical regions, where surface radiation is high and particle scattering is slight due to the sparsity of organisms and particulate matter. Water in the Sargasso Sea, which has a deep photic zone, is clear and well isolated from land effects by the bounding currents; its low particulate content is indicative of relatively sterile regions, or *biological deserts*. These regions are biologically poor because their surface waters have a low nutrient supply and thus cannot sustain abundant plant life to support a large part of the food chain. Coral reef systems in these waters, however, are very productive, forming localized bottom communities of plants and animals that support each other.

The thickness of the photic layer varies due to changes in the clarity of water and available radiation, whereas that of the epipelagic layer is considered nearly constant. The thickness of the other layers is somewhat arbitrary, since they represent zones that separate biological communities as well as environments. Thus their boundaries are not distinct, but overlap regions of diffuse nature.

The distribution of benthic organisms is determined by the type of material forming the sea floor, the food supply in the overlying water column and in the sediments, and the water environment. However, since the organisms are restricted to the sea floor (with the exception of larval stages in the water column), a separate classification system is used for their benthic habitat. The benthic habitat zones are the *intertidal* or *littoral* zone, between high and low water; the *sublittoral zone,* the sea floor extending from low water to the bottom of the epipelagic layer (about 200 m); the *bathyal* zone, 200 to 4000 m; the *abyssal* zone, 4000 to 5000 m; and the *hadal* zone, the still deeper ocean floor (Fig. 14.2).

The classic division of the ocean into the subregions of the benthic and pelagic realms is based on its physical, chemical, and biological environments. As applied to tropical waters, however, these divisions describe quite different environments than they do for temperate or polar regions. This deviation is especially evident in environments near the sea surface; at deeper depths, there is less seasonal and latitudinal control, and thus the geographical distribution of deeper communities of organisms is much greater.

BIOLOGICAL PRODUCTIVITY OF OCEANIC ENVIRONMENTS

The fact that there is considerable seasonal change in surface solar radiation in the temperate zone means that the thickness of the photic zone varies throughout the year. In the spring, when radiation levels increase, plant life grows rapidly, using the nutrients supplied to the surface waters from depth by winter wind mixing. The rapid population increase is also affected by another factor: the heating of the seawater from the surface downward to produce a stable water column. If solar radiation increases but the water column remains neutrally stable, external sources of energy, such as wind and currents, will produce sufficient turbulence and vertical mixing to prevent the suspended phytoplankton from remaining in the photic zone long enough to respond to the increased radiation. This would retard the population increase and diminish the spring bloom. Usually, if salinity variations with depth and time do not control the stability of the water column, surface temperature increases related to increased levels of solar radiation do. Thus surface heating imposes a stability on the water column that reduces vertical mixing and increases the residence time of phytoplankton in the layer above the developing thermocline, which is also in the photic zone. The retention of phytoplankton near the surface along with the increased radiation and available nutrients triggers the spring bloom.

In an estuary, stability may be controlled more by the vertical salinity structure than by seasonal temperature patterns. Thus if stability is maintained

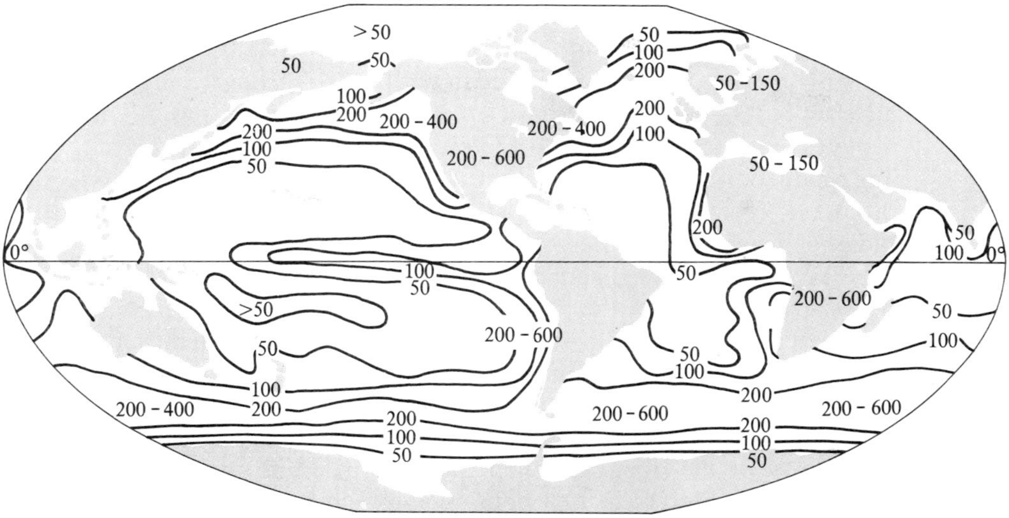

Fig. 14.3 The estimated annual productivity of the oceans, based on production of organic carbon by photosynthesis. Units are grams of organic carbon per square meter of sea surface per year. (After J. W. Hedgpeth, ed., *Treatise on Marine Ecology and Paleoecology.* Ecology, vol. 1. The Geological Society of America, Boulder, Colorado, 1957.

year round due to dilution of the surface water by freshwater, the spring bloom is directly dependent on increasing radiation. If stability is seasonally controlled, the spring bloom may be delayed until the water column gains stability from the higher spring discharge rates of the diluting rivers. In general, both temperature and dilution act to alter the seasonal stability of the nearshore water column, whereas temperature alone controls this process offshore.

Gradually, depletion of nutrients by the plants, natural mortality, and herbivore grazing will act to decrease the rate of increase in the plant population, maintaining it at a level in which uptake of nutrients equals their supply. Thus at midlatitudes both solar radiation levels and nutrient supply control the population. The depth of the photic layer tends to decrease as population and the scattering effect increase, and it increases with increasing radiation. The presence of particulate matter, such as suspended sediment supplied by rivers, would also decrease the photic layer. In general, it may be said that at midlatitudes, radiation and nutrient supply may successively limit the population, while organic production is continually low in the tropics because of low near-surface nutrient concentrations despite abundant radiation. The seasonal restoration of surface nutrients when light levels are low at temperate latitudes does not occur in the tropics (see Fig. 14.3).

In the polar regions the available radiation is always minimal despite the seasonal change from near total darkness to near total light. The summer period brings a large increase in duration of light hours, but the low angle of the incident radiation still keeps radiation levels relatively low. A marked increase in plant population is found in response to the increasing radiation at high latitudes, but this increase in organic production rarely completely depletes the nutrients and is short-lived. The most important factor in controlling production in these high latitude zones is solar radiation. The photic zone, of course, is shallower at higher latitudes than in the tropics due to the low radiation levels and the presence of particulate matter.

The foregoing are generalizations on the organic productivity of oceanic water and the depth of the photic layer in different latitude zones. There are many exceptions, however. As we have stated, the low nutrient supply in the tropics prevents high productivity of organisms; in temperate regions, both light and nutrients may limit production; and in the polar regions, light alone controls productivity. Thus any mechanism that locally supplies nutrients to the photic zone in the tropics or temperate regions where radiation is abundant stimulates organic production considerably and decreases the thickness of the photic zone due to scattering.

Nutrients are supplied to the photic zone for incorporation into organic material by (1) rivers, which supply land materials, and (2) upwelling of water from depths below the photic zone where nutrients have accumulated due to decomposition. The latter process may occur as (a) a direct flow of deeper water upward to replace surface water in regions of horizontal divergent surface currents or flows, and (b) turbulent mixing processes that blend deeper water or, in shallow water, material from the sea floor into the surface waters.

Thus upwelling in nutrient-poor midocean areas will support a large biological population. In the region of the equatorial currents, for example, the Coriolis force acts on opposite and adjacent currents to cause surface divergences and upwelling which create a more productive zone than that found in most tropical areas (Fig. 14.4). Small isolated regions of high surface productivity may also form if the currents create cyclonic gyres. These gyres produce a sea-surface depression, beneath which a dome of denser (deeper) water must rise toward the surface to balance the pressure distribution. This upward movement supplies nutrients to the photic zone, creating an area of high surface productivity. The cyclonic gyre in the Gulf of Alaska, for example, acts to stimulate surface plant production by supplying nutrients to the photic zone.

In the neritic zone, rivers, coastal upwelling, and mixing over the continental shelves combine to promote production of organic material by supplying nutrients from land sources and at depth to the photic zone. The neritic zone therefore is usually much more productive than the oceanic areas and has a shallower photic zone. If the nutrient supply to the neritic zone is such that it is never a limiting factor in organic production, light becomes the most important factor controlling productivity, which is therefore higher in tropical

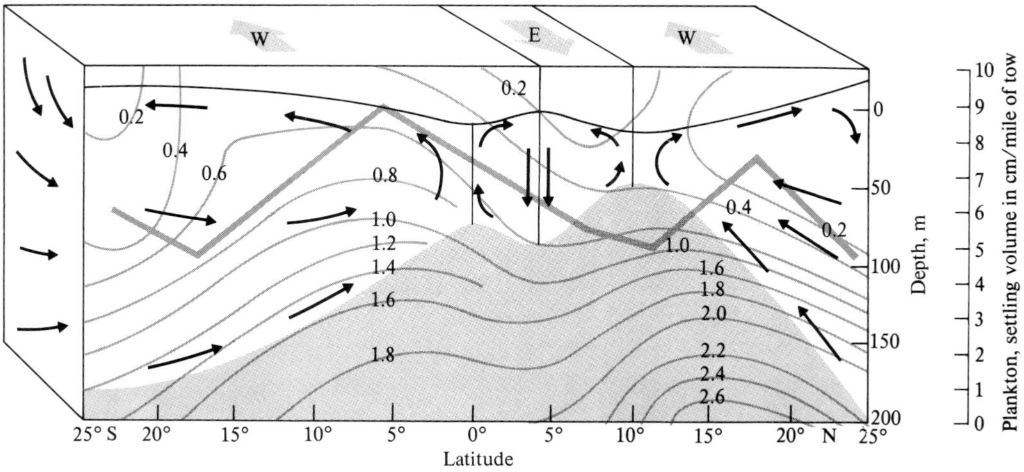

Fig. 14.4 The opposing surface currents found in the equatorial zone and the location of their boundaries relative to the equator produce an upwelling of deeper water that enhances productivity. The vertical circulation associated with the surface divergence caused by the Coriolis forces deflecting the surface current is shown schematically. This vertical circulation supplies nutrients to the photic zone as shown by the isopleths of inorganic phosphate phosphorus, units μg-atoms of phosphate per liter, which act to increase the abundance of plankton. Plankton concentration is given in units of a settling volume of plankton per mile of tow. (Adapted from H. U. Sverdrup, M. W. Johnson, and R. H. Fleming, *The Oceans*, Prentice-Hall, Englewood Cliffs, N.J., 1942.)

regions than in temperate or polar areas. Production in the neritic zone is much more variable than in the oceanic regions because it depends primarily on the processes governing the local supply of nutrients. Although our discussion of pelagic productivity has centered on photosynthesis and plant life in the photic zone, we do not intend to imply that animal life does not contribute to production in the sea. We have emphasized plant life because it is the organic level at which biological productivity starts. In the sea, as on land, plant life must exist in order for animal life to survive; thus regions that produce large amounts of plant life to be consumed by herbivores can also support large carnivore populations.

THE FOOD CHAIN OF THE SEA

The groups of organisms that comprise biological communities depend on a system that allows food energy or biomass to be passed through their community and on to others. The primary source of energy is sunlight, used in photosynthesis by plants, phytoplankton, and other algae to produce organic

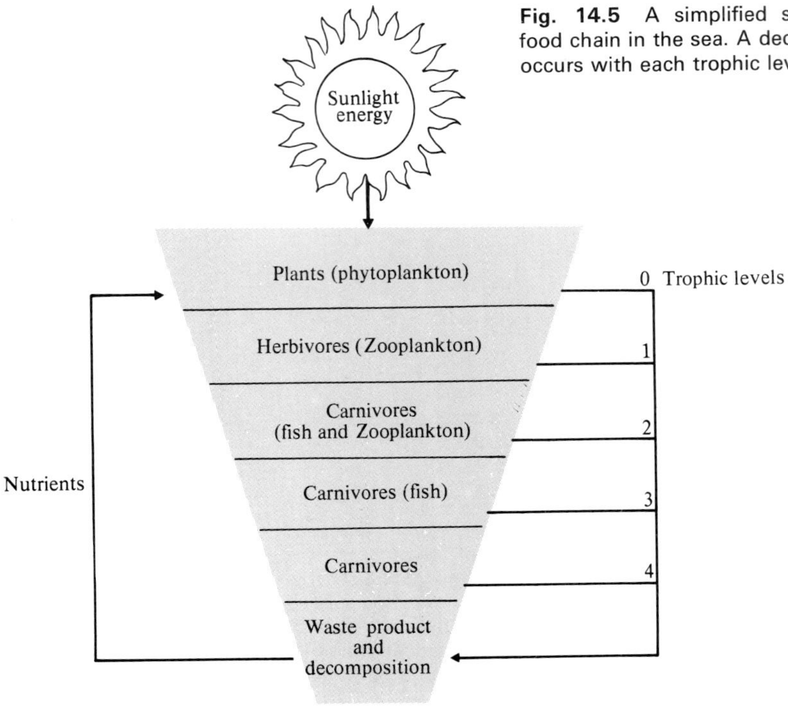

Fig. 14.5 A simplified schematic showing the food chain in the sea. A decrease in organic mass occurs with each trophic level.

material, which is then passed through the *food chain* (or *food web*, as some prefer; see Fig. 14.5). Each step, or trophic level, in the food chain represents a reduction of about 90% in biological mass from the previous level. That is, if there are 100 units of energy in a primary plant level, the herbivores in the next trophic level will use 90 units for metabolic processes and 10 for body materials (increase in biomass); the latter are then passed on to the next trophic level of carnivores. The carnivores then use 9 of the original 100 units for metabolic processes and 1 for body material; this unit is then passed on to the next level.

The 90% decrease, though frequently cited in the literature, is not an accurate indicator of efficiency at all trophic levels or for all species. The decrease in mass may be greater or less than 90%; thus one should consider the cited percentage as an estimate only.

Since food energy is thus attenuated rapidly in the sea, organisms that are in trophic levels remote from the primary source have less total weight (biomass) as averaged over the oceans than those closer to it. At low trophic levels, small organisms are usually very abundant, making up a large biomass by numbers rather than size, while large organisms that feed at a low trophic level are usually few. Both baleen whales, which feed on the small crustacea of the sea, and herringlike fishes are close to the primary level, the former being less numerous than the latter. The transfer of energy and organic matter through the food chain does not stop when material has reached the last predator. In each level, the natural mortality of organisms and their rejection of organic matter as body waste cause material to be liberated in soluble form to the water or to settle to the sea floor for consumption or decomposition. Decomposition or excretion of biogenic materials liberates nutrients back into the water column, where they are recombined with fresh solar energy to produce more organic material and complete the cycle.

As we have used them thus far, the terms "organic production" and "productivity" refer to the *rate* at which organic material is generated in a marine environment, i.e., grams of material per square meter of sea surface per unit time. A region of high productivity, however, does not necessarily have a large organic biomass at any given time. The biomass present at a given time, which is called the *standing stock*, or crop, is the biomass expressed in grams of organic material per square meter of overlying sea surface. The difference here is easily demonstrated. Consider a field that can raise five crops per year with a yield per crop of 2 kg/m^2. Then the standing stock just before harvest is 2 kg/m^2, but the average for the entire year is less, since the field at times is devoid of organic material. The annual productivity, however, is 10 kg/m^2.

A region in the sea may be very productive due to the rate of increase in phytoplankton biomass, but planktonic herbivores may simultaneously deplete the phytoplankton at a high rate. Therefore, the region may have high productivity but a small standing stock.

Productivity may also refer to an increase in population, or the rate of change of a population per square meter of sea surface or of a cubic meter of water. However, in this context productivity does not necessarily indicate the rate of production of organic biomass, since the number of organisms may increase at a faster rate than the population's biomass. If the net population change is zero because natural mortality and predation equal reproduction, the population is in steady state. This equilibrium is not static, however, and there may be considerable increase in biomass as the population density remains constant. The problem here, of course, is one of defining terms and recognizing that a change in body weight is not synonymous with reproduction.

There is a considerable range of organic production rates in the sea. Dr. Eugene Odum, a noted ecologist, estimates that an estuarian environment is about twenty times as productive as the average open sea, seven times as productive as an alfalfa field, and twice as productive as a corn field. Thus the open sea is on the average about one-tenth as productive as a corn field, or not especially good as an environment for food production, except that it is a large field when acreage is considered. However, usable food resources are not uniformly distributed in the sea and thus can be economically harvested in the regions of greatest abundance and in the form of protein rather than carbohydrate. Although the sea may help ease the world food problem by supplying much-needed protein, it is not an infinite resource and, if harvested heavily, will change as a producing environment. The damage that would be done to the food chain by removal of a trophic level is obvious.

OTHER FACTORS AFFECTING THE OCEANIC ENVIRONMENT

We have discussed how solar energy, nutrient supply, depth, and proximity to land affect the biological environment. Obviously, variations in other factors, such as salinity and temperature, also have an effect. The distribution of marine organisms within the marine environment, for example, is determined by food supply, habitat requirements, tolerance to temperature and salinity, and predation. Thus, as often as not, the physicochemical and geological aspects of a region are just as important as depth, proximity to land, and latitude in characterizing an environment. These latter environmental factors, however, are not so regular in their distribution that they can be schematically represented, as in Figs. 14.1 and 14.2.

The environment, which to a large degree controls the biological community, determines which organisms may successfully inhabit an area of the sea. Temperature and its range are the most important physical factors of the near-surface biological environments; they are related, as mentioned before, to the tropical, temperate, and polar subregions. Salinity appears to be less important, since it changes relatively little in the open sea. Yet environmental control may limit certain species groups to a specific water mass which is separated from other water masses only by subtle changes in salinity and temperature. If the association between a marine organism and a specific water mass is strong, the organism may be used as an indicator of the water mass. The reasons for the association between physical environment and organism are difficult to determine and in some cases unknown. In the case of an organism confined to a particular water mass, salinity and temperature factors may not be nearly as important as the current system of the ocean, which tends to confine a water mass and its organisms to a particular geographical location.

Technically, it is relatively easy to measure the physical and chemical properties of a given environment. It is also feasible to determine the composition of the biological community in the same environment. To determine why a specific community is associated with a specific environment is not so simple, however. In many cases we can establish these relationships statistically but may not be able to explain them. Similarly, it is not easy to evaluate the relationship between species forming a community under various environmental conditions. Their interaction, whether beneficial or detrimental, may be as important in determining the species composition of a community as the physicochemical constraints on the environment.

Because they are only partially understood, these relationships are the weak link in our study of the sea's ecology. If they were better known, we could make relatively simple measurements of the environment to determine and even predict the state of the biological population. One need hardly emphasize the impact on open ocean fishing if we could predict where fish would be at a given time.

The symmetrical latitudinal surface environments, tropical, temperate, and polar, are defined by sea-surface temperatures (Fig. 14.1). These zones have movable boundaries that tend to follow the sun in the seasonal traverse between the two hemispheres. The zones also tilt due to surface transport of warm water poleward on the western edges of the major oceans and an equatorward transport of cool water on the eastern sides. The tilting has its maximum effect across the anticyclonic gyres of the central ocean basins.

Beneath these uppermost environments of the sea, the latitude and seasons exert little influence on the environment or the biological community. The horizontal temperature and salinity fields of the sea are nearly uniform in the bathypelagic and abyssal zones, allowing nearly worldwide distribution of many marine forms. The biota of the hadal zone are more likely to be different in the widely separated basins and trenches of the ocean because of an intervening dissimilar environment.

The effect of depth in decreasing latitudinal influence on the oceanic environment is similar to that of elevation on land. On land, an arctic environment, as described by mean temperature and range, is a surface found at sea level above 70° Lat, at an elevation of about 2 km at 45° Lat and at an elevation of about 4 km at the equator. Only a few isolated peaks have elevations that reach or exceed this "temperature surface" at the lower latitudes. When they do, the environment on the mountain at or above this surface will be of the arctic type. The soil types and plant and animal life of these isolated peaks differ, and the isolation of one arctic-alpine environment from another may be quite complete. In subarctic environments the temperature surface is found at lower elevations, and there are therefore many more landforms penetrating it. Hence the isolation of animal and plant life by intervening environments is not as complete as in the arctic-alpine case.

We may reverse these examples and apply them to the oceans. Since temperature may be used to describe the tropical, temperate, and polar areas, a temperature surface within the sea may describe the extension of the polar sea surface zone into lower latitudes. We may trace a surface temperature that is commonly at the sea surface in the arctic regions toward lower latitudes, where it lies deeper. Marine organisms may be widely distributed if they are successful in these cold temperatures and are not restricted by other specific environmental conditions, such as contact with the sea floor, special food requirements, availability of sunlight, or pressure. The limited distribution of species in the deeper hadal zone and the changes in communities in the trenches correspond to similar effects on communities on isolated high mountain peaks.

Other factors enter into this problem. The thermal layer in the sea, representing a constant temperature environment, may have discontinuities which cause abrupt changes in the types of marine organisms. Submarine ridges may rise from the sea floor to interrupt a thermal layer and isolate water masses having different sources and properties but a common boundary. We have already pointed out this effect in connection with the exclusion of Antarctic bottom water from the Angola Basin. A classic example of isolation of both water and biological communities is the Wyville Thomson Ridge, which extends between Scotland and the Iceland-Faeroe ridge system. To the north of the ridge there are arctic-type water, flora, and fauna, and to the south the water, flora, and fauna are related to the central North Atlantic. A benthos survey showed that only 11 of the 48 species found on the ridge lived on both sides of it, in the two different environments. Sharp thermal boundaries between water types do not necessarily depend on topography; they may also be maintained by the currents, as is the cold wall of water forming the inshore boundary of the Gulf Stream off the central east coast of the United States.

Water properties also change sharply due to currents off the coast of Peru. Occasionally a disturbance in the weather pattern affects the currents, thus subjecting the biopopulation to a different set of environmental conditions. This may cause a mass slaughter of marine organisms, which in turn interrupts the food chain and pollutes the water with decaying material. Even sea birds may be killed. This phenomenon has been named the "Callao Painter" because the hydrogen sulfide produced by the decaying material blackens the lead-base white paint on ships in the area.

Compared to temperature, salinity has little control of the environment in the open sea. However, in the neritic zone, especially in the estuaries, salinity may be very important. In this region, organisms may be subjected to large changes in salinity and temperature; sessile benthic organisms in the intertidal zone even suffer dessication. Some organisms spend their entire lives in the rapidly changing estuarian environment. Others, which normally inhabit the open ocean, return to the estuaries or even to freshwater to reproduce. Sea-run trout, salmon, and eels, for example, migrate between freshwater and the sea for spawning and maturation. Organisms that undergo these drastic changes must adapt physiologically or perish. They must maintain their water and salt balance when their tissue fluids are more salty than the surrounding medium at one time and less salty at another. Either they must withstand the changing osmotic pressures or there must be a mechanism for transfer of salt or water to or from their body tissues.

As we have pointed out, estuaries are part of the neritic environment and must have an unspecified net dilution by freshwater. Simply stating that a region is an estuary does not suffice to identify it as a specific environment. We must consider other factors to classify estuaries as subsections of the marine environment and to distinguish between them. First, we might consider the range of salinity over each tidal cycle, which is indicative of the amount of freshwater present. Second, we might take into account such factors as shallowness, the presence of extensive marsh systems as opposed to distinct shorelines, bottom type, tidal range, tidal prism, and flushing rate. Obviously, two estuaries with the same physical characteristics such as shape, tidal volume, mud flats, and so on, are far from identical if one is exposed to large amounts of sunlight so that its seabed and then the water on the rising tide undergo considerable heating. Thus external climatic conditions also act to stipulate an estuarian environment.

In general, the smaller the environment the more factors we must consider to define it and distinguish it from other environments. On the oceanic scale, we can identify large generalized environments using only a few characteristics—light, depth, and distance from shore. For major subareas of the ocean, we must also consider latitude and seasonal effects and possibly water mass and bottom sediment characteristics.

Marine mammals and some free swimmers, such as fish and turtles, migrate tremendous distances across the sea, encountering environmental changes on their way. During their seasonal breeding cycles, organisms on northward or southward migration routes pass from tropical or subtropical environments to polar environments and back. Sea birds leave the regions in which they usually reside for a different environment for nesting.

Unlike the mobile free-swimming organisms (*nekton*), which can readily move to another environment if it is of benefit to do so, benthic animals or plants are restricted to living in or on the bottom. However, if during reproduction and early growth they form swimming or floating larvae, currents may transport the young to a new area. Occasionally, the adult organisms may be transported by rafting as a part of a self-maintaining community. It is not uncommon to find objects such as pieces of wood, logs, coconut husks, and other flotsam drifting on the sea; often, they contain attached communities of plants and animals, some of which would normally be found living on the sea floor in shallow water. Barnacles, nereid worms, crabs, isopods, hydrozoans, bryazoans, crustaceans, algae, and wood-boring mollusks are among the forms that may drift on a piece of flotsam in self-sustaining microcommunities. The flotsam acts as the substrate for attachment, a flotation device to keep the community in the photic zone, and a transport vehicle. As long as the community does not encounter detrimental salinity and temperature changes and maintains a balance of food requirements, it can survive a considerable journey across the sea. Large marine mammals and ships may also play a role similar to that of flotsam in transporting organisms. If the organisms survive the journey and are deposited in a new environment where they can thrive and reproduce, a new colony can be started.

It is small wonder that a newly created, sterile environment, such as a fresh volcanic island, cannot remain free of life long. Airborne and bird-carried seed eventually produce ground cover, while the migrating voyagers of the sea gradually inhabit its beaches and waters.

There is no simple way to describe the oceanic environment. We may think of the factors controlling it in the same way as we do of those on land.

We have already mentioned factors that exercise "climatological" control and those that describe the substrate or sea floor. Although the number of possible combinations of these factors is almost infinite, they represent only a few of the forces at work. For example, one should also consider the effect of the fluid, since the environment of organisms immersed in a solution such as seawater is quite different from that of organisms immersed in a less dense gaseous fluid like air.

The problems of obtaining oxygen for respiration, body support, mechanical abuse by motions in the fluid, heat exchange, and food supply (to mention only a few) differ according to environment. Consider the difficulty that an omnivore (one that eats both plant and animal matter, as opposed to herbivore and carnivore) attached to one place would have obtaining an adequate food supply on land. The sessile organisms of the sea floor do quite well, however, since water currents, the rising tide, or the organic debris falling from the overlying water column continually bring a fresh food supply that can be filtered from the surrounding water or sediments. Very delicate and bizarre organisms which are almost neutrally buoyant also live in the sea. Although they have few or no body structures for support, they survive very well since they move with the fluid rather than resisting its motion; they react as does the water they displace. Populations at the edge of the sea, where turbulent energy and wave forces are expended against solid barriers, must either be mechanically strong enough to withstand the constant battering, or resilient enough to give way rather than absorb the full impact of the expended energy. Sea mammals and reptiles that move with ease and grace in the water are very cumbersome on land; some cannot even support themselves when they become stranded.

Even plants of the sea require environmental conditions that are different from those of their counterparts on land. In the oceans, the photic zone has thickness, whereas on land it is only a surface. Algae derive their nutrient supply and gases for respiration and photosynthesis directly from the liquid that bathes them rather than through special structures (leaves and underground root systems) that land plants use to acquire nutrients from soil and air. To accomplish this, the ratio of surface area to volume must be large; therefore, sea plants must either be very small (like phytoplankton, the unicellular algae) or very large, with many filaments or fronds. The sea plants must also provide themselves with mechanisms for staying in the photic zone. The unicellular plants are aided by the stability of the water column near the sea surface, but they also produce oils within their cells which help to control their buoyancy. Multicellular algae either are attached to the substrate in shallow water or have small, gas-filled bladders that support the fronds on or near the sea surface if their attachment is deep.

Seasonal changes in the sea's temperature are small compared to those on land. Therefore, the coats of sea mammals do not change with the seasons as do those of land animals. However, these mammals must be permanently protected by fat if they dive deep into the sea, thus experiencing rapid temperature changes. Since the latitude effect on sea-surface temperatures is also not as great as it is on land, sea mammals can easily traverse great ranges of latitude. Land mammals, however, may be restricted by latitude; they may even undergo anatomical changes in order to survive in a different environment. For example, compare the arctic hare with the desert jackrabbit. The compact shape and small appendages of the former help to conserve body heat by cutting down the ratio of radiative surface area to body volume. The jackrabbit, on the other hand, has much larger appendages; therefore its surface area is larger, promoting a loss of body heat. The two animals probably would not survive if they changed places.

The primary unit of an oceanic biopopulation is the community, a grouping of organisms with common environmental requirements. A successful community usually consists of organisms that can at least tolerate each other and probably are mutually dependent and/or supportive. Communities may be very small, e.g., the travelers on a piece of flotsam, or they may cover extensive areas of the oceans that support similar planktonic and benthic species.

Land communities are often time-dependent due to the continuous modification of the environment by biological and terrestial processes. A succession of one land community by another in response to these changes is not uncommon. Communities in the sea, however, are nearly powerless to effect semipermanent changes in the environment; they can alter the substrate, but exercise little change in the water's properties. This coupled with the slow rate at which the oceanic environments change through time assures a longer existence for a marine community than a land community. This is reflected in a slower rate of evolutionary change in marine species than in land species.

STUDY QUESTIONS

1 Temperature (average value and seasonal range) is one of the most important parameters in determing the distribution of biological forms in the sea. Discuss why its effect is greater at shallow depths than at mid depths.

2 What physical process in the sea is reduced at low latitudes and tends to restrict the productivity of oceanic waters?

3 In what ways does high primary productivity tend to be self-limiting? Consider both the chemical and available-energy aspects of this problem.

4 Why does a cyclonic ocean gyre at low latitudes tend to have high surface productivity?

5 Why are there large seasonal variations in primary productivity in some regions of the sea? Consider the changes in solar radiation with latitude and time, as well as the effect of changing surface currents due to variation in the atmospheric wind system and stability of the water column.

6 What single environmental factor has a great influence on plant distribution in the sea but little or no effect on the animal life distribution?

7 Under what conditions would you expect anticyclonic current gyres to have reasonably high productivity?

8 What factors may be important in isolating one biopopulation from another in the littoral and bathyal zones?

9 If there is a 90% loss of biomass in each trophic level of the simplified food chain, how is the biomass dissipated? Why doesn't it transmit to the next higher trophic level?

CHAPTER 15

THE BIOLOGICAL POPULATION OF THE OCEANS: BENTHOS, PLANKTON, AND NEKTON

INTRODUCTION

The organisms making up the biological populations of the sea include members of both the plant and animal kingdom. The two groups are clearly different, except for the intermediate dinoflagellates, some of which can ingest their food as well as photosynthesize. The variety in marine biota is enormous, both with respect to size (ranging from whales with a mass several times that of elephants to microscopic bacteria) and to the number of phyla and individuals within each phylum. A comparison of land animals and plants and marine animals and plants shows that the number of animal phyla is greater in the sea than on land, while there is a greater diversity of plants on land than in the sea. From the viewpoint of diversity—not absolute number—the marine environment may be called the animal realm, while the land environment may be regarded as the plant realm.

On the basis of their locomotive ability and mode of living, marine animals and plants are divided into three groups: benthos, plankton, and nekton. To the *benthos*, or benthic biota, belong the organisms that live on, live in, or spend most of their time crawling about the sea floor. The benthos thus includes buried forms, forms attached to hard substrate (sessile organisms, both plant and animal), and organisms that move about, such as crabs, starfish, octopi, and bottom-dwelling crustaceans. Bottom fish which spend considerable time resting on the sea floor are sometimes included. Benthic algae (*phytobenthos*) that are attached to the sea floor need light and can therefore exist only in shallow water, except for the few that are *heterotrophic* (capable of producing plant material chemically without sunlight). Benthic animals (*zoobenthos*), most of which are invertebrates, exist at all depths of the sea. Because of the many variables in the environment—type of bottom, temperature, food—it is almost impossible to subdivide the benthic animals into categories associated with depth. However, they can be grouped according to their mode of living. Those that live buried in the substrate are called *infauna*, while those that are attached to the surface of the sea bed or crawl over it are the *epifauna*. The distinction between these two groups seems clear enough: however, nature abhors stereotypes, and there are animals, such as crabs and shrimps, that will bury themselves in the sediment only to surface and crawl about or swim to some other location.

Animals and plants that float about in the water column because they lack sufficient locomotive power to swim against the ocean currents are called *plankton.* Plants and animals of all types that are not good swimmers fit this category. Note, though, that the criterion of "good swimmer" is somewhat vague, because animals capable of coordinated swimming as well as those incapable of swimming belong to the plankton category. Even the oceanic sunfish, *mola mola,* may be at the mercy of the stronger oceanic currents and thus be considered as planktonic.

The plankton and the benthos are divided according to size into macro, meso, micro, and nanno forms. The largest planktonic forms are called *macroplankton,* with a lower size limit of about 1 cm. *Mesoplanktonic* forms still visible to the unaided eye range from 1 cm to 1 mm. *Microplanktonic* forms range from 1 mm down to 0.076 mm; they make up the bulk of the material usually collected in fine-mesh plankton nets dragged through the sea. For this reason, microplankton are also called *net plankton.* Finally, the smallest plankton are called *nannoplankton,* or *dwarf plankton,* with a size range from 10μ to less than 1μ. To collect these small drifters, one cannot use towed nets, since the pressure of the water flow through the net mesh destroys their delicate forms. Instead one uses special filtration apparatus and centrifuges to gently separate the nannoplankton from bulk water samples brought on shipboard.

Size is only one of the many factors used to categorize plankton. Another is the division into *zooplankton* (animals) and *phytoplankton* (plants). The planktonic community also has a transient component, namely the eggs, larvae, and gametes of benthonic and nektonic organisms. This plankton, which is a stage of the life cycle of nonplanktonic forms, is called *meroplankton.* Its abundance usually follows seasonal reproductive cycles.

Many of the planktonic forms use their limited locomotive ability to perform definite tasks. Jellyfish, for example, pulsate their bells as they are carried along by the currents—slowly rising upward, then relaxing and slowly sinking. During the descent, they try to trap food under their umbrella whence it is conveyed to their gullet for ingestion. The small crustaceans, copepods and euphasids, are able to dart over short distances with considerable speed. The direction of their motion is somewhat random and does not necessarily result in a preferred horizontal displacement.

Some zooplankton are capable of coordinating their locomotive powers to produce vertical displacement. Planktonic organisms and their predators are often found in large concentrations in distinct layers in the sea. These layers scatter sound beams, and hence can be detected by sonic depth-finding equipment; that is, sound waves are partially reflected back to the source as they pass through these dense population layers of organisms. On the fathogram, this partial reflection is recorded as a false bottom layer which has been named the *deep scattering layer* or DSL (Fig. 15.1). This layer can be observed to rise and fall inversely with available sunlight; it lies at depth during the day and rises toward the surface at night. This diurnal migration indicates that the planktonic forms in this layer are capable of coordinated motion in response to the changing light.

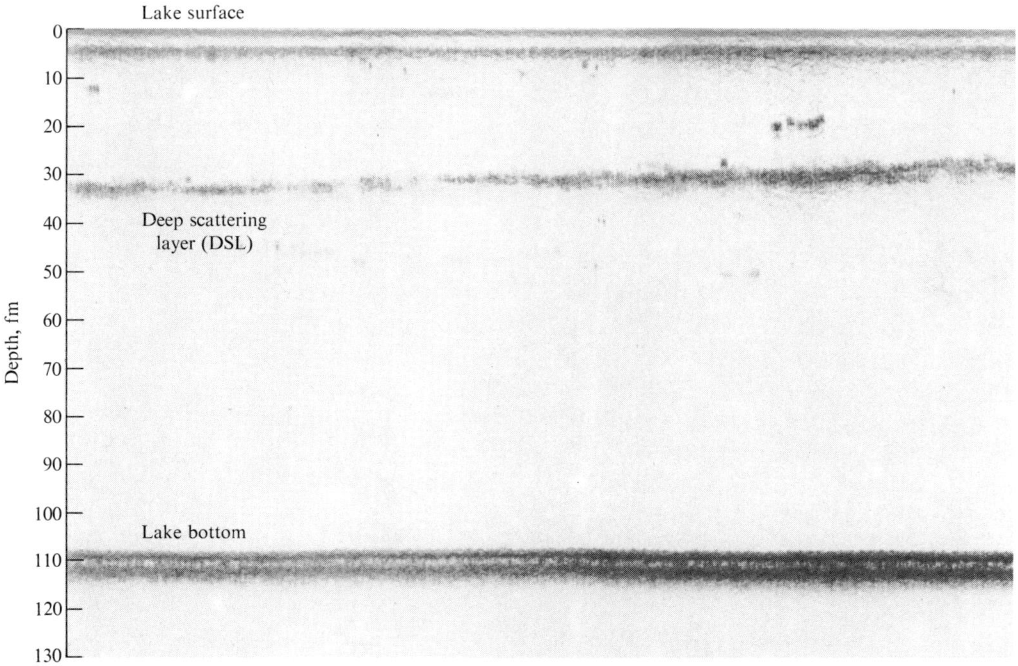

Fig. 15.1 An echo-sounding trace made in Lake Washington showing a partial reflection of the sound beam by organisms forming a deep scattering layer and by individual fish.

The *nekton* of the sea are the strong swimmers which move freely from place to place, unimpeded by the speed of oceanic currents. Nekton are animals: fishes, mammals, reptiles, and cephalopods such as the squid are the major members of this category. These animals can actively pursue their prey, flee if endangered, and traverse great expanses of water in migratory journeys. Compared to the plankton, which contains representatives of practically every animal phylum, the nekton has little diversification. However, the size of the animals in the nekton, their characteristic habit of gathering in schools, and their abundance over discrete areas of the sea make this group the most important marine food resource available for harvesting by man.

The nekton and the benthos cannot exist in large numbers without an adequate food supply. This is provided by the plankton, which appear in the food chain as the primary producers,* or as herbivores and carnivores. Other sources of consumable organics are the benthonic algae and detritus from land; however, these two are not significant on the gross oceanic scale. The

* A primary producer is an organism (plant or protist) capable of manufacturing food from inorganic raw materials.

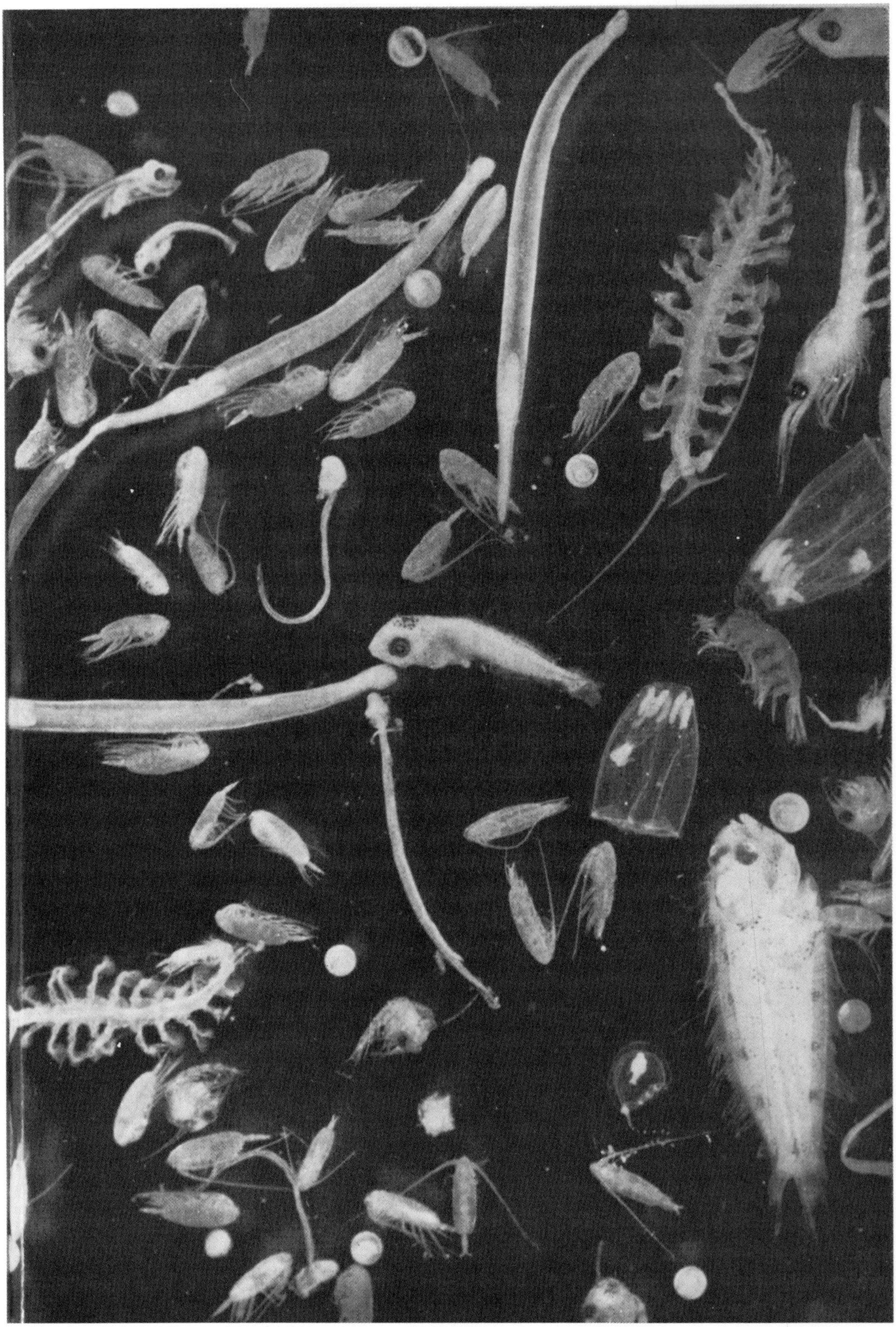

Fig. 15.2 A plankton sample with representatives of the *meroplankton*. (From A. C. Hardy, *The Open Sea*, Houghton Mifflin, Boston, 1964.)

meroplankton, the planktonic eggs and larvae of the benthos and nekton, also are a source of food. However, in spite of the enormously lavish fruitfulness of marine animals, the parent populations could not survive long if they were dependent upon their own reproductive products for food. Hence it seems that both the nekton and benthos of the open sea must have been latecomers to the marine populations, establishing themselves only after a suitable planktonic food base had become available. The law of eat and be eaten is as applicable to the sea as to the jungle.

The sessile benthonic organisms, which cast off sexual stages or spawn their eggs into the water to be fertilized at random, and even the nekton which bear eggs or live young into the plankton are dependent upon the forces of the environment to disperse the young. Currents carry the meroplankton to regions where water conditions and substrate may or may not be conducive to their survival. Thus availability of food, reproduction rate, predation, and natural mortality all combine in a grand scheme to control the populations of the sea (Fig. 15.2).

In some cases the benthos, both plants (phytobenthos) and animals (zoobenthos), exist under wildly fluctuating environmental conditions. Those found in the intertidal zone are subjected to some of the greatest environmental changes on earth. Temperature fluctuations, exposure to air and dessication for differing time spans, pH changes, oxygen tension changes, and variations in the incident solar radiation are just a few of the constantly changing environmental factors. When we consider these factors together with the variety of substrates and amplitudes of the tides found in intertidal regions, the diversity of the benthos in the shallow fringes of the world's seas is not surprising.

THE PHYTOBENTHOS

Despite the number of factors controlling the distribution of benthonic forms, it is possible to make some general statements about the distribution of the phytobenthos because of their dependence on light. However, this is not the case with the zoobenthos, since no single factor, such as solar illumination, exercises a universal control overriding the effect of other environmental conditions.

The phytobenthos of the sea are essentially restricted to the shallow photic layer, where suitable substrate for attachment is also present. Algae, like land plants, contain chlorophyll and are capable of photosynthesis. Unlike land plants, their reproductive structures are unicellular and they lack true

Fig. 15.3
Eel grass.

Fig. 15.4
Fucus.

Fig. 15.5
Kelp.

Fig. 15.6
Sargassum.

vascular systems. [One phytobenthonic plant, eel grass (*Zostera*, Fig. 15.3) is not an alga but a true flowering plant which survives immersed in seawater; it is an important primary food source.]

The algae are divided into groups according to the following criteria: chlorophyll pigments, storage products (carbohydrates, proteins, or fats and oils), and cell wall material (cellulose, pectin, silicon, etc.). The categories we shall describe are based on pigment types.

In the intertidal zone, even the layman will notice that algal type and color change from the high tide mark to below the low water mark. A region of hard substrate with occasional sand pockets and average tidal fluctuation offers the best view of vertical algal zonation. The benthonic algae of such a region are primarily brown algae (Phaeophyta), green algae (Chlorophyta), and red algae (Rhodophyta).

Brown algae are found over a considerable vertical range. It is the fucoxanthin pigment that gives these algae their characteristic color, although it may also produce hues ranging from golden brown to black. The common rockweed (*Fucus*) of the northern hemisphere is a brown alga found attached to rocks high up on the tide line (Fig. 15.4), whereas the brown *Laminaria* and the large kelp (*Nereocystis* and *Macrocystis*), Fig. 15.5, may be anchored to the sea bottom at depths as great as 35 m, with their fronds floating at the surface supported by gas-filled floats. These kelp are the largest of the marine algae. *Sargassum* (Fig. 15.6), which may be attached or planktonic, is also a brown alga.

The green algae are found predominantly in freshwater; only a few are found in the sea. The common sea lettuce (*Ulva* and *Monostroma*) grows at shallow depths in the colder water regions; it is frequently found together with the brown algae over the same vertical range within the intertidal zone and

Fig. 15.7 Ulva.

just below (Fig. 15.7). Sea lettuce is used as food in the Orient but its important role is that of a primary producer in the marine food chain. The seaweed soup served in Chinese restaurants is often made with Ulva.

The red algae are the most diverse of the benthonic algae. Accessory pigments and environmental conditions combine to produce coloration ranging from green to red-brown, bright red, blue, and purple-red to black. Red algae are found over a wide region extending from the high intertidal zone to the lower limit of the photic zone. The variation in color is directly related to the vertical changes in habitat and hence light conditions. The dull green or black forms are found in the high intertidal zone, with purple-red, brown, or rosy-red colors predominating in the lowest portions of the intertidal region. At greater depths, the color changes towards more vivid reds, since it is the red-producing pigments that facilitate photosynthesis at the greater depths, where short-wavelength blue light predominates. Thus the distribution of the benthonic algae shows reds predominating at the deepest photic depths, browns at the shallower depths and intertidal region, and greens at the shallow end of the brown-alga range.

The benthonic algae of the sea are harvested by man. The brown algae are harvested for direct consumption as food (kombu–Japan) or for manufacture of alginates which are used as emulsifying agents in soaps, plastics, dairy products, and films. They also are a source of iodine and fertilizer. The dulse familiar to inhabitants of our maritime states, who eat it in the form of candy, relish, or in soups, is made from the red alga *Rhodymenia* (Fig. 15.8). *Porphyra*, another red alga, is cultivated for food in Japan, where it is known as *nori*. Agar-agar and carageenin are extracted from red algae and find many uses in industry.

There are many other algae in the sea. Some are planktonic; others are both planktonic and benthonic. The blue-green algae (Cyanophyta) occur in uni-

Fig. 15.8 *Rhodymenia.*

cellular colonial and filamentous forms. They color the Red Sea red and form mats of growth in shallow coastal areas. Together with the red algae they play an important role in the coralline reefs of tropical waters. The golden algae (Chrysophyta) are planktonic and of microplankton to nannoplankton size. The centric diatoms and coccolithophores are particularly prevalent in this group (pennate diatoms are considered as primarily benthonic). Another planktonic alga is the little-known yellow-green (Xanthophyta). Motile algae, e.g., dinoflagellates and cryptomonads, belong to the Pyrrophyta and are planktonic. Both the Chrysophyta and Pyrrophyta will be considered when we discuss phytoplankton in greater detail.

THE ZOOBENTHOS

The light-controlled zonation of the phytobenthos, in spite of the considerable overlap between brown, green, and red forms, is a very tidy arrangement compared to the zonation in the zoobenthos. Attached benthonic algae need a hard substrate to cling to. Hence, except for algae that form mats or the benthonic diatoms, algae are nearly absent from sand and mud beaches. The benthonic fauna, however, appear on both hard and soft substrates and hence are found in two distinct substrate environments. In addition, their distribution is not governed by any one controlling factor such as sunlight. The hydrostatic pressure of the overlying water column does not exert much influence either—animals that are more than 90% water and have no internal gas storage chambers are nearly as incompressible as the surrounding seawater.

In soft sediments the region above high tide is inhabited by crabs, isopods, and a variety of crustaceans that scuttle to and fro, seeking moisture by burrowing or traveling to the water. In the intertidal region, the loose sediments support shrimp, crabs, and clams as well as burrowing marine worms. Below

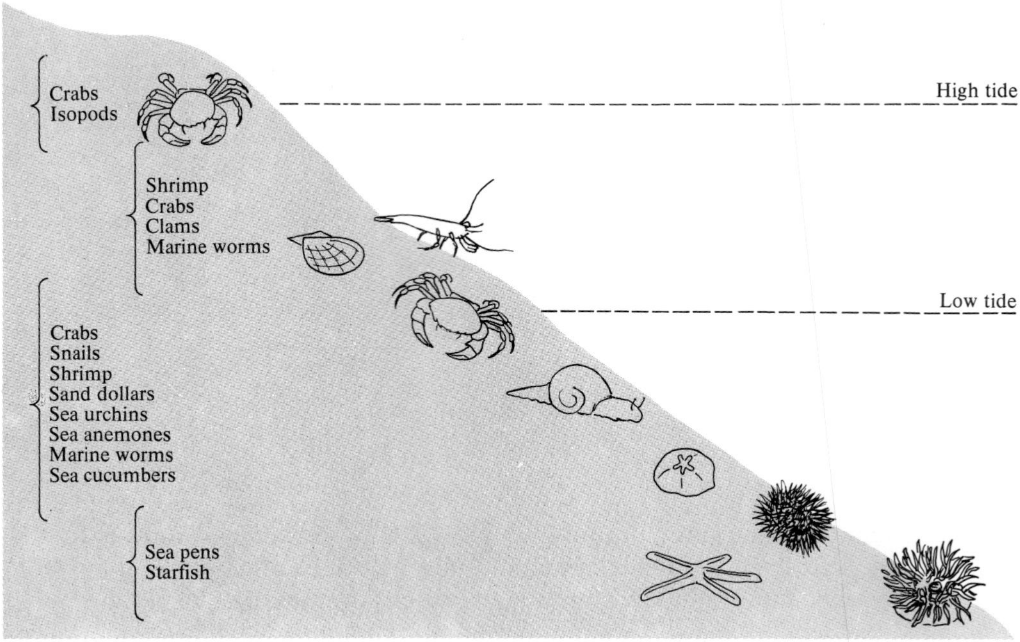

Fig. 15.9 A typical distribution of marine benthic forms found on soft bottom sediments.

the low tide line, we find more crabs, snails, shrimp, and sand dollars. Burrowing worms and those living in tubes, sea cucumbers, sea pens, clams, snails, and brittle stars are found down to depth (Fig. 15.9). The variety is so immense that it almost defies efforts at listing. In reviewing the fauna found in soft sediment, one must also consider environmental variables such as type of soft sediment and the oxygen level in the overlying and interstitial water, because there is a marked difference between organisms living in fine-particle sediment with high organic content, active decomposition, and low oxygen content and those found in coarser clean sand and abundant oxygen supply. Since the deeper reaches of the ocean usually have a substrate of soft sediments, and food and oxygen are present to sustain life, many of the organisms found in shallower water are also found on the deep-sea floor. At present, the abundance and distribution of populations of the deep ocean bottom are not well known and have not been sampled adequately. However, the sparse samples collected from these regions, as well as recent photographic and television records, indicate that worms, sea cucumbers, shrimplike creatures, and brittle stars exist on the bottom at nearly all depths and are more abundant than previously thought.

The benthic fauna inhabiting the hard rocky substrates at the edge of the sea are different from those found in soft sediments. The presence of algae in

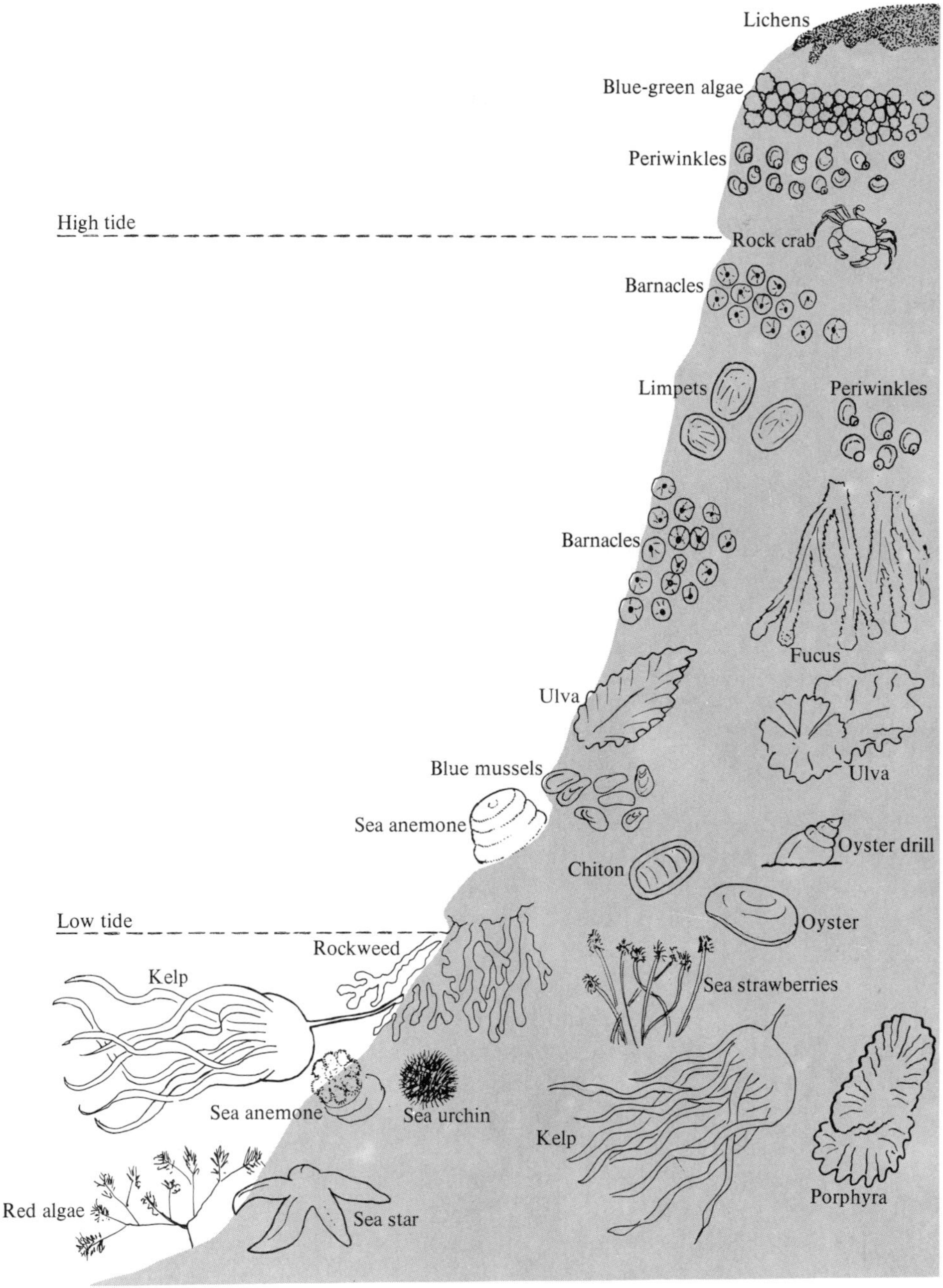

Fig. 15.10 A typical distribution of benthic flora and fauna found on a hard-rock substrate when an appreciable tide range is present. (Adapted from W. H. Amos, *The Life of the Seashore*, McGraw-Hill, New York, 1966.)

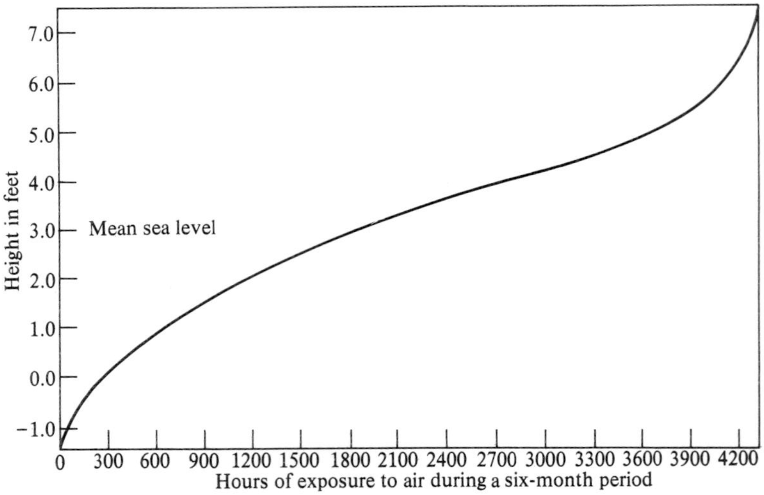

Fig. 15.11 Intertidal benthic forms suffer a degree of exposure to air and dessication that is determined by their location above or below mean sea level and the amplitude of the tide. This example is from San Francisco, California. (From J. W. Hedgpeth, ed., *Treatise on Marine Ecology and Paleoecology*. Ecology, vol. 1. The Geological Society of America, Boulder, Colorado, 1957.)

the rocky environment also acts to modify the fauna types since the algae offer their own structure as still another substrate and a hiding place which attracts certain organisms. Above the high-tide line one finds lichens, a combination of algae and fungi, algae, and the ever-present snail, *Littorina*, that looks like a small dark periwinkle. Crustaceans, such as crabs, isopods, and amphipods, also dwell here. The habitat of the barnacles and moss animals (bryozoans) extends from the upper tide limit down to midtidal and even lower regions. Limpets, more barnacles, mussels, whelks, chitons, oysters, and rock-burrowing mollusks are found at the lower tide level (Fig. 15.10). Rock pools in which seawater is trapped and held above the low tide line contain a conglomeration of animal forms, some of which typically live in the zone below low tide, where dessication does not normally occur. Sea anemones, segmented worms, starfish, crustaceans, mollusks, and sea urchins abound. Below the low-tide mark, we find sea anemones, starfish, sea urchins, the colorful unprotected sea slugs (Nudibranchia), and many other forms. The amplitude of the tide, the exposure of the rocky coast to waves and spray, the duration of its exposure to the air, and the temperature variations caused by the direct solar heating of exposed intertidal rock surfaces produce zonation of the benthonic fauna (Fig. 15.11). Definite zone limits are set for individual species by these environmental conditions, but because of the large variation of species within each phylum and the variation of tolerance within each species, zonation by phyla, class, orders, and even families is not particularly appropriate.

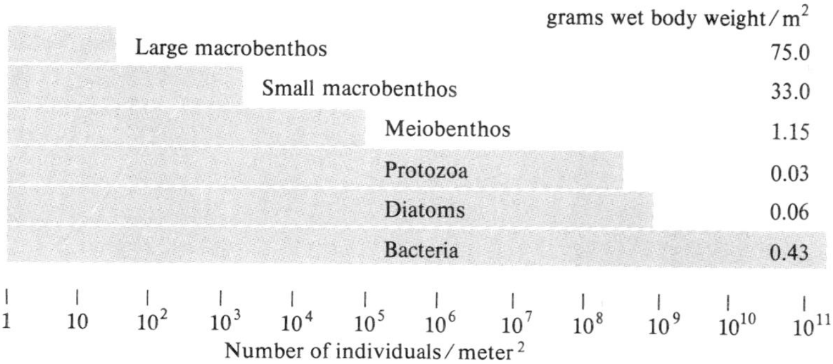

Fig. 15.12 The number of individuals in the benthic community tends to increase as their size and contribution to the biomass decreases. (From J. W. Hedgpeth, ed., *Treatise on Marine Ecology and Paleoecology*. Ecology, vol. 1. The Geological Society of America, Boulder, Colorado, 1957.)

DISTRIBUTION OF BENTHONIC ORGANISMS

We can apply certain general rules to determine the mass distribution of the zoobenthos as a function of depth and average the result over the world's oceans. However, a zonation based on biomass distribution is not necessarily indicative of the number of organisms or the most frequently encountered forms. Nonetheless, we can risk making a generalized statement that the arthropods are found at the highest tidal levels, followed by the mollusks, which in turn combine with and blend into the coelenterates at the lower tide levels and down to greater depth. Below the coelenterates, which usually have very large mass bulk, come the echinoderms. These and the annelid worms then continue down to the deepest oceanic depths. Protozoa and bacteria, of course, are present at all levels.

This crude division of the ocean floor into zones of predominating phyla will no doubt have as many exceptions as examples conforming to the rule. (The reader who has more than a casual interest in the study of benthonic fauna should consult one of the many books focusing on specific environmental areas.) In a shallow water area possessing a variety of substrate conditions, an average sampling of organic benthonic material will usually produce a pyramid of organisms versus mass (Fig. 15.12). It is quite evident from this typical figure that the larger benthonic forms are few in number but constitute the bulk of the biomass.

An interesting aspect is the effect of latitudinal variation of mean temperature and temperature range on the infauna and epifauna. The epifauna are more

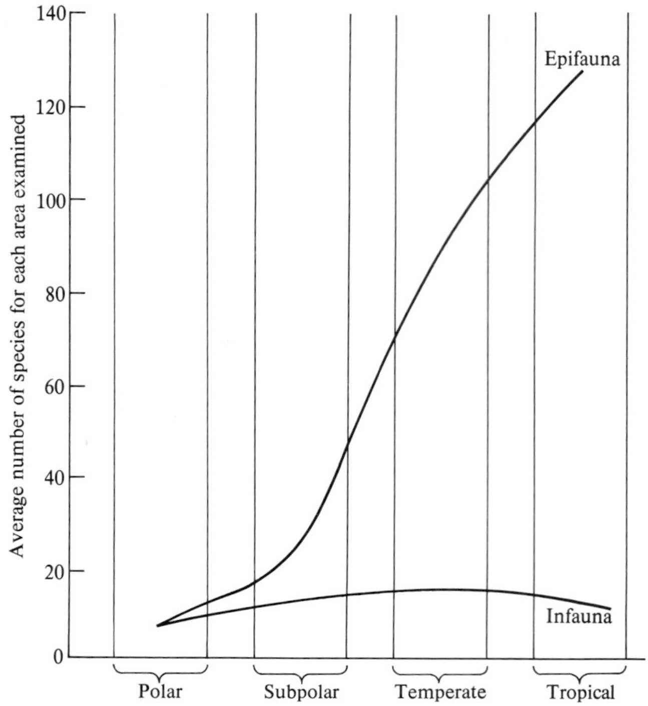

Fig. 15.13 In shallow coastal regions the benthic epifauna are influenced considerably more than the more protected infauna by climatic variation on the earth. (From J. W. Hedgpeth, ed., *Treatise on Marine Ecology and Paleoecology*. Ecology, vol. 1. The Geological Society of America, Boulder, Colorado, 1957.)

Fig. 15.14 This winter flounder resting on a checkerboard pattern shows how versatile is its use of chromatophores for camouflage. (Courtesy of Field Museum of Natural History, Chicago.)

abundant in the very shallow regions of the sea, while the infauna are distributed over a much larger area and depth range and predominate at the greater depths. Since most of the epifauna are in the shallow regions, including the intertidal zone, they are subjected to a large variety of environmental conditions and stresses. The buried infauna in the same region are not as likely to be exposed directly to subaerial climatic stresses since they are protected by the overlying sediment. Thus climatic control on epifauna distribution is much more noticeable than it is on the infauna. The annual range and annual average value of temperature are the most important natural factors controlling the epifauna distribution and abundance in shallow water. In the arctic or polar intertidal zones the epifauna are extremely sparse; they increase in number from polar to tropical environments. The infauna, however, existing in large part below the low-water mark and protected by the sediment, do not show any marked effect of latitudinal climatic control (Fig. 15.13).

Within the sea, as on land, some animals take on shapes and colors that make them appear what they are not. This masquerade is carried on for both protection and attraction. Protective masquerading is accomplished by camouflage (Fig. 15.14). Bottom fishes, for example, change coloration and pattern to imitate the ocean bottom; the carapace of some crabs serves as substrate for both flora and fauna which form a small garden on the crab's back that effectively disguise its form. Animals such as the sea anemones, tube worms, and corals, have assumed the guise of bright flowerlike blossoms either to attract unwary food or to indicate that they are dangerous. Masquerading is most prevalent among the organisms living on the sea floor, where the variations in substrate and the associated animal and plant populations offer considerable scope for changes in patterns of form, color, and texture. The art of hiding in the water column has different problems from those that arise from imitating the ocean floor. In the water column, transparency of body is important, since it makes the organisms nearly invisible. In the near-surface zones, heavier-bodied animals that are not transparent take on a body coloration that gives them a light-colored ventral side and a dark dorsal side. If seen from below, they are difficult to discern against the filtered light scattered in the surface water above them, and if seen from above, they appear dark against the dark background of the deeper water. At depth, where light is scarce or nonexistent, black and dark red coloration predominates.

THE PHYTOPLANKTON

Although the benthos contains many beautiful forms of both animal and plant life, as well as some that are offensive to the human senses, the most bizarre and intricate forms are found in the plankton. The drifters of the sea are both plant and animal. Like the benthos, their number tends to increase as their size and biomass decrease. The role of the plankton in the food economy of the seas is well known. The importance of the planktonic plants is embodied in the simple statement made by Alister Hardy in his *The Open Sea: The World of Plankton*: "*All* flesh is grass."

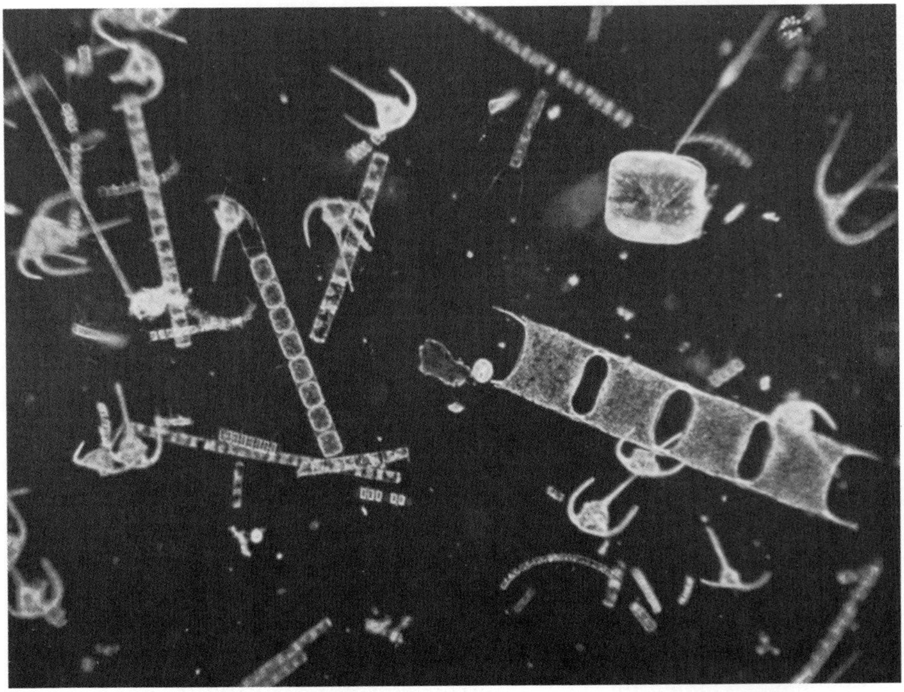

Fig. 15.15 Plant plankton (Phytoplankton). (From A. C. Hardy, *The Open Sea*, Houghton Mifflin, Boston, 1964.)

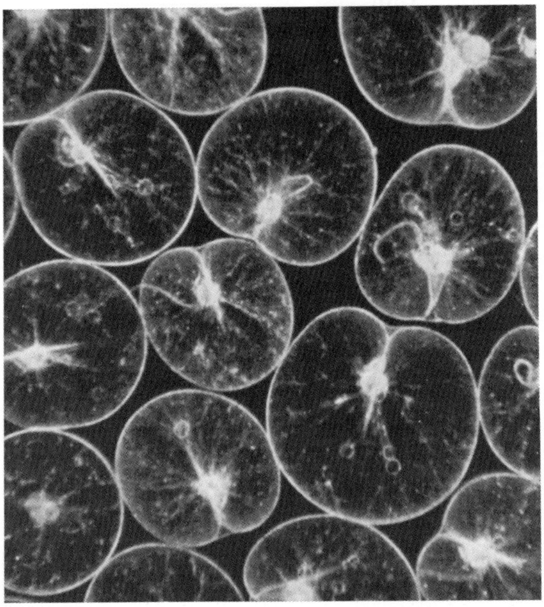

Fig. 15.16 Living dinoflagellates: *Noctiluca scintillans*. Photo by Douglas Wilson. (From A. C. Hardy, *The Open Sea*, Houghton Mifflin, Boston, 1964.)

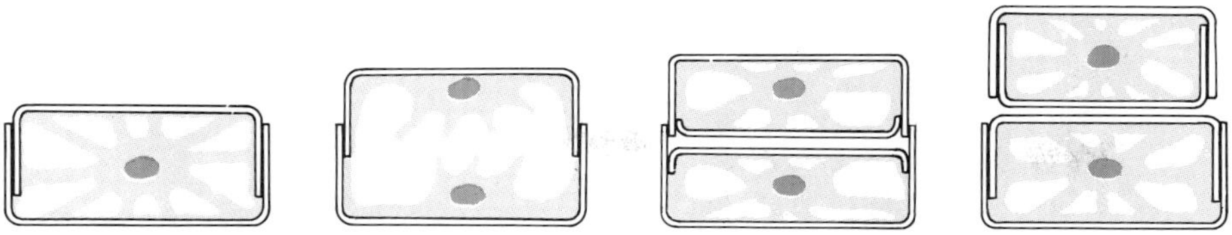

Fig. 15.17 The division of a parent pillbox-like diatom (a) into two daughter cells (b,c) produces one cell that is always smaller than the parent cell and one the same size (d). (From A. C. Hardy, *The Open Sea*, Houghton Mifflin, Boston, 1964.)

If fine nets are drawn through the sea and the trapped contents are microscopically examined, one finds planktonic forms such as those shown in Fig. 15.15. Some of these plankton samples are almost wholly plant while others are wholly animal. Almost all the phytoplankton are unicellular algae. Chains of individual cells as well as single-cell forms are common in the sea. The dinoflagellates, some of which are capable of both ingesting food and photosynthesizing, are also abundant. It is these intermediate (half-plant, half-animal) organisms (Pyrrophyta) that are primarily responsible for the so-called toxic red tides that are detrimental to other marine life and man. However, not all reddish water phenomena are toxic or caused by dinoflagellates. The coloration of the Gulf of California, which once was called the Vermillion Sea, and the Red Sea of the Middle East is due to a blue-green alga. However, another blue-green alga produces a toxic condition in certain areas of the Indian Ocean. *Noctiluca* (Fig. 15.16), a dinoflagellate, imparts the color and appearance of tomato soup to water. This dinoflagellate is nontoxic and is one of the many marine organisms capable of producing "biological light" and luminescence in the sea. This ability is evident from the name *Noctiluca*, which means night light.

The phytoplankton include many representatives of algal groups. It is the unicellular algae in the phytoplankton, as opposed to drifting algae such as the fabled sargassum weed, that make up the bulk of the marine primary producers. These small cells, 1 μ to 1 mm in size, include diatoms, dinoflagellates, coccolithophores, silicoflagellates, cryptomonads, chrysomonads, green algae, and blue-green algae, with diatoms, dinoflagellates, and coccolithophores being the most abundant. The diatoms produce a relatively thick siliceous test that houses the protoplasm of the plant. This rigid test, or frustule, has two halves or valves that fit together in pillbox fashion. Reproduction is vegetative; that is, the diatom forms new valves inside its cell, which then splits apart into two daughter cells (Fig. 15.17). One daughter cell remains the same size as the parent cell, but the other is smaller. On each division, therefore, the size of half the diatom daughter cells becomes smaller and smaller until a limiting size is reached and the diatom forms an auxospore stage. In this stage, sexual differentiation of the cells takes place, and the fertile protoplasm is released

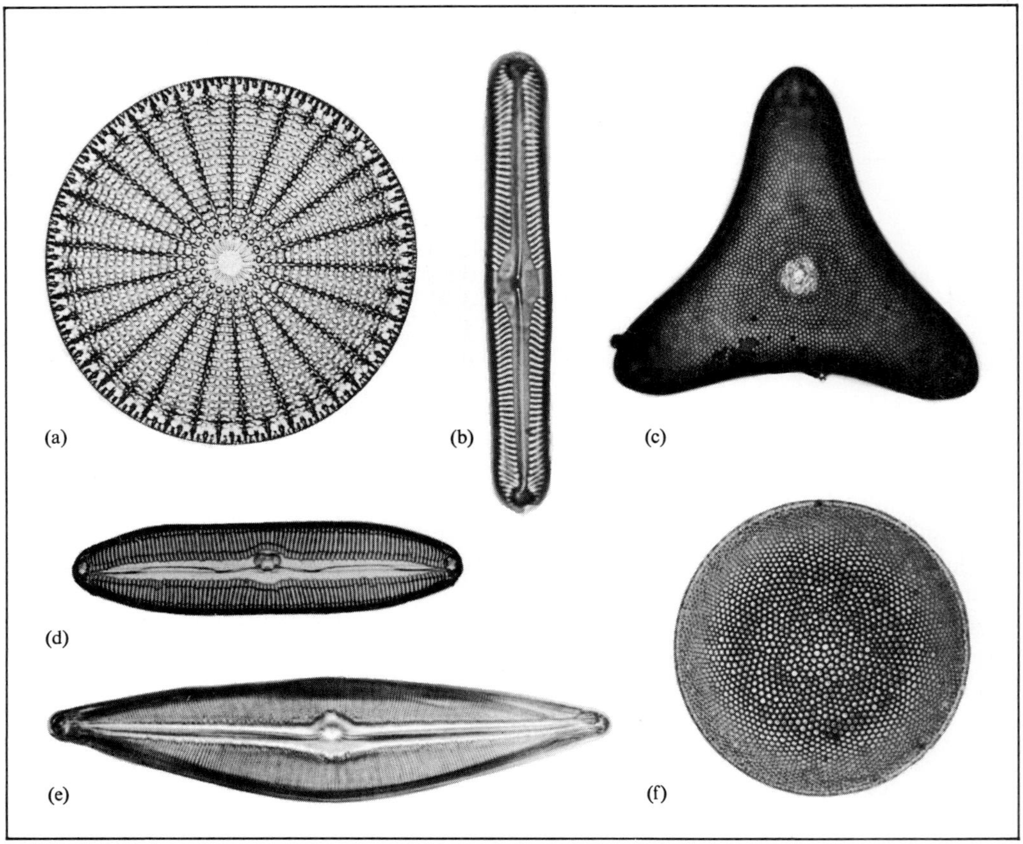

Fig. 15.18 Photomicrographs of diatoms. (a) *Arachnoidiscus*. (b, d, e) *Pinnularia*. (c) *Triceratium*. (f) *Coscinodiscus*. (a, b) Courtesy of G. Dallas Hanna. Other photos courtesy of Hsin Yi Ling.

from its casing to enlarge to adult size, form a new test, and begin to repeat the process of vegetative production.

The diatoms are divided into two groups based on shape. Those that are radially symmetrical are called *centric* diatoms. These float freely in the water column. Diatoms with bilateral symmetry are called *pennate* (Fig. 15.18). Pennate diatoms do not readily float and are principally confined to the sea floor (benthonic diatoms). However, they are also found adhering to the surfaces of flotsam and sea life in the water column. Figure 15.19 shows several diatoms whose frustules, or tests, have ornately perforated structures. The perforations, called *areolae*, allow the transfer of nutrients, gases, and wastes in and out of the protoplasm. Some diatoms have siliceous spindles and protoplasm threads that link individual cells together into chains (Fig. 15.15). The chain lengthens as each member cell undergoes binary fission to produce linked daughter cells until mechanical action fractures the chain or an auxospore is formed.

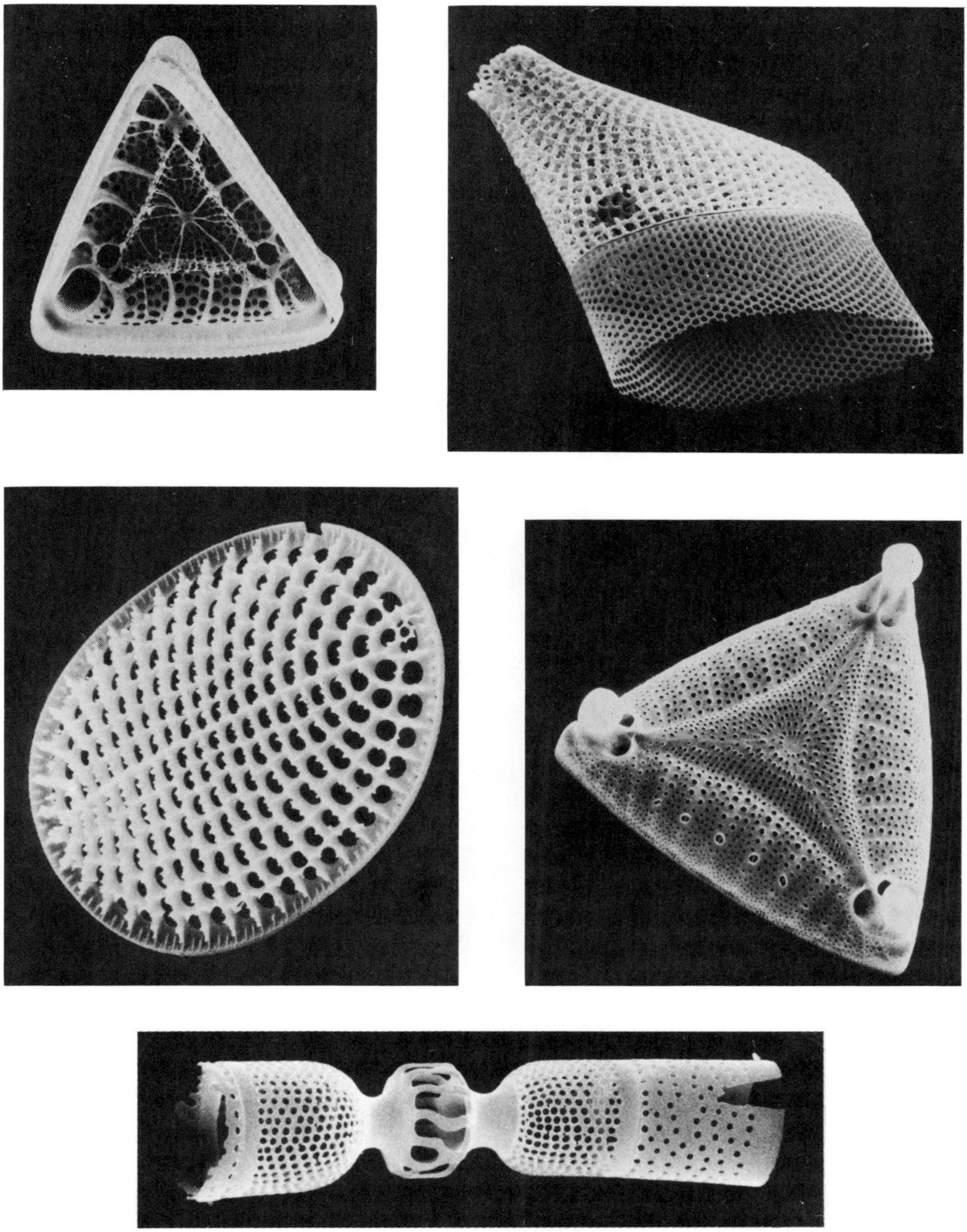

Fig. 15.19 Diatoms with areolae. (From G. Dallas Hanna, "Nature's Opaline Gems," and G. Dallas Hanna and A. L. Brigger, "Stereoscan Microscopy of Diatoms." *Pacific Discovery*, October 1968. Photos courtesy of Engis Equipment Co., Morton, Grove, Ill.)

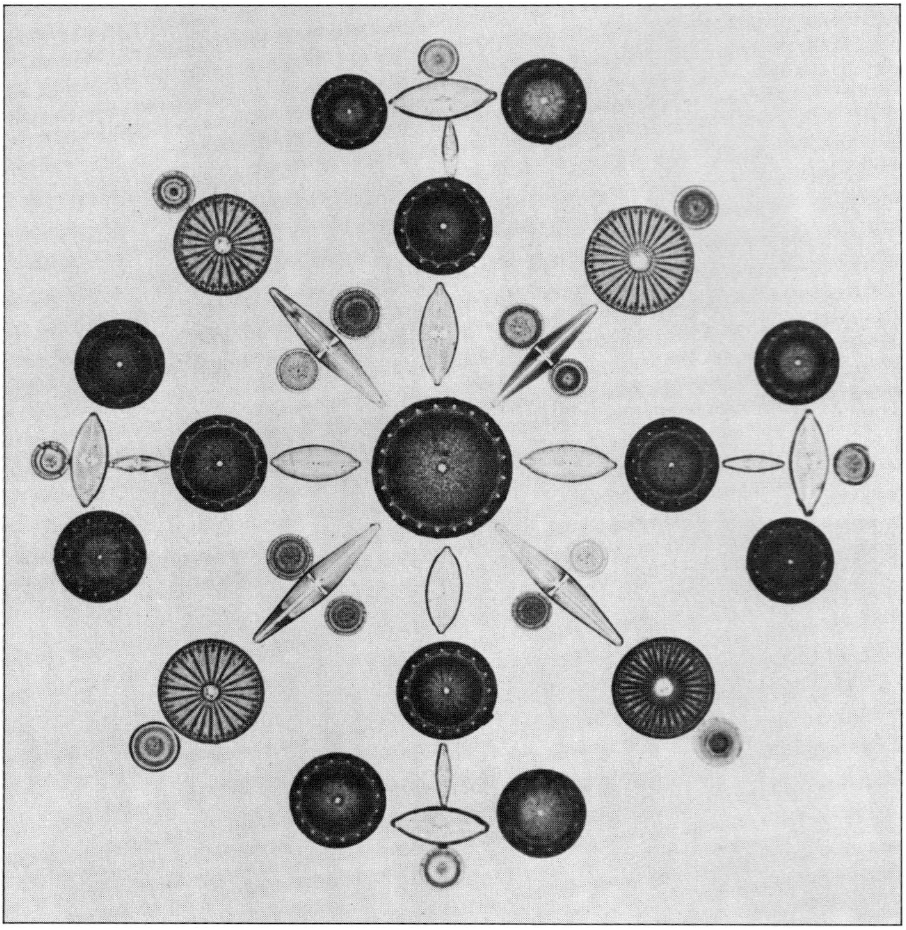

Fig. 15.20 A series of different species of diatoms arranged in a geometrical pattern to illustrate the skill developed in preparing such minute objects. Prepared by W. M. Grant. Courtesy of G. Dallas Hanna.

Diatoms were used in the past to test the quality of microscope lenses. The frustules of diatoms were painstakingly mounted on glass slides (Fig. 15.20) to create symmetrical patterns which when viewed through a microscope could be used to test the quality of the lenses.

Pennate diatoms are not as passive as the centric forms and are capable of limited motion by means other than the adjustment of their buoyancy. They have a flexible structure outside their frustule, called *raphe.* In order to move they pump cytoplasm through this structure to produce undulations that weakly propel the cell. Diatoms have existed on the earth since the Triassic Age and are found in fresh and salt water. Their silica tests, which are resistant to chemical changes, accumulate at the bottom of the oceans, forming siliceous oozes. In some cases, these deposits have, millions of years later,

Fig. 15.21 Diatom mine of Johns-Manville Corp. at Lompoc, California.

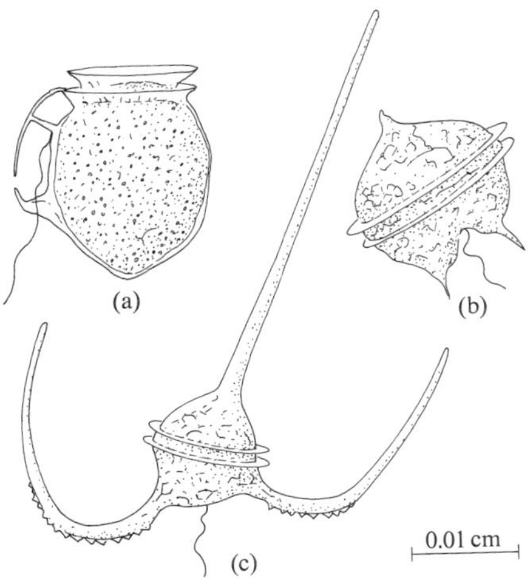

Fig. 15.22 Dinoflagellates. Examples shown in drawing are (a) *Dinophysis acuta*, (b) *Peridinium granii*, (c) *Ceratium macroceros*. (From A. C. Hardy, *The Open Sea*. Houghton Mifflin, Boston, 1964.)

been lifted above sea level by large-scale geologic upheavals. This has occurred, for example, in California, where the deposits are mined and marketed as diatomaceous earth (Fig. 15.21).

The dinoflagellates, which have both plant and animal properties, are the most ecologically diversified group of the phytoplankton. They are also more mobile than the pennate diatoms, since they are equipped with flagella, small whiplike appendages that beat away in the water to propel the individual (Fig. 15.22). Although they are mobile enough to vertically migrate in response to light, they are nonetheless planktonic. As in the case of the diatoms, there are two principal subgroups of dinoflagellates: armored cells and naked cells. The armored forms are covered with small platelets of cellulose that form the cell wall. The naked forms have a thin flexible membrane (*pellicle*) that has no differentiating substructure. There are usually two flagella set in grooves in the cell wall, one at right angles to the other. Reproduction is accomplished by binary fission, as in the case of diatoms. Reduction in size of successive daughter cells does not occur, and a sexual stage such as the auxosphore of diatoms is not known.

The bioluminescence of the sea is caused primarily by dinoflagellates. This phenomenon, often referred to incorrectly as "phosphorescence," occurs when the cell itself or the water containing the luminescing cells is disturbed. The disturbance touches off a chemical reaction in these organisms, and the energy released in the process is given off as light. However, in addition to the dinoflagellates, there are many other luminescing marine organisms. Since dinoflagellates are part of the food chain, the chemicals involved in the light-producing reaction are passed up the trophic levels to other animals. Jellyfish often glow when disturbed, and the flesh of fish and some sea birds luminesces because of accumulated enzymes. Other marine creatures, such as the middepth fishes, have light-producing organs that can be flashed at will. (The common firefly familiar in central, eastern, and southern sections of the United States has a similar ability.)

Even though dinoflagellates are next to diatoms in terms of abundance in the phytoplankton, their remains are not accumulated in the sediments except as consumable organic matter, since cellulose forming the pellicle or armored cell wall is not resistant to decomposition. The two dinoflagellates that are usually involved in toxic red tides, which in extreme cases can seriously injure many marine forms, are *Gymnodinium brevis* in the Gulf of Mexico and *Gonyaulax* of the more temperate waters.

The abundance of phytoplankton in the sea is quite variable, depending on many environmental factors. Cell counts may range from 10^3 cells per liter of water in the biological desert areas of the sea to 10^8 cells per liter in more productive regions. Rapid population increases can occur that tend to make phytoplankton populations appear "overnight." The sudden blooms stimulated by light and nutrients form rapidly, with diatoms undergoing as many as three cell divisions per day per cell and dinoflagellates as many as six or more per day.

The phytoplankton group next in abundance to diatoms and dinoflagellates is made up of the coccolithophores, which, on the evolutionary scale, predate the diatoms and dinoflagellates. The coccolithophores, like the dinoflagellates, are equipped with two flagella. Unlike the dinoflagellates, they having a covering formed of joined calcareous plates called *coccoliths* (Fig. 15.23). These tiny forms also reproduce by binary fission, although there is also some sexual reproduction. The coccolithophores are not as widely distributed as

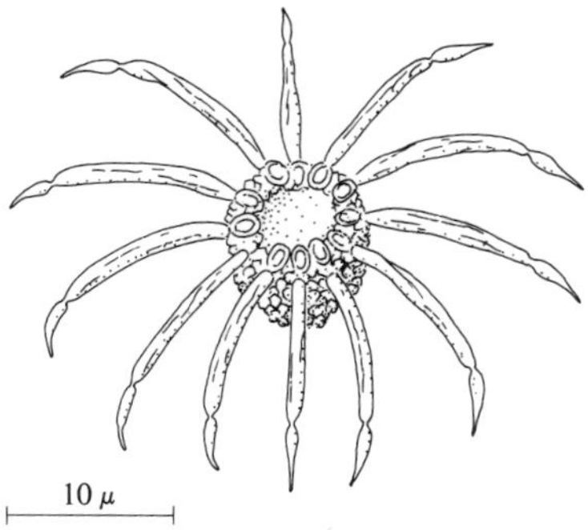

Fig. 15.23 A coccolithophore, *Michaelsarsia aranea*, reconstructed from electron-microscope photographs. (From R. W. Fairbridge, ed., *Encyclopedia of Oceanography*, © 1966 by Reinhold Pub. Corp. Used by permission of Van Nostrand Reinhold Co.)

diatoms and dinoflagellates are. They are confined to warm-water regions and survive nicely in regions of low nutrients. Since their coccoliths are predominantly calcite, they are found in the biogenic sediments of the shallower regions of the sea at the lower latitudes.

THE ZOOPLANKTON

The zooplankton contains representatives of nearly every animal phylum and as a result is extremely diversified. Some of the zooplanktonic organisms are permanent forms while others, meroplankton, are transient (the egg and larval stages of benthonic and nektonic organisms). The most abundant types in the zooplankton group are the small crustaceans, such as the copepods and euphausids. Arrow worms (*chaetognaths*), the voracious carnivores of their world, are also abundant (Fig. 15.24). Other groups, foraminifers, radiolarians, tunicates, mollusks, cnidarians, amphipods, ctenophores, and pteropods, are also well represented in the zooplankton. However, these typically make up less than about 50% of the total average zooplanktonic biomass. The

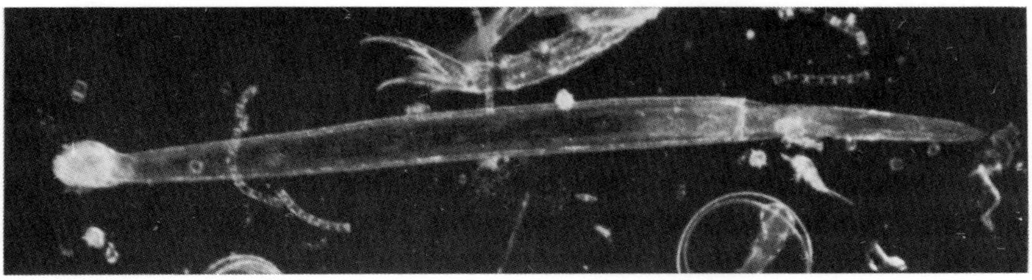

Fig. 15.24 A living arrow worm in the plankton. Photo by Douglas Wilson. (From A. C. Hardy, *The Open Sea*, Houghton Mifflin, Boston, 1964.)

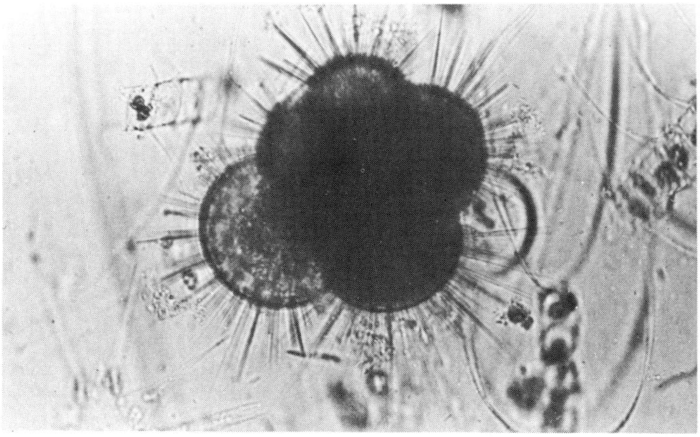

Fig. 15.25 Globigerina, a pelagic foraminifera. Photo courtesy of Ralph Buchsbaum.

crustaceans of the plankton are primarily small copepods and euphausids. Some euphausids are reasonably large and occur in sufficient numbers to form an adequate food source for baleen-type whales. The large mixed populations of crustaceans that are found in subpolar waters and called *krill* or *whale food.* The krill of the circumpolar waters in the southern hemisphere is abundant enough to support a blue whale population that was many times larger than that of today. The food supply is still there, but the number of whales has dwindled due to man's unwillingness to temper his rapaciousness with rational conservation measures.

Foraminifers and radiolarians are amoeba-like protozoans. One of the most common foraminifers is *Globigerina.* In regions in which these organisms are abundant the calcareous tests of dead cells produce a sedimentary ooze (Fig. 15.25). These encased protozoans have *pseudopods* (false protoplasmic feet) which flow in and out through the reticules of the test to help in attachment, movement, and food gathering. The radiolarians have silica or strontium-sulfate tests. They are beautiful creatures whose skeletons take on the form of ornate sunbursts with delicate radiating spines. They are also an important source of biogenic sedimentary material.

Mollusks, which include heteropods and pteropods, are widely distributed in the plankton. Pteropods (meaning "wing-footed") and heteropods belong to the class of mollusks called Gastropoda and are related to snails. These animals are bizarre, nearly transparent creatures; some species are protected by shells. The hard remains of pteropods also contribute to the buildup of sedimentary ooze.

Ctenophores, the comb jellies, occupy a phylum of their own. They have eight rows of exterior cilia that propel them through the water, trailing their small tentacles in quest of food. The round forms (Fig. 15.26) look like gooseberries and hence are called sea gooseberries. A flattened form, Venus's girdle,

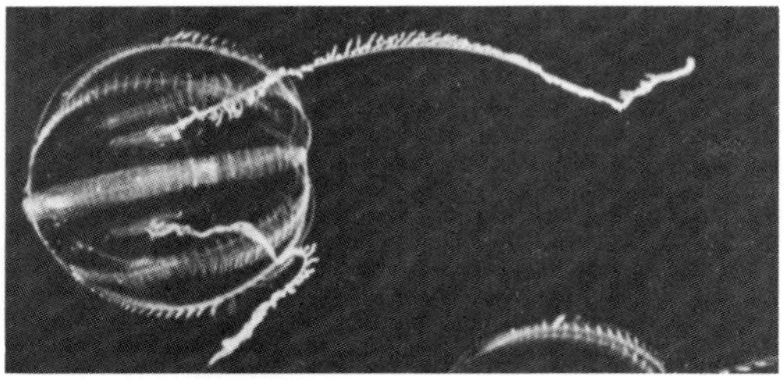

Fig. 15.26 Ctenophore ("comb-bearer"). Photo by Douglas Wilson. (From A. C. Hardy, *The Open Sea*, Houghton Mifflin, Boston, 1964.)

may grow to a length of more than one foot. These elongate irridescent shapes catch the sunlight in their beating cilia and offer a spectacular sight to observers on a drifting ship.

The tunicates (phylum Chordata) exist in both planktonic and benthonic forms. The sessile benthonic form produces free-swimming, tadpole-shaped larvae which form part of the meroplankton and disperse the species to new locations. In the process of maturation the larvae take on the tubular or irregular shape of the adult sessile form. *Salpa*, a common pelagic tunicate, is widely distributed throughout all the oceans. When it occurs in abundance, it is the bane of the planktologist, whose nets it clogs with a gelatinous mass. However, it also contributes to research: its intestinal system contains minute organisms that can be collected for further study.

The jellyfish are all planktonic and form two classes of the phylum Cnidaria (or Coelenterata): Hydrozoa and Scyphozoa. Most members of the hydrozoans produce two body types: the polyp, a sessileform, and the medusa, which floats or swims freely in the water and thus aids in the dispersal of the species. (Despite its different appearance, the medusa is essentially an upside-down polyp.) *Gonionemus*, a small medusa common in the Atlantic Ocean, and *Aequorea* are common jellyfishes of the hydrozoan class. To this same class belongs the colonial Siphonophora. The Portuguese man-of-war, *Physalia*, and the small by-the-wind sailor, *Velella*, are widely distributed colonial hydrozoans. These plankton are a collection of different hydrozoan polyps, each having a specific function to assure survival of the colony. Thus one group is responsible for developing the float, another for feeding, the third for protection and food-gathering, and the fourth for reproduction.

The class Scyphozoa includes the larger jellyfishes and marine medusae. *Aurelia*, the common jellyfish, belongs to this class, as does the magnificently colored *Cyanea*. A third class of cnidarians, Anthozoa, includes the benthonic sea anemones and corals. Members of this class do not possess a medusa body form.

The planktonic forms and their debris make up the base of the food pyramid of the open sea and are the material that sinks forever downward as the "eternal snowfall" to carpet the sea floor with organic substances and biogenic sedimentary materials.

THE NEKTON

The nekton, the free swimmers of the sea, are at the top of the food pyramid. Important nekton are the mammals, the bony fishes possessing internal bony skeletons (class Osteichthyes), and the cartilaginous fishes whose skeletons are made of cartilage (class Chondrichthyes).

The mammals—whales, seals, and sea otters—today play only a minor role in the economy of the ocean. At one time, whaling constituted a major industry of maritime states. For example, it is estimated that in the 19th century Yankee whalers alone killed approximately 100,000 whales. Although the romance-shrouded Captain Ahab and his crew have been replaced by today's highly mechanized factory ship and speedy hunter craft equipped with explosive harpoons, and although some of the markets for whale products have disappeared, whaling fleets are still plying the oceans, primarily in Antarctic waters. Whale oil is used as a low-residue lubricant for high-precision instruments and as a base for margarine production. The sperm whale is the only source for ambergris, a base used in making expensive perfumes. Thousands of tons of whale meat are consumed in Japan, and there are many other industrial uses for whaling products.

Seals, sea lions, elephant seals, and walrus have been harvested in the past for food and pelts but now live in a state of semiprotection. However, they are still an important source of food and clothing for the people of the Arctic regions. The crews of fishing vessels will also occasionally kill these animals if they endanger their nets or feed on the fish the fleet has set out to catch.

The luxurious pelts of sea otters used to be actively traded between Europe, the Orient, and the Americas. They were in such demand that the species was nearly exterminated. Governmental intervention and protection have saved this mammal from complete extinction, and it is now beginning to repopulate the shores of the northeastern Pacific Ocean.

Dolphins, porpoises, and blackfish (several types of small toothed whales) also populate the oceans in considerable numbers. However, they are not commercially harvested except for display in aquatic parks.

Reptiles are rare in the open sea. One of them, the sea snake of the tropics, is poisonous. Turtles (green, tortoise shell, and leatherback) are numerous but barely affect the sea's economy. These lumbering members of the nekton are important only as a source for meat, oil, and eggs for the shore dwellers of the tropics. Like the whales, they serve as a moving substrate transporting attached organisms from one place to another.

The cartilaginous fishes (sharks, skates, and rays), because of their scavenging propensities and occasional attacks on man, are usually thought of as the main predators of the sea. Yet this group contains filter feeders as well as members equipped with sinister jaws and teeth. The largest fish of the sea, the whale shark, is a filter feeder unable to tear flesh apart. Sharks are found in all latitudes, from the tropics to the polar regions. Most prey on the easy victim, the infirm and struggling. This is one of the reasons that man in water is susceptible to attacks by sharks. Water simply is not man's element. His swimming movements on the sea surface are inefficient and out of rhythm with those of marine creatures of similar size. To the shark, the human swimmer appears as a weakling struggling for life and therefore fair game.

Sharks are the sanitation crews of the seas. They are able to digest nearly every type of natural organic material and thus keep the water clean of unhealthy and decaying matter. In the past, before the development of synthetic processes, sharks were fished commercially as a source of vitamin D, which was extracted from their disproportionately large livers. The flesh of sharks and skates is edible and marketable if processed properly to remove the urea. At present, the harvest of cartilaginous fish is not commercially important. The fish are, however, a significant link in the food chain and economy of the sea.

The bony fishes of the nekton play the greatest role in the economy of the oceans and man's marine fisheries. The bottom fishes (cod, flounder, and halibut) have long been an important food source for man. The herringlike fishes contribute the greatest bulk to the annual oceanic crop harvested by man. The tunas of the warmer water regions are also important. Although they make up a smaller portion of the annual catch, the price per pound is higher than that of herring and thus they are an important commercial item. Another commercially important fish is the troutlike salmon, caught primarily in coastal waters rather than in the open sea.

Scientists have become increasingly aware of the need to estimate the annual harvest potential of ocean fish in order to determine the sea's share of the world's food resources. For all practical purposes this food yield potential is almost entirely limited to the nekton, although shellfish, shrimps, and crabs should be and often are included. Current estimates of potential harvest yields range from 55 to 2000 million metric tons per year, with a yearly mean value of about 85 million metric tons. These estimates are based on both extrapolation of catch trends and energy flow through the food chain of the sea. In describing the food chain, we mentioned that only a fraction—perhaps 10%—of the matter in one link of the food chain is passed on to the next trophic level. Based on this percentage and the assumption that the phytoplankton as primary producers generate about 1.9×10^{10} metric tons of carbon, the annual yield potential at the third trophic level of the oceans (second-stage carnivores) can be estimated to be 19.0 million metric tons.

Let us consider the actual yield of the oceans' fisheries broken down according to geographic areas (Table 15.1). Although our breakdown is somewhat arbitrary and the available data scanty, it is possible to estimate that about 53% of the yield comes from the Pacific Ocean, 25% from the Atlantic Ocean, and 22% from the Indian Ocean. It is quite evident from Table 15.1 that the present yields do not compare favorably with the average estimated yield potential of 85 million metric tons per year.

Table 15.1 Oceanic fisheries production, 1965 (millions of metric tons)

Region	*Production*
Africa	3.
North America	4.4
South America	8.9
Asia (except USSR)	19.9
Europe (except USSR)	10.8
Oceania	0.1
USSR	4.9
Total	52.0

Data courtesy of James Crutchfield, University of Washington.

About 90% of the world yields shown in Table 15.1 is taken from the 7.6% of the world's oceans overlying the continental shelves. In the open sea beyond the shelf area, the productivity drops to about 1/80 of that of the ocean regions covering the shelves. However, the area of open oceans is so great that it is estimated to contain perhaps as much as 40% (or more) of the projected total annual ocean yield. The harvesting of the open sea populations is much more difficult, since the problems of locating, concentrating, and economically harvesting the fish stock increase with distance from the shore.

The growing world population and the protein deficiency in the average diet make it imperative to exploit the nekton of the ocean to its maximum sustaining yield. At present, many of the marketable fish are being harvested in reasonable numbers. It is the unmarketable fish, the trash fish, which will have to be exploited further to produce a protein supplement in the form of fish flour. This can be used directly as food supplement in the diet or as feed to increase the protein yield of stock animals. Worldwide acceptance of fish meal and fish protein concentrates is expected to increase. This will in turn stimulate the fisheries that provide this foodstuff.

The oceans should not be thought of as an infinite reservoir of food. Some fisheries have already reached the point at which harvesting nearly offsets production. If all fisheries are developed to the same extent, a limiting yield will soon be reached. Since not all fish in the sea are on the same trophic level, harvesting does indeed affect the food chain. Thus to ensure optimum yield levels, men will have to carefully study the impact of harvesting on the dynamics of food production in the sea and establish limits on the harvesting at each trophic level.

Books about fish usually focus on dangerous, colorful, or bizarre representatives of the species. They contain photographs of sharks, strange-looking seldom-seen fish of the deep ocean levels, and colorplates of the delicate and colorful fish of the tropical reefs. These fish may be photogenic but they do not play an important role in the economy of the seas. The food fishes do, of course, as noted above; these, however, are seldom shown because they are commonplace and not particularly attractive. The mid-depth pelagic fishes, such as the lantern fish, hatchet fish, and gapers, are also represented in print, as are the relics of the past, the coelacanths. These mid-depth fishes with their greatly enlarged mouths, light organs, photophores, and lures protruding from their foreheads to attract prey are the strange monsters from the deep capturing the imagination of the reader. Unfortunately, as monsters these fish fail to be impressive, because for the most part they are very small, only a few inches long. However, even the poor goose fish of the northeastern seaboard of the United States is good for a story and picture in the local paper when it has the misfortune to be caught.

It should be remembered that on the oceanic scale, the commonplace is usually the most important, for being common frequently is indicative of being abundant. The nekton are no exception here. The more abundant the individuals in the nekton, the more important they are in the biological realm of the oceans.

STUDY QUESTIONS

1 What are the classifications of the biological marine populations with respect to locomotive ability?

2 What are the classifications of the planktonic community with respect to size?

3 What is meant by the term "meroplankton" and to which general groups do these organisms belong?

4 Why does the benthos tend to decrease in population density and diversification as depth and distance from shore increase?

5 What do you conclude from the gradual change with depth in the pigmentation of the benthonic algae if you relate this phenomenon to the quality and quantity of available sunlight?

6 How important are the benthonic algae compared to phytoplankton from the standpoint of gross productivity of the world's oceans?

7 Why does the benthonic community supported by a soft-substrate littoral zone differ from that found in a hard-substrate region?

8 Why are the major coral-reef communities limited to tropical and subtropical regions of the seas?

9 Are all red tide or red-water phenomena usually toxic?

10 Why do marine plants and animals require structures that are different from those required by land forms?

11 Describe two ways in which sea plants may be shaped to permit efficient extraction of nutrients from seawater and exposure to sunlight for photosynthesis.

12 Why is it difficult to classify dinoflagellates as either plants or animals?

13 If marine plants produce 4 g of carbon per square meter of sea surface per day, what is the daily maximum rate at which organic carbon can be harvested at the third trophic level, given that harvesting is to just keep pace with primary productivity? Assume a 10% efficiency at each trophic level.

CHAPTER 16

WAVES

INTRODUCTION

Surface waves are the most frequently observed and easily recognized form of motion in the sea. Not all waves are visible, however; even when the surface appears still, there are waves not perceptible to the human eye. Surface waves of long period or length, such as storm surges, tides, and subsurface internal waves undulate across the sea hidden from the casual observer or unrecognizable as wave forms.

Waves are generated by many sources, the most common of them being the wind. As the wind blows over the water's surface, friction between air and water exerts a viscous drag which stretches the surface as if it were an elastic membrane. One may recall the children's trick of floating a needle on the still surface of a cup or glass of water to demonstrate this elastic quality (surface tension). As the wind distorts and stretches the surface, its elasticity attempts to return it to equilibrium. This process—distortion by the wind and restoration toward equilibrium by surface tension—causes the water to undulate and produces tiny wavelets or ripples. As the wavelets grow and gravity takes over as the restoring force, they move forward before the wind and form a series of surface wind waves. Eventually the waves may outdistance the wind that generated them and move across the sea as free waves which are no longer forced by the wind, to decay and die out at sea or on a distant shore.

Tides, which are waves of oceanic dimensions in length, never escape their generating force to become free waves; they are constantly forced to orient their crests to the moving line of centers between the earth and the tide-raising bodies, the sun and moon. Yet tides are similar to wind waves because they obey some of the same mathematical laws of harmonic motion.

Rarely do waves resemble the simple forms a child draws. If we could freeze the sea surface at a given instant and inspect its profile, we would find it nearly impossible to recognize a simple sinusoidal wave form. Waves of different length, period, amplitude, and direction of travel combine to produce a highly irregular sea surface. Only if we could find an area of the sea with a single series of waves of one length, period, amplitude, and direction could we detect a simple waveform surface.

In order to understand water waves and their properties, we shall work with a simple wave system. We shall then assume the complex state of the sea surface as the sum of many such systems.

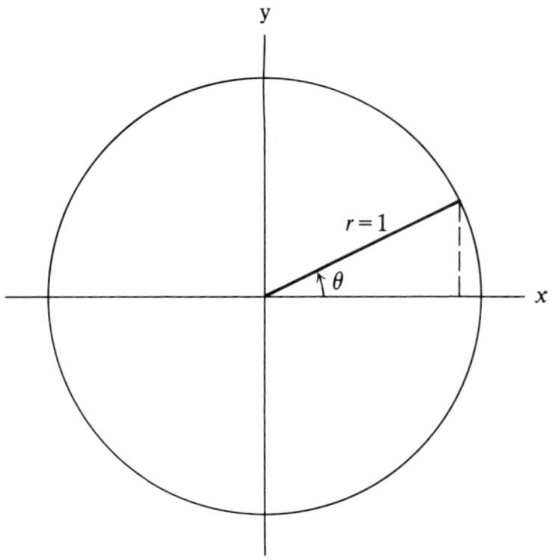

Fig. 16.1. The unit circle of radius one.

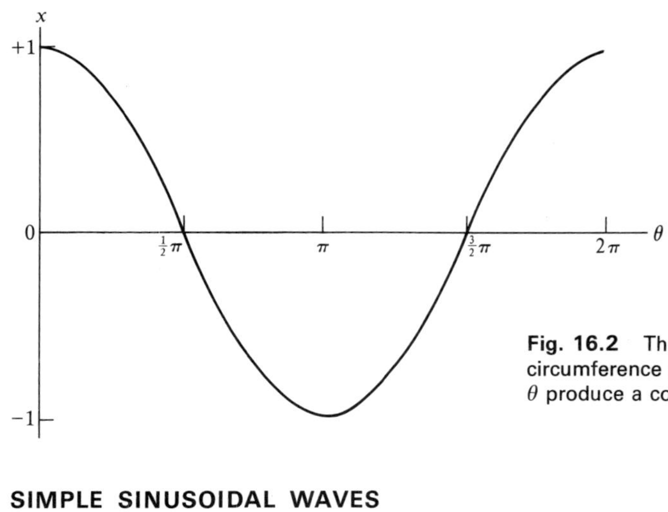

Fig. 16.2 The values of x at positions on the circumference of the unit circle described by θ produce a cosine curve.

SIMPLE SINUSOIDAL WAVES

The simplest waveform that approximates a sea-surface wave is a sinusoidal curve, which is produced by circular motion and plotted as a sine or cosine curve. Consider a circle of unit radius centered on the origin of an x, y coordinate system (Fig. 16.1). Since the radius rotates counterclockwise, beginning at the x-axis, the line dropped from the terminus of the radius to the x-axis describes how x varies along the circumference of the circle as angle θ increases. In one full rotation, the radius of angle θ goes from zero to a maximum of 2π (360°). Thus we can plot x versus θ as shown in Fig. 16.2. The curve will continue if the radius is rotated a second time so that its maximum angle is 4π. This sinusoidal curve is a cosine curve; its counterpart, the sine curve, is depicted by a plot of y versus θ. We can calculate the displacement of the cosine curve above or below $x = 0$ for any value of θ from

$$x = A \cos \theta. \tag{16.1}$$

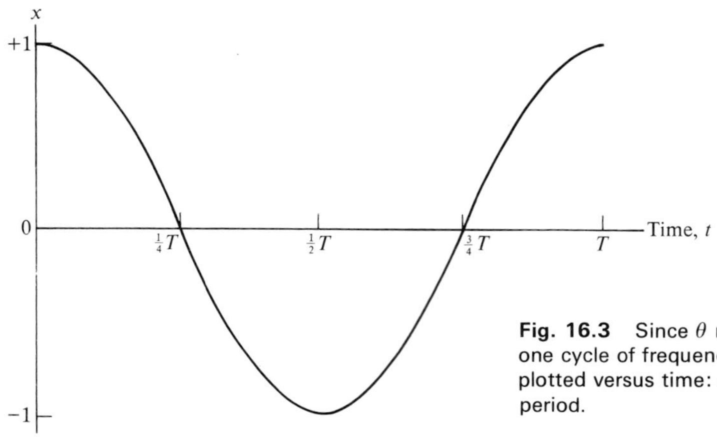

Fig. 16.3 Since θ may be expressed as fractions of one cycle of frequency, the cosine wave may also be plotted versus time: $\theta = (2\pi/T)\,(t)$, where T is the period.

In the depicted case, A, the amplitude of the curve, equals 1; we could increase A by enlarging our unit circle. For the cosine or sine curve, A is the maximum value of x.

We shall now consider whether the angle θ can be expressed in other ways. If the radius rotates at a uniform rate, time can describe angle θ. Assume that the time required for a full rotation is T, the *period.* The rotation rate of the radius is then $2\pi/T$, and θ is

$$\theta = (2\pi/T)t, \tag{16.2}$$

where t is any fraction or multiple of T. We may now replot the cosine curve as x versus time (Fig. 16.3), so that

$$x = A \cos\,[(2\pi/T)t]. \tag{16.3}$$

Since we are dealing with a waveform, we may also assume that the cosine curve between $0 \le \theta \le 2\pi$ describes one wavelength L when one rotation is completed; thus angle θ is a function of L in the same manner that it is of T:

$$\theta = (2\pi/L)l, \tag{16.4}$$

and

$$x = A \cos\,[(2\pi/L)l] \tag{16.5}$$

(see Fig. 16.4).

We can now use a cosine curve as an approximation of a simple water wave. Let us consider a wave as a displacement of the sea surface above and below its equilibrium position that extends in one direction at a given instant

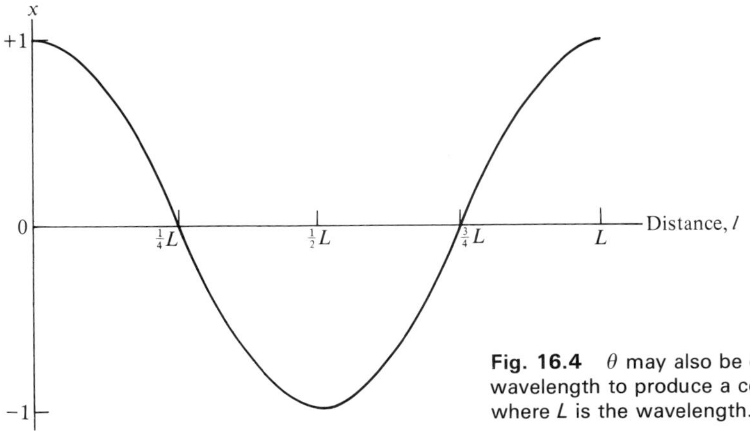

Fig. 16.4 θ may also be expressed as fractions of one wavelength to produce a cosine curve, $\theta = (2\pi/L)\,(l)$, where L is the wavelength.

in time or appears as a variation in water level with time if we observe only one point in space. We shall denote the displacement of the water surface above or below its equilibrium value by η, with a maximum value of A (the amplitude). The waveform is changing in time and space, both of which affect angle θ. Thus the equation for the elevation of the water surface is

$$\eta = A \cos\,[(2\pi/L)x - (2\pi/T)t], \tag{16.6}$$

where x is the direction that the waveform is traveling.

If we assume that the sea surface freezes when $t = 0$, we can look at the waveform and its variation in the x-direction; if we set x equal to zero, we can see how the water surface undulates with time at a fixed position in space (i.e., $x = 0$). To simplify the description of the sinusoidal sea surface, $2\pi/L$ and $2\pi/T$ are referred to as the *wave number* (k) and the frequency (σ) respectively. The peak of the wave is called the *crest*; the lowest point is the *trough*. The distance between the crest and trough is the wave height H, equal to twice the amplitude ($H = 2A$). As before, T and L are the period and length of the wave respectively.

Water waves are transverse; that is, the oscillation of the sea surface is at right angles to the direction in which the waveform propagates. In the simple sinusoidal approximation, the waveform progresses from the area of generation to its point of eventual decay or dissipation without creating a net displacement of water in the direction of wave propagation. As most of us know, energy which is propagated by the waveform can be quite awesome. Let us now consider what we can learn about water waves and what restrictions are imposed when we approximate water waves by using a simple wave theory. Although we shall need some mathematics, we can make approximations that yield simple and easily used equations.

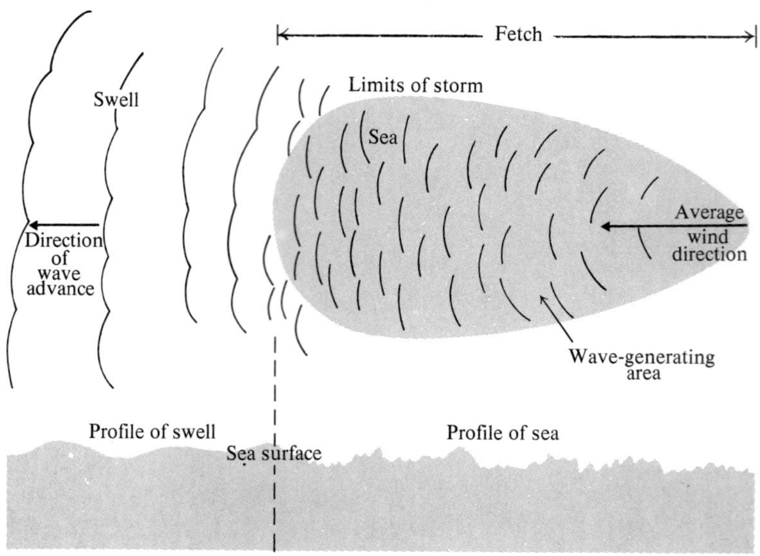

Fig. 16.5 Waves sort themselves out by virtue of their differing speeds of propagation and form a regular wave pattern known as swell when they leave the area of active wind generation. (After M. G. Gross, *Oceanography*, Charles E. Merrill Books, Columbus, Ohio, 1967.)

SINUSOIDAL WATER WAVES

We shall first consider the relationship between the uniform cosine wave and waves that occur naturally at sea. As we have stated, the smallest waves are created by distortion of the sea surface by wind and its restoration by surface tension.

As wavelets or ripples form, they roughen the surface of the sea, allowing more energy to pass from the moving air to the water. The waves gradually grow higher and longer until all sizes and frequencies are present in the area of active wind generation; gradually gravity takes over as the restoring force as the waves increase in size. The wave generation area appears confused and disordered, with no pattern resembling a cosine curve (Fig. 16.5). The confused state of the water's surface in an area being actively stirred up by storm winds is called a *sea*. The use of this term is evident from the expression, "There is a high sea building in the hurricane." Wave measurements made in a sea would show that the water surface is actually a composite of many cosine curves, each representing waves of a certain period, height, and length; the eye, however, cannot distinguish individual wave types. The waves radiate outward from the area of generation, propagating at speed C, the *celerity*, which is the wavelength divided by the period (L/T). Waves of all sizes and frequencies are generated, each traveling at a speed determined by its particular L/T. Thus water waves are dispersive; i.e., the faster ones propagate through the slower ones, leaving the area of generation more rapidly and causing a gradual sorting of wave types as they move away from the area of wind generation. If there are no winds to impart further energy to the waves, they continue as free waves, gradually decaying and losing their energy. Decaying waves that have outraced their generating storm appear more regular in form due to natural sorting by dispersion. When observed at a distance from their source, these well-sorted, uniform waves, usually called *swell*, look more like idealized cosine waves. They tend to have long crests and repeat themselves in regular sequences called *wave trains* and their heights are small compared to their lengths.

At sea, waves from one area of wind generation may traverse many miles and be further modified by new storms or by wave trains from other storm areas. In this discussion we shall assume that, once generated, waves become free and appear as regular undulations at the air-sea boundary. How and where they are formed or how many times they are modified is of little consequence to our analysis. Free waves can travel large distances over great circle routes in the ocean. Recently, for example, researchers have been able to identify and follow groups of free waves from their source of generation south of New Zealand to the beaches of Alaska.

The regular rise and fall of the sea surface as described by simple wave motion theory requires that the water particles under the traveling wave form move in a closed orbit with an orbital velocity composed of components in both the horizontal and vertical directions, designated as u and w respectively (the particle velocities). The vertical particle velocity at the sea surface is determined by the rate of its rise and fall; that is, $w = \Delta\eta/\Delta t$, where Δ (Greek delta) indicates the change of a variable and $\Delta/\Delta t$ is the rate of change. We can easily observe the motion imposed on water particles by a passing wave form if we suspend neutrally buoyant particles in a glass-walled wave tank and observe their motion from the side. However, we must use mathematics to understand what is happening.

The mathematical analysis of simple water waves is only as good as the approximations we employ. To use a simple cosine wave as an approximation of a water wave, we must require that (1) the amplitude be small in comparison to the wavelength ($A/L < 1$) and (2) the waves have very long crests that are perpendicular to the direction of propagation. We must also establish a coordinate system. In this case, we shall assume that Z is zero at the equilibrium sea level and is positive upward, while X is the horizontal direction and is positive in the direction of propagation. The sea surface, displaced from equilibrium by a small-amplitude wave, lies a distance η from the line $z = 0$. The sea floor is a distance d below the equilibrium sea level and is therefore located at $z = -d$. We can now stipulate boundary conditions at both the sea surface and the sea floor:

$$w = (\Delta\eta/\Delta t) \quad \text{at} \quad z = 0, \tag{16.7}$$

$$w = 0 \quad \text{at} \quad z = -d. \tag{16.8}$$

A third boundary condition is that pressure at the sea surface equals atmospheric pressure P_a, and is constant.

In Eq. (16.7), $w = \Delta\eta/\Delta t$ at the actual sea surface where $z = \eta$, but since η is small for small-amplitude waves we shall accept the approximation at $z = 0$ (and see later that it is useful). Equation (16.8) implies that no water can move vertically through the sea bottom. In other words, the seabed is impermeable. We assume also that the depth of the water is uniform.

The three boundary conditions given above are required for a simple evaluation of the dynamic conditions that determine the motion of water particles in the water column. In a two-dimensional problem, we need the x-equation of motion to describe u, the horizontal particle velocity, and the z-equation of motion to describe w, the vertical particle velocity:

$$\frac{\Delta u}{\Delta t} = -\frac{1}{\rho}\frac{\Delta P}{\Delta x}, \tag{16.9}$$

$$\frac{\Delta w}{\Delta t} = -g - \frac{1}{\rho}\frac{\Delta P}{\Delta z}, \tag{16.10}$$

where P, g, and ρ are the pressure, acceleration due to gravity, and density of the water, respectively. We need two additional equations to describe continuity of flow and the restriction to irrotational motion. For an incompressible fluid, the continuity equation in two dimensions is

$$\frac{\Delta u}{\Delta x} + \frac{\Delta w}{\Delta z} = 0. \tag{16.11}$$

That is, for a given change in velocity (flow of water) across an assumed box in the x-direction, there must be an equal and opposite change in velocity in the z-direction so that the amount of water remains constant in the box—what goes in must come out at an equal rate. The equation for irrotational motion in two dimensions,

$$\frac{\Delta u}{\Delta z} - \frac{\Delta w}{\Delta x} = 0, \tag{16.12}$$

states that if u, the horizontal velocity, changes in the z-direction to produce a velocity shear on a particle of water and thus a torque, then w, the vertical velocity, must change in the x-direction to produce an equal but opposite shear so that no turning moment is generated on a water particle. In other words, a velocity structure involving u and w does not force the particle to rotate on its center.

Equations (16.9) through (16.12) and the boundary conditions form a mathematical linear problem which can be satisfied by a waveform solution of the type

$$\eta = A \cos (kx - \sigma t). \tag{16.13}$$

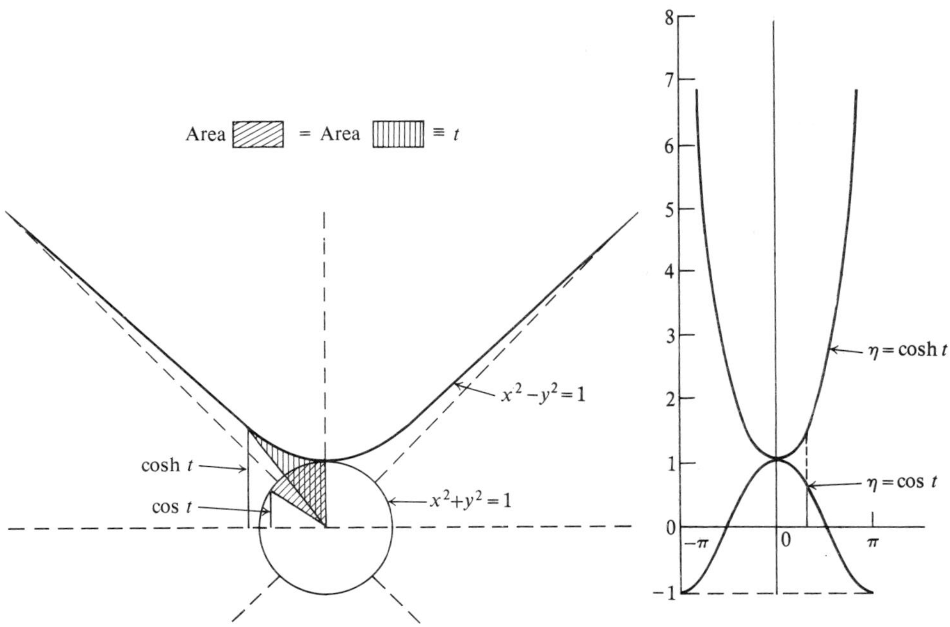

Fig. 16.6 A comparison of hyperbolic cosine function and the cosine. (From B. Kinsman, *Wind Waves*, © 1965. By permission of Prentice-Hall, Inc., Englewood Cliffs, N. J.)

Using this solution, we can show that the following relationships hold for Eqs. (16.8) through (16.12):

$$u = A\sigma \frac{\cosh k(z + d)}{\sinh kd} \cos (kx - \sigma t), \tag{16.14}$$

$$w = A\sigma \frac{\sinh k(z + d)}{\sinh kd} \sin (kx - \sigma t), \tag{16.15}$$

$$P = Pa - \rho gz + \rho gA \frac{\cosh k(z + d)}{\cosh kd} \cos (kx - \sigma t). \tag{16.16}$$

These relationships also satisfy Eq. (16.7) and the third boundary condition if σ is defined by

$$\sigma^2 = gk \tanh kd. \tag{16.17}$$

The functions introduced in these equations—cosh, sinh, and tanh—are the hyperbolic sine, hyperbolic cosine, and hyperbolic tangent respectively (Fig. 16.6). Although we shall not explain their special properties here, their values, like those of the sine or cosine of an angle, are expressed numerically and may be found in any standard table of trigonometric functions.

Equations (16.14) and (16.15), which define the particle velocities of water under the waveform, are dependent on depth z, distance x, and time t. If the products of velocity times increments of time are summed for all values of x and z, the totals equal the distances the particles are displaced for all values of x and z. This summation process is called *integration*. The displacements derived for the x- and z-directions are:

$$\xi = -A \frac{\cosh k(z + d)}{\sinh kd} \sin (kx - \sigma t), \tag{16.18}$$

$$\zeta = A \frac{\sinh k (z + d)}{\sinh kd} \cos (kx - \sigma t). \tag{16.19}$$

These two distances (ξ and ζ) combine to describe the circular path in time that a water particle takes about a point with particular values of x and z.

We now have all the equations we need and may simplify them for use. We have stated that C, the celerity or speed of propagation of a wave, equals the wavelength divided by the wave period:

$$C = \frac{L}{T} = \frac{\sigma}{k} = \frac{2\pi/T}{2\pi/L}. \tag{16.20}$$

Since Eq. (16.17) determines σ^2 from the acceleration due to gravity, wave number, and depth of water, it follows that

$$C^2 = \sigma^2/k^2 = g/k \tanh kd. \tag{16.21}$$

This expression is the *fundamental* equation for determining the speed of a water wave under the conditions stipulated thus far. We may now make approximations to simplify the determination of C.

THE SHORT-WAVE APPROXIMATION

If the product kd, $2\pi d/L$, is large enough, $\tanh kd$ is approximately unity and

$$C_0^2 = g/k = (g/2\pi)L_0 , \tag{16.22}$$

where the subscript zero identifies the approximate celerity and length of short waves. For $\tanh kd$ to be approximately unity, *d/L must be greater than $\frac{1}{2}$* or *kd must be greater than π*. Waves that meet this condition are called *short waves* because their lengths are short in comparison to the depth of the water. Reversing this statement, we say that the depth of the water must be great compared to the length of the waves; therefore, short waves are also known as *deep-water waves*.

At the beginning of this chapter we stated that the wind distorts the sea surface and forms very small ripples or surface-tension waves. These waves are very short with lengths measured in inches or fractions of inches; when working with them, we must consider a surface-tension term as well as a gravity term:

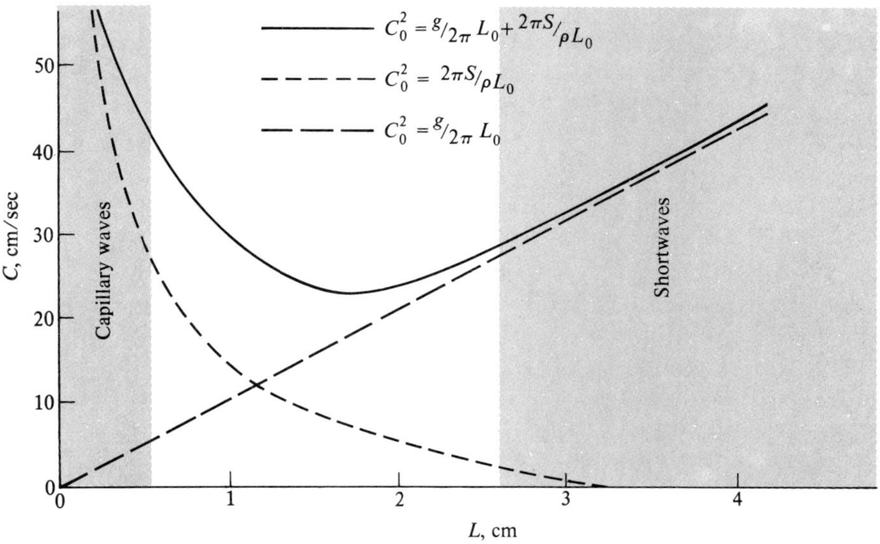

Fig. 16.7 The propagation speed and restoring force for waves with length of less than 0.5 cm is governed principally by surface tension. When waves grow to lengths greater than 2.5 cm, they act principally as gravity waves.

$$C_0^2 = (g/2\pi)L_0 + (2\pi S/\rho\, L_0), \tag{16.23}$$

where S is the surface tension (about 74 dynes/cm for water at 20°C) and ρ is the density of the water. The graph in Fig. 16.7 shows the region in which either term or both describe celerity as a function of length. A wave does not have to be very long to become a pure-gravity wave with a celerity adequately described by the first term in Eq. (16.23).

From Eq. (16.22), which shows celerity squared as a function of length only, we can derive another relationship involving the wave period T:

$$C_0^2 = L_0^2/T^2 = (g/2\pi)L_0 \tag{16.24}$$

or

$$L_0 = (g/2\pi)T^2; \tag{16.25}$$

that is, the length of short waves is determined by T, which in turn determines, C. Thus we find directly from T:

$$C_0 = (g/2\pi)T. \tag{16.26}$$

Equations (16.22) through (16.26) completely define celerity under the short-wave approximation. Both C and L are related to T, which for our purposes is determined by the system generating the wave. We further assume that once T is stipulated by the generating source, it is conserved and does not change as long as the wave exists.

We may also simplify Eqs. (16.14) through (16.16), which define the particle velocities u and w and pressure P, using the assumptions for either short or long waves. For short waves, we have

$$u = A\sigma e^{kz} \cos (kx - \sigma t), \tag{16.27}$$

$$w = A\sigma e^{kz} \sin (kx - \sigma t), \tag{16.28}$$

$$P = Pa - \rho gz + \rho gAe^{kz} \cos (kx - \sigma t). \tag{16.29}$$

Note that e^{kz} has replaced the hyperbolic functions. This term represents an attenuation factor; the constant e (the base of the natural logarithim) = 2.71828. Since a number raised to a negative power decreases as the value of the exponent increases, e^{kz} for a given k becomes smaller as z increases, taking on larger negative values. The rate at which e^{kz} decreases depends on k: the shorter the wave length, the larger the wave number and the faster e^{kz} decreases with depth. Thus u, w, and P all decrease or are attenuated rapidly with depth if L is small; in fact, the effect of short waves on the movement of water particles nearly disappears when z equals $\frac{1}{2}L$. At this depth the variations in hydrostatic pressure due to wave trains passing above is nil, since the total pressure is determined solely by atmospheric pressure and the hydrostatic head. Thus a sensor located at great depth on the sea floor cannot detect short waves passing across the sea surface.

The water particle velocity equations and Eqs. (16.18) and (16.19), their integrated forms describing the displacement of water particles, show that particles move in circular orbits under the passing wave form. The equations for u and w, as well as for ξ and ζ, are similar to each other but out of phase; the attenuation factor causes them to decrease as z increases negatively, producing weaker particle motions and reducing the size of the orbits. The radius of the orbit under deep-water conditions (short-wave approximation) is

$$r_0 = \sqrt{\xi^2 + \zeta^2} = Ae^{kz}. \tag{16.30}$$

THE LONG-WAVE APPROXIMATION

If waves are very long compared to the depth of the water, we may simplify the basic celerity equation (16.21) to obtain the *long-wave approximation* for a *shallow-water wave.* In this case kd is small, so that tanh kd approaches the value of kd; specifically, d/L *must be less than* $\frac{1}{20}$ or kd *must be less than* $\pi/10$. When this condition is met, Eq. (16.21) is approximated by

$$C^2 = (g/k)(kd) = gd. \tag{16.31}$$

Wave celerity is therefore controlled entirely by the depth of the water, since g, the acceleration due to gravity, is for all practical purposes a constant. We stipulated above that T, the period of a wave, once given does not change. Thus, since

$$C = L/T = \sqrt{gd} \tag{16.32}$$

or

$$L = T\sqrt{gd}, \tag{16.33}$$

and C varies with depth, L must also vary with total depth of water. Wind waves generated on the open sea are short waves. As they run away from the storm area and approach a coast, they begin to *feel bottom*; i.e., they pass through short-wave conditions to intermediate conditions (Eq. 16.21) to shallow-water or long-wave conditions. When depth decreases, causing the wave speed to slacken, the wavelength must gradually decrease since T is conserved; thus the waves are said to *shorten up*. Because the ratio of H to L becomes larger as L decreases, the waves also steepen gradually until they become unstable and break. At this point, the wave steepness is often given as $H/L = \frac{1}{7}$; the crest angle becomes less than 120°.

We have made two approximations of the wave celerity equation on the basis of deep-water (short) or shallow-water (long) waves. These terms are frequently confusing, since some long waves in midocean are sufficiently long to qualify as shallow-water waves by definition. The mean depth of the oceans is approximately 4000 m; therefore, a wave that is 80 km or longer meets the criteria for long (shallow-water) waves. Tides, waves caused by atmospheric pressure disturbances, and *tsunamis* (*seismic sea waves*) may exceed these lengths, in which case they travel at speeds determined by the water depth. From the approximate depth of the ocean, we find the characteristic celerity for these midocean long waves as

$$C = \sqrt{(9.8\ \text{m/sec}^2)\ 4000\ \text{m}} = 200\ \text{m/sec}, \tag{16.34}$$

or approximately 400 miles per hour.

There are also approximate forms for u, w, and P for long wave conditions. Equations (16.14) through (16.16) become

$$u = \frac{AC}{d}\cos(kx - \sigma t), \tag{16.35}$$

$$w = \frac{AC}{d}k(z + d)\sin(kx - \sigma t), \tag{16.36}$$

$$P = Pa - \rho gz + \rho gA\cos(kx - \sigma t). \tag{16.37}$$

Note that the hyperbolic functions have again been replaced by simpler relationships. In the case of long waves, u and w do not appear to have a symmetrical equation form; u is unaffected by z, while w has a maximum value at the surface, where $z = 0$, and decreases to zero at the bottom, where $z = -d$. (Recall that we stipulated earlier that w must be zero at the sea floor.) Because

w varies with z but u does not, the water particles move in elliptical rather than circular orbits. The orbits flatten vertically as z becomes more negative, until at the bottom they have only an unmodified horizontal axis and their motion (u) is back and forth, oriented with the direction of wave travel. The fact that the horizontal particle motion is unaltered with depth until bottom friction is encountered, while vertical motion is continually reduced with depth, helps to explain why the waves "feel bottom."

The hydrostatic pressure effect of passing long waves, unlike that of short waves, can be detected by a sensor placed on the sea floor. The total hydrostatic pressure is the sum of atmospheric pressure at the sea surface, the hydrostatic head of the water from equilibrium sea level to the bottom, and $\rho gA \cos(kx - \sigma t)$, the oscillatory pressure generated as the waves pass.

THE USE OF WAVE EQUATIONS

We now have enough information to solve some rather sophisticated problems. Consider first the prediction of the arrival times of a tsunami. When an earthquake is recorded in the ocean (Fig. 16.8), a forecasting group is alerted to the possibility that a seismic wave has been generated. The task is first to draw a map with lines radiating from the quake center out across the ocean; then the ocean depths are determined from charts and the speed of a possible seismic

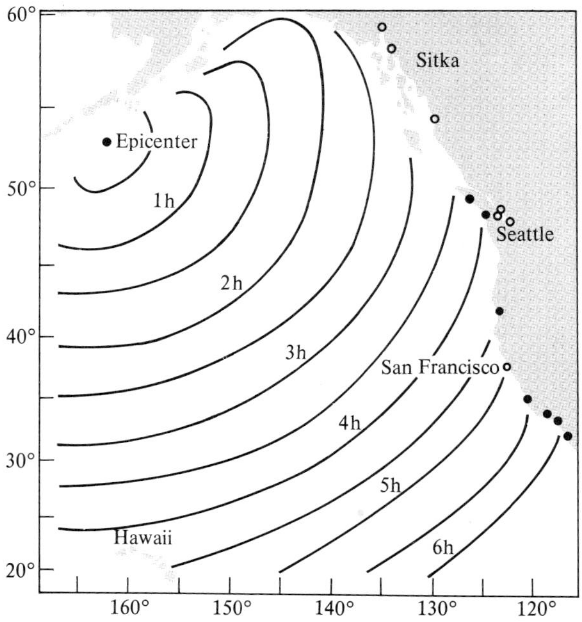

Fig. 16.8 Positions of the April 1, 1964, tsunami in hours after the Alaska earthquake. Open circles mark tide stations where little effect was noted. Closed circles mark locations of large observed waves. (From R. W. Fairbridge, ed., *Encyclopedia of Oceanography*, © 1966 by Reinhold Pub. Corp. Used by permission of Van Nostrand Reinhold Co.)

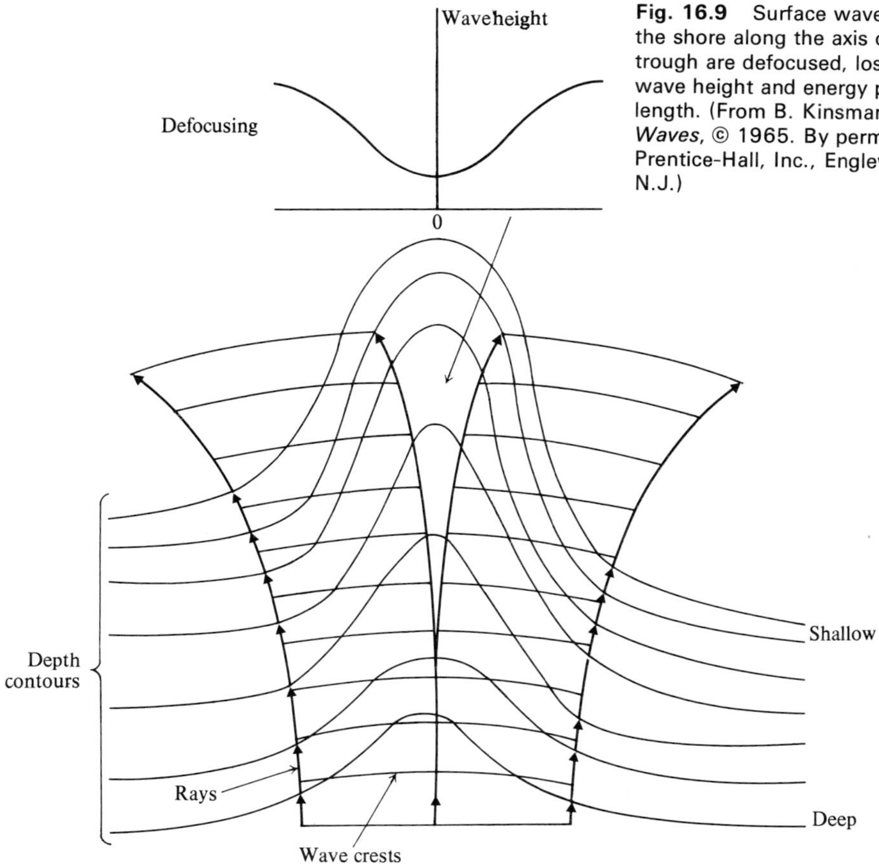

Fig. 16.9 Surface waves approaching the shore along the axis of a sea floor trough are defocused, losing both wave height and energy per unit crest length. (From B. Kinsman, *Wind Waves*, © 1965. By permission of Prentice-Hall, Inc., Englewood Cliffs, N.J.)

wave from $\sqrt{gd}$ along the radiating lines is calculated. It is then a simple matter to estimate the distance the wave will travel in each of the hours after it has been generated, and to draw corresponding position rings around the quake center. The anticipated arrival time is thus determined from the intersects of the time rings and the shore. Shore stations are then alerted to watch for the wave. If the early arrival stations report a wave, then all shore points are alerted and given the approximate time of arrival. Although this method is satisfactory for estimating arrival time, it does not yield any information on the height of the anticipated tsunamis at a particular location.

We may also use the wave equations to solve problems dealing with coastal water depths and bottom topography. We have shown that as a wave moves into shallow water, its celerity and length decrease. In other words, one part of the crest of a wave approaching a beach obliquely is in shallower water, slowing down, while the offshore end is in deeper water maintaining its speed. The crest will therefore tend to curve parallel to the beach line (Fig. 16.9).

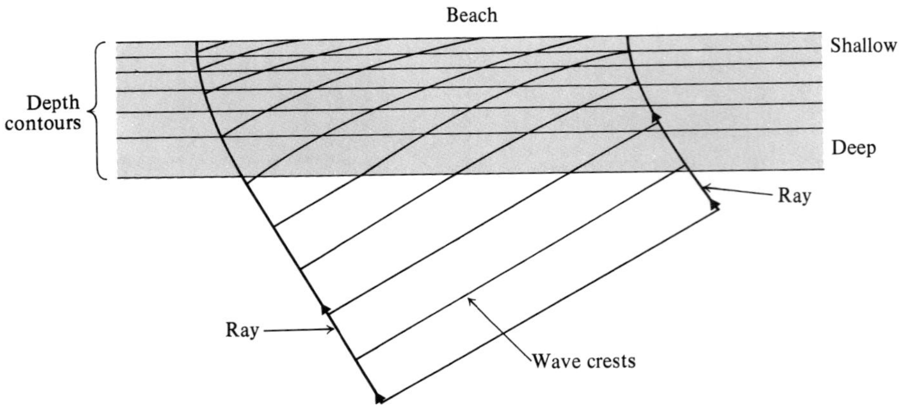

Fig. 16.10 Waves approaching a uniform beach obliquely are refracted as the end of the crest in shallow water propagates more slowly than the seaward end. (From B. Kinsman, *Wind Waves*, © 1965. By permission of Prentice-Hall, Inc., Englewood Cliffs, N.J.)

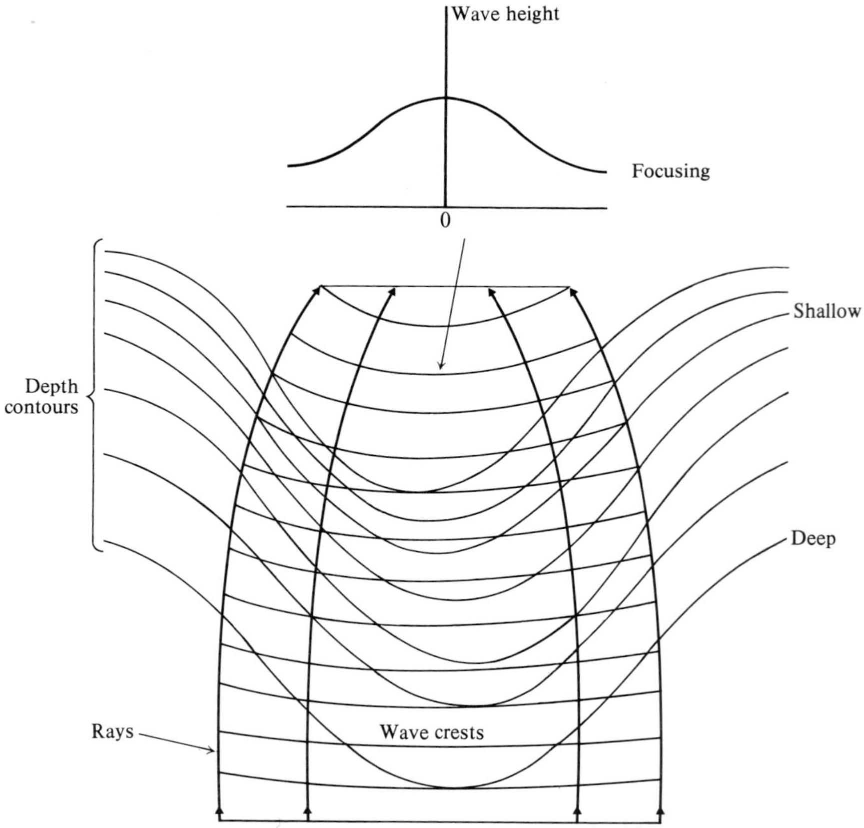

Fig. 16.11 Surface waves approaching the shore along the spine of a submerged ridge are focused and gain both height and energy per unit crest length. (From B. Kinsman, *Wind Waves*, © 1965. By permission of Prentice-Hall, Inc., Englewood Cliffs, N.J.)

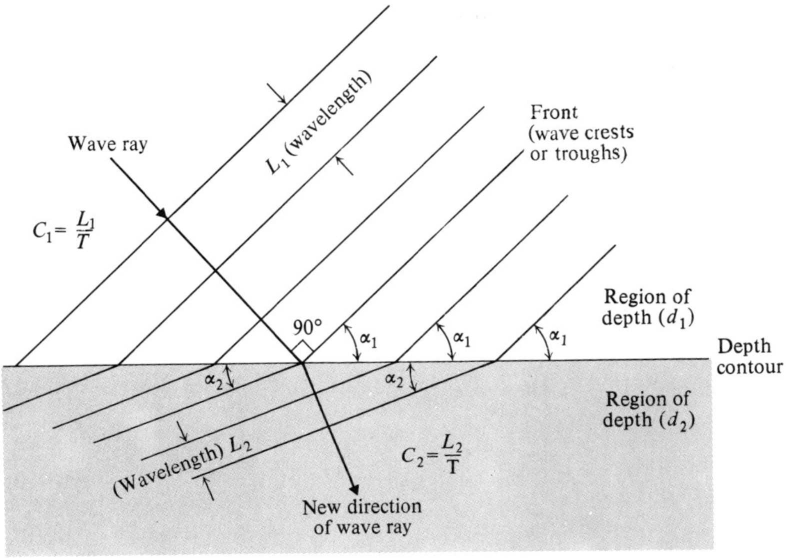

Fig. 16.12 Waves are refracted in accordance with Snell's law as depth acts to alter their speed of propagation.

The degree of this bending, or *refraction*, is given by Snell's Law (see Chapter 8). If we calculate the change in celerity from the degree of curvature and the decrease in the length of successive waves, we may determine the water depths and the bottom topography along the coast (Figs. 16.10 and 16.11). The speed of a long wave as it moves through different depths of water (Fig. 16.12) is given by

$$C_1 = gd_1 = L_1/T, \tag{16.38}$$

$$C_2 = gd_2 = L_2/T. \tag{16.39}$$

Since T is a constant,

$$L_1/\sqrt{gd_1} = T = L_2/\sqrt{gd_2} \tag{16.40}$$

and

$$L_1/L_2 = \sqrt{d_1}/\sqrt{d_2} = C_1/C_2. \tag{16.41}$$

We can calculate the change in depth from changes in L taken from aerial photographs.

We may sometimes fix approximate locations of storm centers (areas of wave generation) from wave observations alone. Since short wind waves travel away from their generating area at a speed determined by their periods, longer period waves (with longer wavelengths) will arrive at a remote point before shorter ones from the same storm center. For example, assume that waves of period T_1 arrive at a given location one day earlier than waves of period T_2. If the waves are from the same storm, we may find the distance D traveled by both as follows: Since

$$D = C_1 \times \text{travel time} = (g/2\pi)\,T_1 \times \text{travel time} \tag{16.42}$$

and

$$D = C_2 \times (\text{travel time} + 1\text{ day}) = (g/2\pi)\,T_2 \times (\text{travel time} + 1\text{ day}) \tag{16.43}$$

where the travel time is that required by the longer period waves to complete the journey, then we have

$$(g/2\pi)\,T_1 \times \text{travel time} = (g/2\pi)\,T_2 \times (\text{travel time} + \text{one day}), \tag{16.44}$$

or, simplifying,

$$(T_1/T_2) - 1 = 1\text{ day/travel time}, \tag{16.45}$$

where

$$\text{travel time} = 1\text{ day}\,[T_2/(T_1 - T_2)]. \tag{16.46}$$

We can now calculate from Eqs. (16.42) and (16.46) the distance to the storm center and determine its direction from the direction of the incoming waves. We can obtain a more accurate fix of the storm center by backtracking the waves from several shore positions.

Waves occur at sea in all sizes, lengths, and periods, ranging from ripples to forms so long that the human eye cannot distinguish them. A wave becomes essentially invisible, in fact, if its period exceeds 40 sec; we can see the rise and fall of the water surface only against a reference line. Even a tsunami traveling at 200 m/sec passes without notice on the high seas, since the eye cannot detect a wave that is 1 to 2 ft high and about 300 km long. The tides, which are also wave forms of long length and period, cannot be seen from the deck of a ship at sea. Seamen focus their attention instead on shorter wind waves, which may expend their energy against a ship and can cause serious trouble.

WAVE ENERGY

All water waves possess two kinds of energy: kinetic, by virtue of the motion of the water, and potential, created by the displacement of the sea surface from equilibrium. The energy of a wave is not concentrated in a given location, but distributed under the sea surface and over the entire wave length. We shall now explain how to determine the wave energy contained in a volume of water extending from $z = \eta$ to the bottom, from $x = 0$ to L, and one unit of width along the crests.

Consider first the potential energy (PE) of a static column of water extending from sea surface to seabed, of unit cross-sectional area. The PE is $\rho g d(d/2)$, where d is the depth of water. If a wave distorts the sea surface, the PE becomes $\rho g(d + \eta)[(d + \eta)/2]\Delta x$, where Δx is the unit width of the cross-sectional area oriented to the direction of wave travel. The summation of the PE for each column over the wavelength L, of which Δx is a small increment, yields the total PE of the water per unit width perpendicular to direction of travel over the entire wavelength:

$$\text{total PE} = \tfrac{1}{2}\rho g \sum_{0}^{L} (d + \eta)^2 \,\Delta x. \tag{16.47}$$

To find the total PE in the presence of waves, we must determine both the PE of the static water column without waves and then the additional PE that they contribute, as follows:

$$\text{PE} = \tfrac{1}{2}\rho g \sum_{0}^{L} \eta^2 \Delta x + \rho g d \sum_{0}^{L} \eta \,\Delta x. \tag{16.48}$$

However, since η has as many positive as negative values, all of equal magnitude over one wavelength of a simple sinusoidal wave, the second term, $\rho g d \Sigma_0^L \eta \Delta x$, becomes zero and only that part of the wave's contribution to PE involving η^2 remains significant. The PE due to the presence of a simple sinusoidal wave is therefore

$$\text{PE} = \tfrac{1}{2}\rho g \sum_{0}^{L} \eta^2 \,\Delta x. \tag{16.49}$$

We can also determine the kinetic energy (KE) of a wave over a depth range of η to $-d$ and one wavelength 0 to L per unit width:

$$\text{KE} = \tfrac{1}{2}\rho \sum_{0}^{L} \sum_{-d}^{\eta} (u^2 + w^2)\,\Delta z\,\Delta x, \tag{16.50}$$

where u and w are the orbital velocities of water particles beneath the wave form. The total energy E of the wave is the sum of PE and KE. We can simplify the calculation of PE and KE by substituting Eqs. (16.13), (16.14), and (16.15) for η, u, and w as functions of x, t, and z into Eqs. (16.49) and (16.50) and then summing over the depth and wavelength. We thus find for the averaged energies, where A is the amplitude and H is the wave height,

$$\overline{\text{PE}} = \tfrac{1}{4}\rho g A^2 = \tfrac{1}{16}\rho g H^2, \tag{16.51}$$

$$\overline{\text{KE}} = \tfrac{1}{4}\rho g A^2 = \tfrac{1}{16}\rho g H^2. \tag{16.52}$$

The average total energy per unit width of a wave due to the wave is therefore the sum of $\overline{\text{PE}}$ and $\overline{\text{KE}}$, or

$$\overline{\text{E}} = \overline{\text{KE}} + \overline{\text{PE}} = \tfrac{1}{2}\rho g A^2 = \tfrac{1}{8}\rho g H^2. \tag{16.53}$$

We note from this equation that the higher the wave, the greater its energy; however, exceptionally high waves are uncommon in the sea. If we averaged all waves over the oceans over a long period of time to determine the energy distribution with wave type, we would find that the greatest amount of energy is associated with rather small wind waves, since they are the most abundant and frequent. A plot of $\overline{\text{E}}$, the average energy, versus wavelength, or period, or frequency σ is called an *energy spectrum*. It may be used to identify waves produced from a specific generating source or storm area, even if they move thousands of miles across the ocean.

PROPAGATION OF WAVE ENERGY

Waves carry their energy with them as they propagate, transferring it not at the celerity of the wave, but at a speed at which the group of waves moves, or *group speed* (V). To illustrate, let us consider the following simple experiment: A large rock is dropped into a pool, producing a ring of small waves that move outward from the center of disturbance. Water remote from the center is not affected until the outer edge of the wave-train ring reaches it. The ring moves at the group speed V, which is the speed at which the energy is propagated away from the disturbance center. Close inspection will show that individual waves which form inside and disappear outside the ring move faster than the ring itself. If we measure C and V, assuming that the waves are gravity waves, we can show that for these short waves

$$V = \tfrac{1}{2}\, C. \tag{16.54}$$

If we perform the same experiment in very shallow water, so that the ratio of L to d meets the conditions for long waves, we find

$$V = C. \tag{16.55}$$

Under these conditions, the lengths of the individual waves are sufficiently large that their speed is determined by the water depth. Since in our example the length of the wave group is twice the width of the wave train forming the ring, the group speed is also determined by d. In general, however,

$$V = nC \tag{16.56}$$

where

$$n = \tfrac{1}{2} + (kd/\sinh 2kd). \tag{16.57}$$

The $\frac{1}{2}$ and 1 approximations of n hold if the conditions for short and long waves respectively are met. In the case of intermediate waves, we must evaluate n by using Eq. (16.57).

We have discussed briefly the refraction of waves as they come under the influence of shoaling conditions, i.e., approach a beach. Using our knowledge of wave energy, we can determine the energy transported per unit width of crest as averaged over one wavelength. We have also stated that as two successive waves approach shoal water, their crests are initially parallel. If we draw two lines parallel to each other and perpendicular to the two crests, the wave energy of the area between the perpendiculars, or *wave rays*, is equal to $\overline{E}$ times the number of unit widths separating the rays. As the crests approach the shore and gradually bend due to refraction, the wave rays begin to either diverge or converge. Since ray convergence creates more energy per unit width than does divergence, more energy is expended on headlands and shoals than in bays and on the beaches at submarine canyon heads. Waves are also higher in regions of converging rays, since $\overline{E}$ is related to A (see Figs. 16.9 and 16.11).

OTHER TYPES OF WAVES

Waves other than those formed at the surface are found in the sea. *Internal waves*, which propagate between layers of water with dissimilar densities, move within the sea in much the same manner as do their surface counterparts; the former, however, tend to have longer periods and lengths and slower speeds of propagation than the latter. In regions in which the total depth d is large compared to either L or h, the thickness of the surface layer, the celerity of an internal wave is given by

$$C = \sqrt{gh\,[(\rho_2 - \rho_1)/\rho_2]}, \tag{16.58}$$

where ρ_1 is the density of the surface layer and ρ_2 is the density of the deeper layer. When L is large in comparison with d,

$$C = \sqrt{gh\,[(d - h)/d]\,(\rho_2 - \rho_1)/\rho_2}. \tag{16.59}$$

High atmospheric pressure creates depressions of the sea surface, while low atmospheric pressure creates bulges. These surface irregularities also travel as single waveforms, often keeping pace with their generating sources as forced waves. Another type of wave is the *standing wave* or *seiche*. Although seiches cause the surface to rise and fall, they do not progress across the surface of the water as do other waves. For example, winds may exert stress on a water surface, causing water to accumulate against the shore; when the wind ceases, the effort of the water to return to equilibrium may create an oscillation at a period unique to the region of occurrence. We may visualize these seiches, frequently found in lakes, by tilting a basin partially filled with water and then immediately setting it on a level surface. The water will slosh back and forth at a period determined by the size and shape of the basin. The maximum amplitude of a seiche in a closed basin occurs at the ends, or *antinodes*; the *node*, or point of no vertical change, is located near the center. Since standing waves do not progress horizontally, they have no celerity. In effect, they are the result of a progressive wave being reflected back on itself, with the celerities of the two opposing waves cancelling each other.

We may determine the period of oscillation in a rectangular basin of depth d (d is smaller than the basin length l so that shallow-water wave conditions hold) from

$$T = 2l/\sqrt{gd}. \tag{16.60}$$

We note from this equation that the fundamental standing wave (only one node being present) has a wavelength twice that of the basin ($L = 2l$).

Under normal conditions, large natural closed water basins are seldom tilted, but their surfaces may be. If the force causing the tilt is removed, a free standing wave may result. Every natural basin has a period of resonance for each direction in which a seiche may be formed.

Standing wave oscillations also may occur in a basin with an open end connected to the sea. In this case the node of the fundamental wave is located at the basin entrance, such that the basin length is one-fourth the wavelength. In the Bay of Fundy, one of the most famous of such basins, standing tides in excess of 50 ft occur at the head, while a 6- to 8-ft tidal range is normal at the mouth. The bay's resonant period is nearly the same as the tidal period at the mouth; thus the input of tidal energy in rhythm with the natural period of oscillations produces an extremely large tidal range.

There are many theoretical mathematical approximations of water waves besides our very simple sinusoidal one, which is crude but allows us to identify most of the wave properties. If one observes free water waves carefully, he will note that they are actually not sinusoidal, but trochoidal, with flat troughs and steep crests. He will also find that water particles move not in closed orbit but in orbits that are continuously displaced in the direction of wave propagation. Thus real water waves produce a small transport of water in the direction of wave propagation which can be supplied to a beach, where it is then carried off by a longshore or rip current (see Fig. 5.3).

Using more sophisticated mathematical theories for investigating waves, we may determine additional properties of waves and depict their shapes and behavior more accurately. Unfortunately, however, the acquisition of this 5 to 10% increase in knowledge about waves requires a 100 to 200% increase in mathematical complexity.

We may also study waves without using theoretical approximations by statistically recording sea-surface elevations and their change in time and space. For example, we may analyze a wave record to determine the maximum wave height, average height of the highest 10% of the waves, variety of wave periods, average height or energy associated with each period, etc. These characteristics may then be empirically related to the characteristics of the wave-generating region, the wind strength, the time the wind has blown, and the distance over which the wind blows in one direction (*fetch*).

Relationships between wind, time, and fetch and the generated sea state are used to forecast wave conditions that might pose a danger to vessels at sea. However, it is extremely difficult to make accurate predictions. Observations from a ship indicate local conditions, but the waves may have been generated in an entirely different region for which the ship has no data. It is easier to hindcast wave conditions from past weather data than to predict them. Atlases give monthly averages for wave occurrence, height, direction, and swell for the oceans, but since most ships are continually moving from place to place, this information is of little help. There seems to be no substitute for direct observations of the sea state, although the advent of satellite monitoring of the sea should aid forecasters considerably.

STUDY QUESTIONS

1 What ratio of d to L must hold in the deep- and shallow-water approximations for wave celerity?

2 Why doesn't $H/L = \frac{1}{7}$, the limiting wave steepness ratio, describe the slope of a wave near its crest with a crest angle of 120°? The crest angle is also considered limiting and if exceeded the wave will break.

3 The celerity of a free wave is always L/T. Derive the relationships that relate C to L alone and to T alone under deep-water conditions.

4 At what depth will the horizontal particle velocity of a deep-water wave be one-tenth its maximum surface value, given that the wave period is 7 sec and the wave height is 3 m?

5 If we consider the midocean tide as a shallow-water progressive wave, then the surface tidal current in midocean is described by the horizontal particle velocity. What is the maximum tidal current if the amplitude of the tide wave is 0.5 m and the depth of the ocean is 4 km?

6 If we mount a pressure recorder on the sea floor at 2-km depth, what is the minimum wave amplitude that it can measure if it cannot detect pressure fluctuations less than 10^7 dyn/cm^2? Assume the length of the wave to be 100 m.

7 A displacement-type ship hull generates a water wave whose maximum length approximates 150 m, the waterline length of the vessel. The ship can move efficiently through the water at speeds up to that which the maximum-length wave propagates. What is its maximum efficient hull speed?

8 Assume that the draft of the vessel in Problem 7 is 7 m. If the ship passes over a shoal 12 m deep, what happens to its speed, and why?

9 The length of a wave in water of 50 m depth is 85 m. What is its length at 18 m depth as it progresses shoreward? If its celerity at 85 m is C, what is its celerity at 18 m?

10 An atmospheric disturbance at sea generates storm waves with a period of 8 sec. How long will it take the wave energy generated by the storm to reach a beach 800 km away?

11 If a seismic disturbance generates waves with a period of 15 min, how long will it take their energy to traverse 800 km? Assume the mean depth of the water to be 4 km.

12 A wind blowing along the axis of a lake 50 km long gives the water a surface slope; if the wind suddenly ceases, an oscillatory seiche results. What is the period of the fundamental oscillation if the lake is 50 m deep?

CHAPTER 17

TIDES

INTRODUCTION

The tides are the great waves of the ocean that slowly beat against the shore, producing a rise and fall in sea level with astonishing regularity. Their wavelengths and periods are much larger than those of the ordinary wind waves discussed in the last chapter. Tide waves are not free because they cannot escape their generating force, which acts on them continually. Thus they are compelled to maintain a rhythm governed by the motion of the earth and the principal celestial tide-raising bodies, the moon and sun.

In other respects, however, tide waves act similarly to wind waves. Since the tide may be thought of as a simple progressive long wave traveling across or around the sea, we may consider the horizontal particle velocity of the water particles under the tide wave as the tidal current associated with the rise and fall of sea level. Like wind waves and swell, tide waves may be reflected and refracted; they reflect off of land barriers and pass back through the advancing wave, or oscillate in an ocean basin to create a standing wave. Unlike wind waves, however, tide waves are so long and their periods so great that water set in motion by them is influenced by the Coriolis force; thus they may rotate in an ocean basin with the turning of the earth.

Since tides are forced waves, we shall first study some of the conditions that cause them. The gravitational attraction between mobile seawater particles and the moon and sun creates the tidal force. The movement of these two celestial bodies relative to the earth thus determines the distribution of the tide-raising force. If we can calculate the magnitude of this force and its variation and movement in time, we can describe a theoretical *equilibrium* tide for the earth that will give us insight into the mechanics of the actual tides.

THE TIDE-RAISING FORCE

In Chapter 1, we showed that two masses exert an attractive force on each other that is described by Newton's law of gravitation,

$$F = G\frac{M_1 M_2}{R^2}, \qquad (17.1)$$

where G is the universal constant of gravitation, M_1 and M_2 are the masses, and R is the distance between their mass centers. The attractive force F exists between the two principal tide-raising bodies and any unit mass located on or in the earth. The force per unit mass, F/M_1, is equal to GM_2/R^2, where M_2 is the mass of the tide-raising body and R is the distance between the unit mass on the earth and the center of the tide-raising body.

Let us first consider the tide-producing force (or attractive acceleration) acting on a unit mass located at the earth's center. We can calculate this acceleration for the sun and the moon, using tabulated data and the law of gravitation (Table 17.1).

Table 17.1 Gravitational attraction on a unit mass at the earth's center due to the sun and moon

Body	*Mass,* g	*Distance between earth and celestial body, cm*	*Force/unit mass at earth's center, dyn/*g
Sun	1.971×10^{32}	149.5×10^{11}	5.876×10^{-2}
Moon	7.347×10^{25}	384.4×10^{8}	3.317×10^{-3}

The accelerations acting on the unit mass at the earth's center are small in comparison to the earth's acceleration due to gravity, g, and that caused by the sun is greater than that of the moon due to the former's larger mass. However, the attraction of a unit mass at the earth's center alone does not produce tidal activity on the earth; rather it is the difference between the attractive acceleration at the earth's water surface and at the earth's center. Therefore, we must now compute the tide-raising force per unit mass on the earth's water surface caused by the sun and moon and compare it to that found at the earth's center. If the magnitudes of the two forces were equal, a unit mass of surface water would be pulled toward the sun or moon with the same force as is a unit mass at the earth's center; thus there would be no *differential tidal force* and no tides. Since the force per unit mass at all points of the earth's surface is not equal to that at its center, there exists a distribution of differential tidal force that produces a distortion of the water surface, creating the tides.

Let us now consider how this differential force is distributed on a spherical earth model that is uniformly covered with water, and see how it produces a tide.

The earth's radius, 6371 km, is small compared to the distance between the earth and sun centers. Therefore, the attractive acceleration on a unit mass located at the intersection of the sun-earth line of centers and the earth's surface varies only slightly from 5.876×10^{-2} dyn/g, the value given in Table 17.1. The differential force available to produce tides is correspondingly small. The earth's radius is a considerably larger fraction of the distance between the earth and moon centers, however. Thus the attractive acceleration on a unit mass located at the intersection of the earth-moon line of centers and the earth's surface differs more from 3.317×10^{-3} (Table 17.1), indicating that the moon's differential tide-raising force is greater than the sun's. The small changes in R produced by the nonspherical shape of the earth and eccentricity

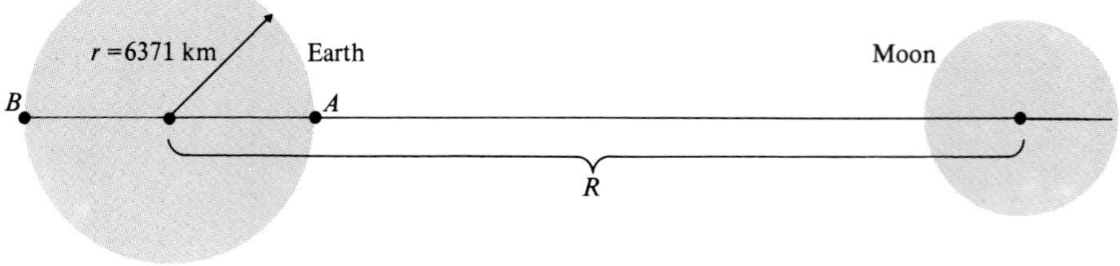

Fig. 17.1 The earth-moon system. R is the distance between the earth and moon centers. The separation distances between the moon's center and a unit mass at point A on the earth and between the moon's center and point B on the earth are $R - 6371$ km and $R + 6371$ km, respectively.

of the orbits of the sun, earth, and moon also affect the differential tide-raising forces, whereas G and the mass of the tide-raising body are for all practical purposes constants.

We shall now show that the differential tide-producing force at the water-covered earth's surface is proportional to $1/R^3$. We can determine the differential force by subtracting the force per unit mass exerted at the earth's center from that exerted per unit mass at the earth's surface. Consider the problem as follows: Let r be the earth's radius, R the distance between the earth's center and the center of the sun or moon, and M the mass of the sun or moon (Fig. 17.1). The attractive acceleration at the earth's center is GM/R^2. At point A the acceleration acting on the celestial body along the line of centers is $GM/(R - r)^2$, while at point B it is $GM/(R + r)^2$. Since $(R - r)^2$ is smaller than R^2, which in turn is smaller than $(R + r)^2$, the attractive acceleration at A must be greater than that at the earth's center, which in turn is greater than that at B. We can now determine the relative effective tide-raising force per unit mass at A and B by subtracting the acceleration at the earth's center from that at A and B respectively. For point A, we find

$$\frac{F}{m_1}(\text{relative}) = \frac{GM}{(R - r)^2} - \frac{GM}{R^2}, \qquad (17.2)$$

and for point B,

$$\frac{F}{m_1}(\text{relative}) = \frac{GM}{(R + r)^2} - \frac{GM}{R^2}. \qquad (17.3)$$

With simple algebraic manipulation, we can rewrite these equations as follows:

At point A:

$$\frac{F}{m_1}(\text{relative}) = -\frac{GM}{R^2}\left[1 - \frac{1}{(1 - r/R)^2}\right] = -\frac{GM}{R^2}\left[\frac{-2r/R + (r/R)^2}{(1 - r/R)^2}\right]; \qquad (17.4)$$

at point B:

$$\frac{F}{m_1}(\text{relative}) = -\frac{GM}{R^2}\left[1 - \frac{1}{(1 + r/R)^2}\right] = -\frac{GM}{R^2}\left[\frac{2r/R + (r/R)^2}{(1 + r/R)^2}\right]. \quad (17.5)$$

We shall now make an approximation that simplifies the right-hand members of Eqs. (17.4) and (17.5). The ratio r/R is about 1/60 for the moon and considerably smaller for the sun. Thus to a first approximation we shall assume that $(r/R)^2$ is negligible, that $(1 + r/R)$ and $(1 - r/R)$ are both approximately equal to 1. The right-hand expressions now become, for point A,

$$\frac{F}{m_1}(\text{relative}) = -\frac{GM}{R^2}\left(\frac{-2r/R}{1}\right) = \frac{2GMr}{R^3}, \quad (17.6)$$

and for point B,

$$\frac{F}{m_1}(\text{relative}) = -\frac{GM}{R^2}\left(\frac{2r/R}{1}\right) = \frac{-2GMr}{R^3}. \quad (17.7)$$

The relative tide-producing accelerations at points A and B have opposite signs, indicating that at point A the acceleration is directed toward the tide-raising body, while at point B it is directed away from the tide-raising body (negative sign). Thus a mass on the side of the earth facing the tide-raising body is pulled away from the earth's center, which simultaneously in turn is subject to a pull from the opposite side. Thus relative to the force on the earth's center the side of the earth opposite to the tide-raising body has a differential force acting away from the tide-raising body. We can now readily calculate the magnitude of $2GMr/R^3$ for both the sun and moon. It should be evident that the ratio $1/R^3$ reduces the relative attractive acceleration due to the sun such that it is less than that due to the moon, although the latter has the smaller mass.

The relative accelerations at points A and B are of equal magnitude but opposite sign. Had we completed our evaluation using the full equation rather than the approximation that r/R is small, we would have found the magnitude of the relative acceleration at A to be slightly larger than that at B. (The reader should verify this assertion algebraically.)

With a little trigonometry, we can now calculate the magnitude and direction of the relative accelerations due to a tide-raising body for all points on the earth's surface. We can further resolve each acceleration into two components, one perpendicular and one tangent to the earth's water-covered spherical surface. The former component is maximum at intersections of the line of centers with the earth's surface; at these points it is directed away from the earth's center and acts against the earth's acceleration due to gravity. At points on the circle formed by a plane perpendicular to the line of centers and passing through the earth's center, the vertical component is negative and directed toward the earth's center, acting in concert with the earth's acceleration due to gravity.

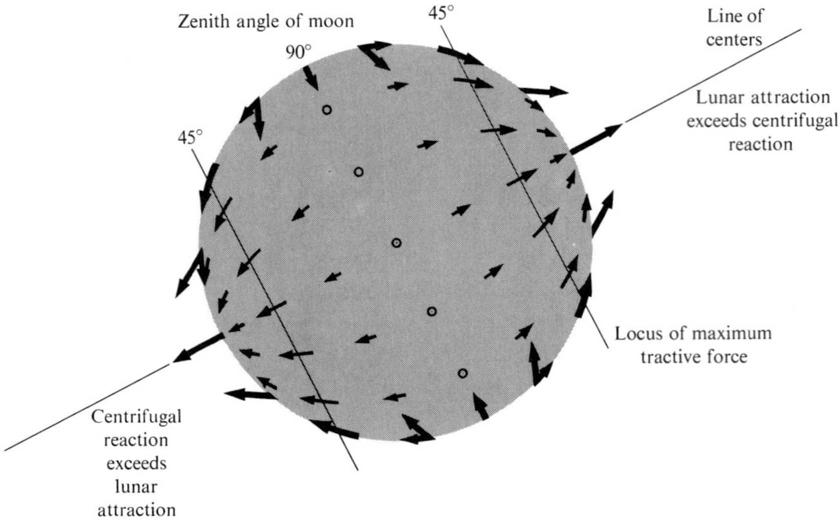

Fig. 17.2 The distribution of the lunar tractive tide-raising force on the earth relative to the earth-moon line of centers. The gravitational forces on the earth caused by the mass of the moon are oriented toward the moon center on the near side, with a mirror image on the far side. The component of these forces tangent to the earth is the lunar tractive force. (From W. S. von Arx, *Introduction to Physical Oceanography,* Addison-Wesley, Reading, Mass., 1962.)

The tangential component is zero at intersections of the line of centers with the earth's surface; it is maximum at positions on circles that are perpendicular to the line of centers and pass through points that lie 45° and 135° relative to that line and to the earth's center. The tangential acceleration is also zero on the circle described by the plane perpendicular to the line of centers and containing the earth's center. This circle divides the earth in two, as the direction of the tangential component is negative in one half and positive in the other. Figure 17.2 shows the orientation and magnitude change in the two acceleration components relative to the line of centers.

The vertical component acts to alter the earth's acceleration due to gravity. If the moon is the attracting body, the vertical component of acceleration causes an object on the earth at the intersection of the line of centers to lose only about 1×10^{-7} of its weight. Although also small, the tangential component acts at right angles to the earth's gravity vector and thus, since it is unopposed, has an effect on the water-covered earth. This horizontal component, often called the *tractive force* per unit mass of the tide-raising bodies, is important in tidal analyses. With it we can formulate the equilibrium tidal theory and study how it produces tides on a uniformly water-covered model earth. We may then apply this theory to the actual earth in order to better understand the tidal phenomena.

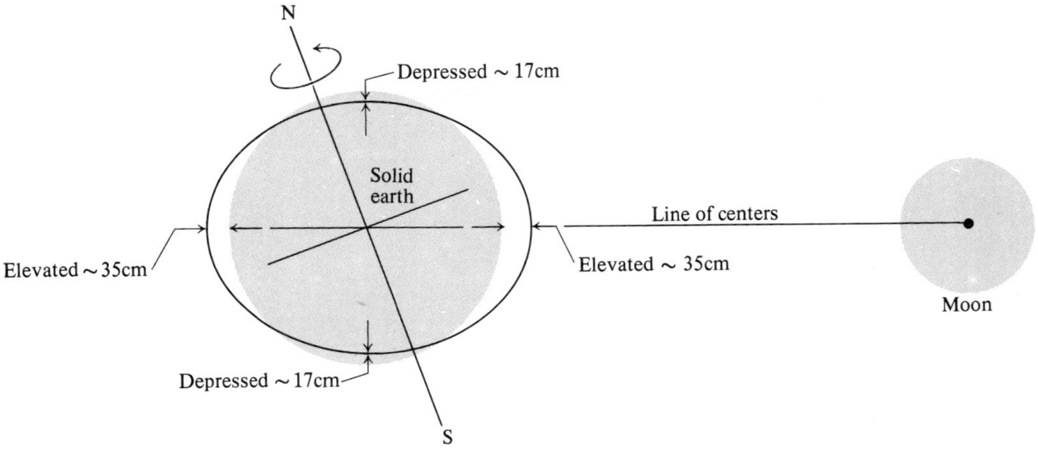

Fig. 17.3 Distortion of a uniform water envelope surrounding a smooth earth by the lunar tide-raising forces, as determined from equilibrium tidal theory.

THE EQUILIBRIUM TIDE

Tangential tractive forces displace water laterally toward intersections of the earth with line of centers. The accumulated water rises to form a high tide bulge, which builds in height until the downslope gravity force balances the tractive force (Fig. 17.3). The maximum elevation of this bulge above the equilibrium spherical surface is about 35.4 cm if the moon is the tide-raising body and 16.2 cm for the sun. The depression of the water level at the intersection of the earth's water-covered surface and a plane perpendicular to the line of centers and containing the earth's center is 17.7 cm for the moon and 8.2 cm for the sun, which shows that the sun is about 46% as effective as the moon in producing tides.

The distorted water envelope of the equilibrium tidal model is considered stationary, with its bulge centered on the line of centers while the earth spins within it. Thus the height of the water covering a point on this model varies as the earth rotates. The changing height of water level as the earth rotates within the water envelope represents the rise and fall of the tides. The eastward rotation of the earth within the water envelope as the latter is held fixed in space makes the direction of rotation of the envelope relative to the earth appear to be opposite, or westward. The movement of the water envelope relative to the earth is similar to the progression of a wave having a length of about one-half the circumference of the earth, the distance between the high-water crests in the water envelope covering the earth.

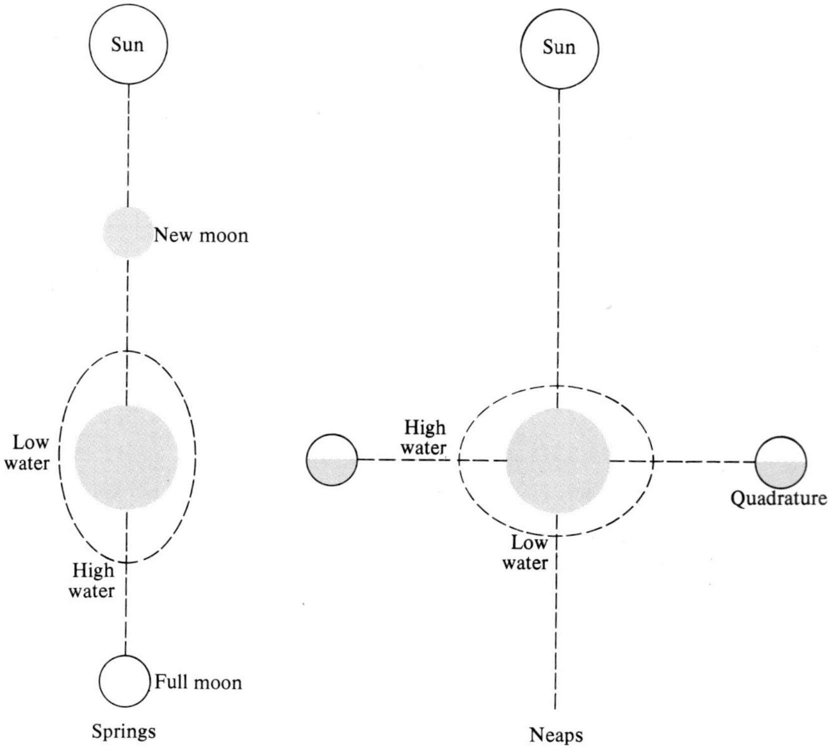

Fig. 17.4 When the moon and sun act along the same plane, the lunar and solar tidal bulges combine to produce large amplitude spring tides. When the lunar and solar tide bulges are 90° out of phase, their combined effect produces smaller amplitude neap tides. (After Russell and Macmillan, *Waves and Tides*, 2nd ed. rev., Hutchinson, London, 1954.)

The principal tide-raising bodies, the sun and moon, exert their forces on the water envelope independent of each other. If the lines of centers between earth and moon and sun coincides, the maximum elevation of the distorted water surface is about 51.6 cm and the maximum depression 25.9 cm. If the lines of centers are perpendicular to each other, the sun and moon tidal bulges are 90° out of phase and tend to compensate each other; that is, a moon-raised low tide occurs simultaneously with a sun-raised high tide, and vice versa. In the first case (sun, moon, and earth in line), *spring tides*, the largest-amplitude tides, are formed; the latter set of conditions produces *neap tides*, with the lowest amplitudes. Since the change from neap tides to spring tides (and vice versa) takes approximately two weeks, they are often called *fortnightly tides* (Fig. 17.4.). The positions of the earth-moon and earth-sun lines of centers are related to the moon's phase; thus spring tides occur during new and full moons and neap tides accompany quarter moons. Obviously, variations in

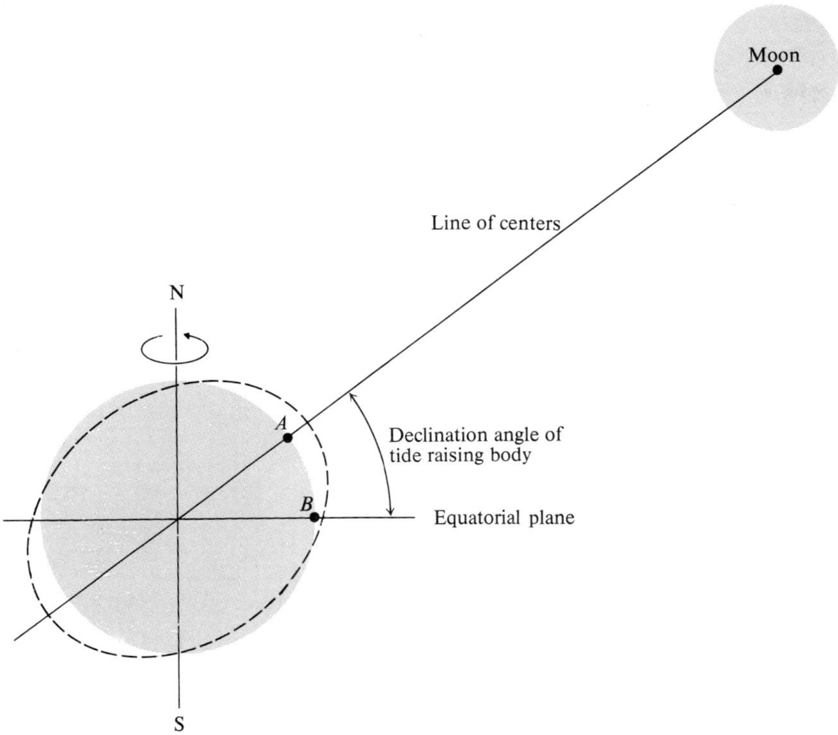

Fig. 17.5 When the moon is at a large declination, the tidal bulge is inclined toward the earth's rotational axis.

amplitude of the springs and neaps depend on how closely the lines of centers coincide.

The variation of water level with time and its magnitude at a given earth point are determined by the point's location with reference to the line of centers. If the line of centers is inclined to the earth's equatorial plane, the high-water positions are displaced north and south of the equator (Fig. 17.5). As point *A* traverses under the water envelope with the rotating earth, it will initially register high water, which will be followed by a low water about 12 hours later, after 180° of rotation; when it has rotated 360°, about 24 hours later, *A* will return to the high-water position. The rotation of point *A* within the inclined tidal envelope produces a *diurnal* (daily) *tide*—one high and one low per day—at point *A*. Since diurnal equilibrium tides are associated with the inclination of the line of centers to the equatorial plane caused by the declination of the tide-raising body, they are also called *declinational tides.*

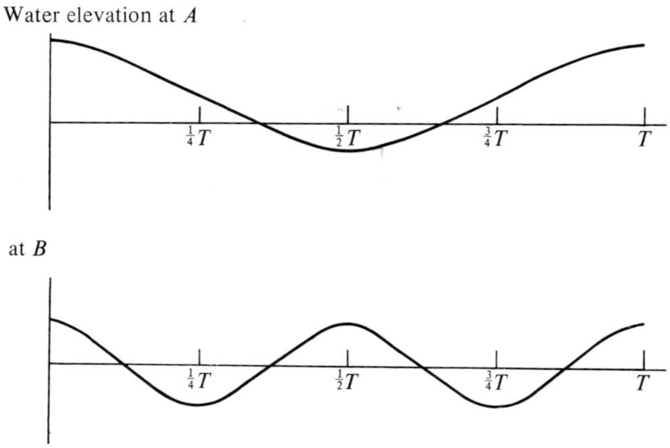

Fig. 17.6 The rotation of the earth within the inclined tidal shape shown in Fig. 17.5 produces a diurnal tide at *A* and a semidiurnal tide at *B*. The period of one rotation is *T*.

Both the moon and sun have angles of declination; therefore both produce diurnal tides. The earth's axis of rotation is tilted $66\frac{1}{2}^\circ$ to the plane of the ecliptic, causing the declination of the sun to change from $23\frac{1}{2}^\circ$S to $23\frac{1}{2}^\circ$N and back annually. The moon is inclined at an angle of 5° to the plane of the ecliptic to make its declination range from $28\frac{1}{2}^\circ$ to $18\frac{1}{2}^\circ$ with a period of 18.6 years.

If the earth point in question is on the equator at *B* (Fig. 17.5), or at *A* when the declination of the tide-rising body is small, the 24-hour sequence of water levels for a 360° rotation will be high water, low water, high water, low water, and back to the initial high water. The earth point will thus be exposed to a *semidiurnal tide*—two highs and two lows per day. If the 2 highs and 2 lows of the semidiurnal tides do not have the same magnitude, the tide is said to possess *diurnal inequality* (Fig. 17.6).

We have stipulated that the model earth rotates on its axis relative to the water envelope, which is fixed in space throughout the 24-hour period (or, more specifically, the sidereal day). However, the real earth's water envelope, oriented on the line of centers, is not fixed in space.

When the sun acts as the tide-raising body, the earth's displacement along its orbit moves the line of centers about $360^\circ/365\frac{1}{4}$ in the direction of the earth's rotation. The earth must rotate $360^\circ + (360^\circ/365\frac{1}{4})$ to complete a solar tide sequence. The time required for this rotation is the *solar day*, which, as we have learned, varies annually; only its mean value is equal to 24 hours.

In the moon-earth system, the moon moves approximately $360^\circ/27.3$ along its orbit around the earth in a sidereal day. This displaces the earth-moon line of centers by $360^\circ/27.3$ as the earth completes one revolution relative to fixed space. The earth must rotate this additional angle to complete a lunar tidal sequence, $360^\circ + (360^\circ/27.3)$. The time required for one rotation relative to the moon is a *lunar day*, which is about 24 hours, 50 minutes long. The lunar day is commonly used as the standard tidal day on which the determination of diurnal and semi-diurnal tides is based.

Whether the solar tides are diurnal or semidiurnal at a given earth point depends on its location relative to the line of centers. The mean values of the solar diurnal and semidiurnal periods are 24 hours and 12 hours respectively. Lunar tides may also be diurnal or semidiurnal at a given earth point, depending on its position and the moon's declination. The respective mean periods for the lunar diurnal and semidiurnal tides are about 24 hours, 50 minutes and 12 hours, 25 minutes. Lunar tides occur about 50 minutes later than solar tides each day, and because of the moon's greater tractive force, they tend to advance the combined lunar and solar tide by 50 minutes each day.

Because of the difference in periods between the lunar and solar tides, their phasing is continually shifting. The matter is further complicated by the fact that R, the distance between the earth and sun or earth and moon, is also continually changing, causing the magnitude of the tractive forces to vary. However, despite the complexities arising from the nutation of the rotational axis, the variation in R, and the changes in the declinations of the moon and sun, we can still readily determine their combined effect in altering the distribution and magnitude of the tide-producing force on earth. From the equilibrium tidal theory, we can also see how these forces act to distort the water envelope of our model earth.

THE EARTH TIDES

We may now apply what we have learned from the equilibrium theory on a water-covered earth to tides on the real earth. We shall find that although the tide-producing forces are the same in both cases, the earth's waters do not respond in exactly the same way as do those of the model. We can attribute this difference to the following factors:

1. The earth's oceans are discontinuous, whereas the model is completely and uniformly covered with water.
2. The celerity of the tide wave relative to the real earth is given approximately by the shallow-wave speed equation, $C = \sqrt{gd}$. Thus the tide wave does not progress at a speed corresponding to the linear speed of a point on the earth's surface, which is determined by the rotation of the earth beneath the water surface. At high latitudes in deep water, the tide wave moves faster than points on the earth's surface, which move at a rate determined by the rotation of the earth beneath the water surface. At high latitudes in deep water, the tide wave moves faster than points on the earth's surface, while at low latitudes and typical oceanic depths it moves slower. Local propagation speed is also affected by ocean depth, which is not constant on the earth as it is on the model.

3 As mentioned in the introduction, water set in motion by the earth's tide waves are affected by the Coriolis force, which was neglected in the model.

4 The earth's ocean basins have natural periods of oscillation which affect their response to tidal forces.

5 The presence of land masses on the real earth can cause reflection of tide waves.

An understanding of the earth's real tides is desirable in that it enables us to predict them so man can plan his activities in response to their behavior when required. We have learned how to determine and predict the tide-generating potential of the sun and moon, but this knowledge is insufficient for the accurate prediction of actual earth tides. We must first measure the rise and fall of the sea surface at the location for which we wish to predict the tides. From these measurements, we should be able to calculate the local mean sea level, mean low water, and mean high water. We must then analyze the tidal record and extract the various sinusoidal components from the tide curve, which are determined by the magnitude and frequency of the known tide-generating forces. When added together, these components yield the water level η at any time t of the theoretical tide:

$$\eta = D + a_1 \cos(\sigma_1 t + \delta_1) + a_2 \cos(\sigma_2 t + \delta_2) + a_3 \cos(\sigma_3 t + \delta_3) + \ldots, \tag{17.8}$$

where D is the difference in height between the mean sea level and mean low water, a_n is the magnitude of the component (i.e., amplitude), σ_n is the frequency, and δ_n is the phase angle. The number of tide-raising components in this series equation depends on the accuracy of η desired.

After extracting the known components from the observed tide record, we will probably discover a residual record that is not related to the frequencies of the tide-raising bodies. We must then describe this component of the local tide mathematically before we can proceed with our predictions.

In order to predict η, we first assume that the tidal curve is generated from a given point in time t_0. We may then set the amplitudes, frequencies, and phase angles of both the astronomical tide-raising components and the locally determined residual component to a single value at time t_0. We can then determine the local η as a function of time at any time after t_0, using an equation of the type:

$$\eta = D + \Sigma\, a \cos(\sigma t + \delta). \tag{17.9}$$

We may also find η mechanically by using a *Kelvin tide-predicting machine* (Fig. 17.7). In this procedure, slotted disks that are eccentric to their rotation points are linked together by pulleys such that each turns at a rate described by the frequency of a different tidal component. Each disk is inside a yoke that converts its eccentric rotation to reciprocal linear motion. The total linear

Fig. 17.7 Kelvin tide-predicting machine. Photo courtesy of National Oceanic and Atmospheric Agency, U.S. Dept. of Commerce.

motion for all the components is then summed on a single wire and recorded as the tidal curve by a graphing mechanism. Since the disks may be rotated by hand within their yokes, they can be adjusted to the phase angles of their respective components before they are locked into place. The magnitude of their contribution is set by their eccentricity. Large tide-predicting machines handle as many as 63 components for high accuracy. However, there are only seven major components of tide generation; their periods and magnitudes relative to the M_2 component are given in Table 17.2. (Their frequencies are 2π/period.)

Table 17.2 Principal harmonic components of the tide

Component	*Period, hr*	*Relative amplitude (% of M_2)*
Semidiurnal components		
Principal lunar, M_2	12.42	100.0
Principal solar, S_2	12.00	46.6
Larger lunar elliptic, N_2	12.66	19.2
Luni-solar semidiurnal, K_2	11.97	12.7
Diurnal components		
Luni-solar diurnal, K_1	23.93	58.4
Principal lunar diurnal, O_1	25.82	41.5
Principal solar diurnal, P_1	24.07	19.4

After H. J. McLellan, *Elements of Physical Oceanography,* Pergamon Press, New York, 1965.

Tides on the earth are usually classified as follows:

1. Diurnal: once daily with amplitudes that vary with the declination of the moon and sun.
2. Semidiurnal: twice daily with amplitudes that vary with the moon's phases.
3. Mixed: diurnal and semidiurnal combinations.

Tides may also be classified in relationship to the moon, since it has the greatest effect on them:

1. Synodic: tides whose variation in range is primarily associated with the moon's phases, e.g., springs and neaps.
2. Anomalistic: tides whose variation in range is primarily associated with the variation in the moon's distance from the earth.
3. Declinational: tides whose variation in range is primarily associated with changes in the moon's declination.
4. Mixed: a combination of synodic, anomalistic, and declinational tides.

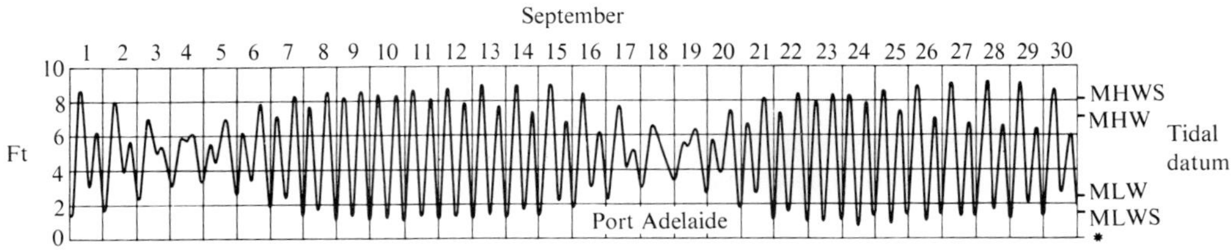

Fig. 17.8 A semidiurnal tide, synodic type, with a large variation in range associated with changes in the moon's phases. (From N. Bowditch, *American Practical Navigator*. H. O. Publication No. 9, U.S. Naval Hydrographic Office, Washington, D.C., 1962.)

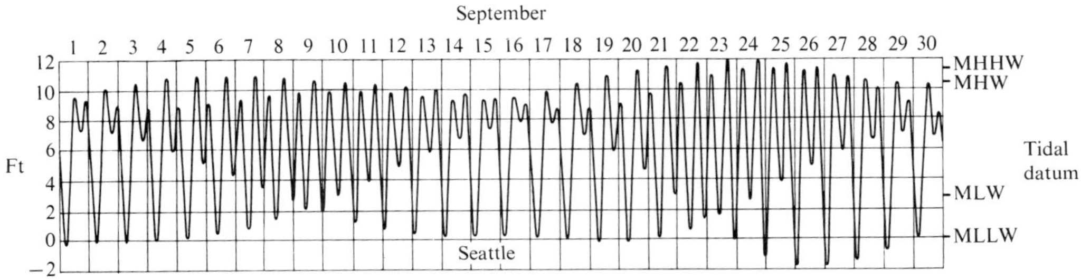

Fig. 17.9 A mixed-type tide having both semidiurnal and diurnal characteristics. When the sun and moon are on the equator, the semidiurnal component is strong. When the moon is at large declination, the tide becomes more diurnal (tropic tides). (From N. Bowditch, *American Practical Navigator*, H. O. Publication No. 9, U.S. Naval Hydrographic Office, Washington, D.C., 1962.)

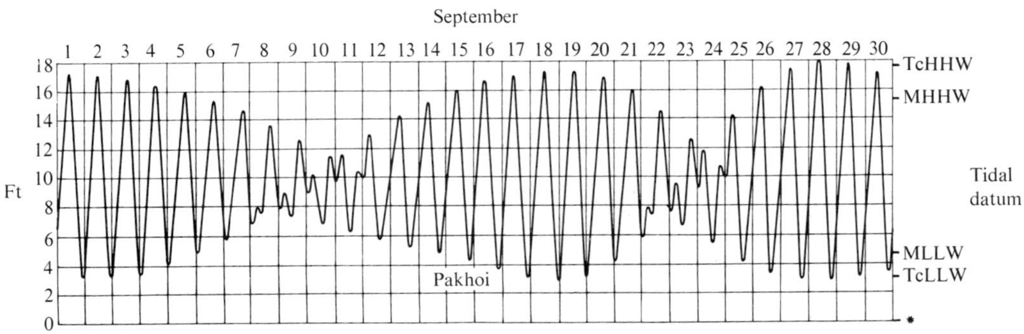

Fig. 17.10 A diurnal-type tide. The changes in amplitude are keyed to the changes in the moon's phases. (From N. Bowditch, *American Practical Navigator*. H. O. Publication No. 9, U.S. Naval Hydrographic Office, Washington, D.C., 1962.)

Figures 17.8 and 17.9 show records of two *synodic semidiurnal* tides. The Port Adelaide, Australia, record (Fig. 17.8) indicates only a small change in amplitude due to the moon, while the Seattle, Wash., record (Fig. 17.9) shows definite spring and neap tides.

In the case of a mixed tide (Fig. 17.10), it is necessary to identify and find the averages of the different highs and lows, since their values can vary appreciably. The average of the highest high-water levels is referred to as the mean higher high tide. The other mixed-tide averages are: mean high water (average of all highs), mean lower high water (average of all lower high waters), mean higher low water (average of lesser low-water levels), mean low water (average of all lows), and mean lower low water (average of greatest low waters); the last is usually the reference datum for regions having mixed tides with a considerable range. Navigational charts give depths relative to mean lower low water rather than to mean sea level to ensure a margin of safety against the accidental grounding of ships.

Mixed tides may show considerable declinational effect. At Pakhoi, on the Gulf of Tonkin (Fig. 17.10), the change in declination of the moon produces a diurnal inequality in the tide. When the moon and sun are at their extreme northern declinations, the tide becomes nearly diurnal, with a prolonged high and short-lived low. This is sometimes called a *tropic tide*. When the celestial bodies are above the earth's equator so as to produce *equatorial tides*, the mixed character is more evident. At Victoria, B.C., the tide is often almost purely diurnal, even though Victoria is located between the Straits of Georgia and Seattle, which have mixed tides.

The small range of the tide at Port Adelaide (Fig. 17.8) is typical of open-sea regions, whereas larger amplitudes are commonly associated with embayment systems whose entrances to the sea are not overly constricted. However, this generalization is not always true, since range is only partly determined by the increase in height/wavelength as a wave moves into shallow water and steepens as its celerity decreases. Another factor controlling the range is the generation of a standing wave tide in some basins. The equilibrium tidal theory yields tidal amplitudes of about $\frac{1}{2}$ m, which are comparable to those of the real earth at midocean, where the tide approximates the equilibrium tide.

Casual observation of tides at the earth's continental margins would lead one to conclude that tide range increases with latitude. However, this assumption does not hold for all areas of the ocean. There is, of course, some increase in range with latitude due to the declination of the tide-raising bodies and to the presence of the deep elongate inlets that predominate at mid- to higher latitudes. However, tides at the high latitudes of the Arctic Ocean are relatively small, while elongate sea inlets of the subarctic and temperate regions may have very large tides. The natural period of oscillation in the Bay of Fundy, for example, nearly matches that of the tide; thus the resulting resonance produces a very large tide range—about 17 m. We can determine whether a tide is of progressive or standing type at a given location by comparing its current (*stream*) to the water level. The maximum tidal current of a progressive-type tide wave occurs nearly simultaneously with the low- and high-water stands, while that of a standing-wave tide occurs halfway between the periods of high and low water.

Although we can learn a great deal about tides from the equilibrium tidal theory and from measurements made with water-level recorders at the shore and sensitive pressure gauges on the sea floor, the tides remain a phenomenon that is neither fully understood nor quantitatively predictable. At best we can use empirical relationships to predict the tides, but we are a long way from a full interpretation of their behavior both in the ocean's basins and relative to the tide-raising forces.

Oceanographers have made some advances in this area, however, and they are incorporated into the modern theory of tides. Primarily, we now have a better understanding of:

1 The behavior of tides as standing waves in enclosed or semienclosed basins, e.g., the free oscillation of a liquid in a basin after the water surface has been tilted and allowed to oscillate about its equilibrium position.
2 The effect on the tides of resonance between the periodic tide-producing forces and basins with similar periods of fundamental oscillation.
3 The deflection of water set in motion by the tractive forces due to the earth's rotation.

TIDES AS STANDING WAVES

If tides behave as standing waves, we may study them as such using the classical approach. As shown in Chapter 16, a rectangular closed basin has a fundamental oscillation period described by

$$T = 2l/n\sqrt{gd}, \qquad (17.10)$$

where T is the period, l is the length of the basin in the direction of oscillation, g is gravity, and d is the equilibrium water depth (Fig. 17.11a). A node indicates the position of no change in level, while antinodes mark the maximum high- and low-water level in the basin. Equation (17.10) gives the period of a basin with only one node. If a basin generates multiple standing oscillations with more than one node, the period of oscillation is

$$T = 2l/\sqrt{gd}, \qquad (17.11)$$

where n is the number of nodes in the system. The length L of the fundamental wave is $2l$, or twice the basin length. Thus Eq. (17.10) becomes

$$2l/T = L/T = \sqrt{gd}; \qquad (17.12)$$

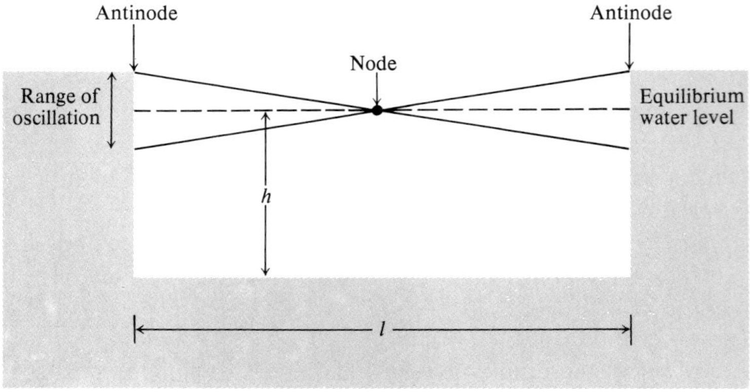

Fig. 17.11a The fundamental standing wave in a closed basin has a central node where no vertical motion of the free surface occurs. Maximum water-level change occurs at the antinodes. The basin's length (*l*) is $\frac{1}{2}$ of the wavelength.

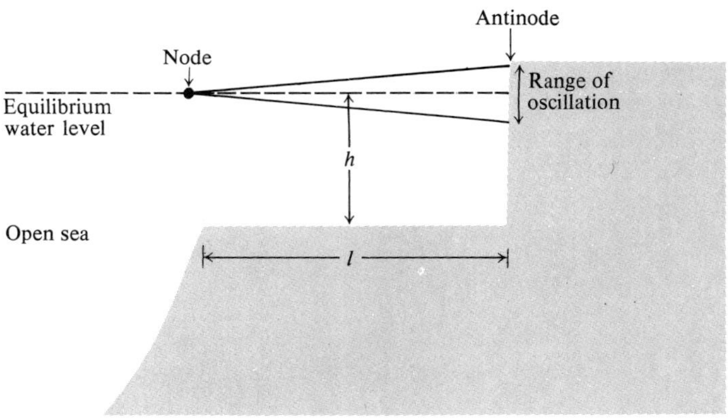

Fig. 17.11b The fundamental standing wave in an open-ended basin has a node at the basin's mouth. Maximum change in water level occurs at the antinode. The basin's length (*l*) is $\frac{1}{4}$ of the wavelength.

that is, the speed of progression of the waveform is governed by the shallow-water wave equation. This speed is typically about 200 m/sec in midocean areas having a depth of about 4 km. Thus if an ocean basin has a period of resonance approximating that of a major tide component—say 12 hours or 24 hours—the basin length is about 4.3×10^3 km or 8.6×10^3 km respectively. These values, of course, are only approximations, since natural basins are not rectangular or of constant depth, and their waves do not oscillate in only one direction but rotate with the turning of the earth.

Unidirectional standing oscillations may occur in elongate gulfs or arms of the sea, however, since their narrowness reduces the rotational capacity of the waves. These embayment systems are usually open-ended rather than closed basins. The node of an open-ended basin is usually located at its mouth, while the antinode is at its head (Fig. 17.11b). Its fundamental period is

$$T = 4l/\sqrt{gd}. \tag{17.13}$$

Resonance can occur in an open basin if its length is about one-fourth of the tide wavelength. Thus the respective lengths of basins with 12-hour and 24-hour resonances are 2.15×10^3 km and 4.3×10^3 km, given a depth of 4 km. We should note that in most gulfs, since $d \ll 4$ km, the wave speed ($\sqrt{gd}$) is much less than it is in the open ocean.

The dimensions of natural basins and gulfs are well within the resonance requirements. Thus standing-wave tides can be generated within a basin by the periodic input of energy; they need not be part of the continuous bulge of water that moves relative to the earth as depicted on the equilibrium model.

The Bay of Fundy in New Brunswick is in near resonance with the periodic tidal pulsations near its mouth. The tidal amplitude in this gulf builds due to the continuous input of energy in phase with the natural oscillations until friction dissipates the energy at the same rate at which it is added. As mentioned earlier, the tidal range at the head of the Bay of Fundy approaches 17 m. Since its average depth is about 80 m, its length must be 302 km for resonance to occur; the actual length is 310 km. Thus despite the oversimplification in the rectangular basin approximation, it is a useful tool for determining first-order resonance conditions.

THE ROTARY MOTION OF THE TIDE

Earlier, we assumed the water envelope of the model earth (Fig. 17.3) to be fixed in space, with the peak of the bulge symmetric about the line of centers. We also noted that the earth rotated within this envelope to produce the rise and fall of the tide at earth points. However, we did not discuss the effect of the earth's rotation and the migration of the line of centers on the distribution of the tractive force at a given earth point, although it should be obvious that this force changes in both magnitude and direction. Figure 17.12 shows the variation of the lunar tractive force at a point on 30°N Lat when the moon's declination is 15°N. A water particle located at the origin of a N-S-E-W axis system is attracted in the direction indicated by the arrows and at a strength proportional to the arrow length.

The gyroscopic motion of the earth tends to make tides rotate in the ocean basins. This rotation occurs about nodal or *amphidromic* points in the sea at which there is no significant vertical change in water level due to the tides. Radiating from these points are cotidal lines, which join all positions at which high water occurs simultaneously. Positions of common tidal range cut across the cotidal lines, forming approximately concentric circles of equal

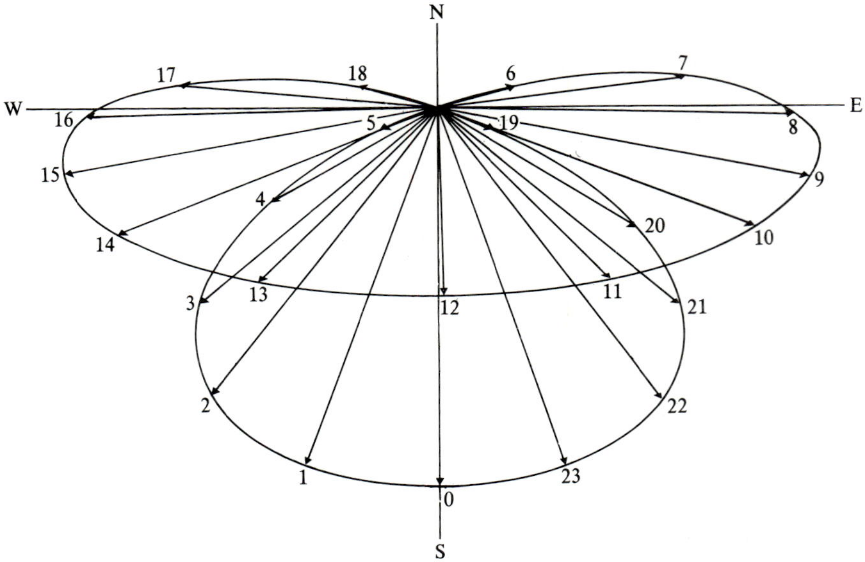

Fig. 17.12 The direction and relative magnitude of the lunar tractive force acting hour by hour on a unit mass of water located on the 30° N parallel when the declination of the moon is 15° N. (From British Admiralty, *Manual of Tides,* 1941.)

tide range which increases with distance from the amphidromic point (Fig. (17.13). The rotation of the tide crest about the amphidromic point is indicated by the movement of the tide crest from one cotidal position to the next. We shall now consider why the standing-tide wave in an ocean basin rotates as it does and why this motion produces a nodal point rather than a nodal line across the ocean basin.

In a simple standing wave, water moves from a high water position at one end of a basin to a low at the other and vice versa; a line node across the basin indicates where the elevation of the water does not change. Water must thus flow horizontally from one end to the other through the nodal line section to permit a change in level. The maximum value of this flow, which is the equivalent of the tidal current in a standing-wave type tide, is directed from the end in which the water level is dropping to the end where it is rising, and is maximum when both ends are at the same level. When one end is at its high mark and the other at its low mark, the current is zero. Thus the current is said to be out of phase with the tide level for a standing-wave tide.

If the ocean basin as part of the earth is made to rotate in the manner that the earth rotates about a vertical axis through the center of a basin [e.g., rotation rate = (the earth's polar rotation rate) (sin ϕ), where ϕ is the latitude of the basin center], the process of wave formation becomes more complicated. Assume that our rectangular basin is in the Northern Hemisphere and has

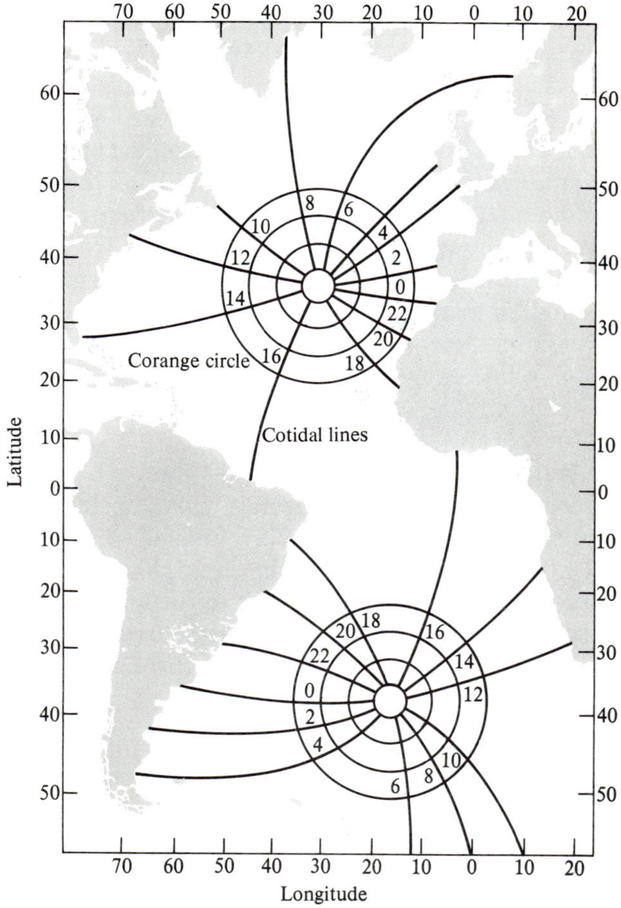

Fig. 17.13 The rotation of an oceanic basin about a vertical through the node of its standing wave makes the wave appear to rotate about the basin. This rotation relative to the basin reduces the node to a point, the amphidromic point. The range of tide increases with distance outward from this point, as indicated by the corange circles. The cotide lines indicate the location of the high-water crest as the wave rotates about the basin. (From Russell and Macmillan, *Waves and Tides*, 2d ed. rev., Hutchinson, London, 1954.)

north, west, south, and east sides. If at time zero it is high water on the north side and low on the south, the following will occur in sequence: As the water level starts to drop in the north and rise in the south, a flow of water (current) must be directed southward. This flow undergoes Coriolis acceleration and is deflected toward the west side of the basin. At the point of maximum north-to-south flow there will be a cross-channel surface slope to the water with a high elevation on the west and a low on the east. At this instant the north and south elevations equal each other and the elevation at the node; therefore the high-water level has migrated counterclockwise from the north side to the west side. As the elevations of the south and north ends rise and drop respectively, both the north-to-south flow and the cross-channel slope decrease. When the south end is at its high mark and the north end at its low mark, the north-to-south flow is zero, requiring the cross-channel slope to be zero also. The high-water crest has now moved counterclockwise an additional one-fourth of the way around the basin to reside at the south end. When the south end starts to drop and the north end rises, the current is from south to north, again causing the cross-channel slope to build. In this case, however, the maximum elevation appears on the east side, when the flow is maximum. Thus the tide crest, or region of high water, is seen to rotate counterclockwise relative to the earth in the Northern Hemisphere about a node which is reduced to a point or, in the actual oceans, a small area.

The flow required to move the standing-wave crest about the basin is the oceanic tidal current. In the case just described, the flow must transfer water in both the north-south and east-west directions to allow one side of the basin to have a higher water level than another. If the basin is as wide as it is long, the magnitude of the current in the two perpendicular directions must be comparable. If the basin is considerably longer in the north-south direction than it is wide in the east-west direction, the north-south currents are correspondingly greater than the east-west flows.

TIDAL CURRENTS

The rotation of tidal currents, like that of tide-wave crests, changes direction with the earth's turning. Since the successive highs and lows in a basin are not always of the same elevation (as is the case for mixed tides), the flow strength is not constant. We can plot a hodograph (a successive current vector diagram) of typical tidal currents for a given location if we measure them over many tidal cycles, separating the rotational tidal current from other long-term nonperiodic currents. Figure 17.14 shows a hodograph of the tidal current at the Columbia River Lightship. The strength and direction of the current are indicated respectively by the distance between the curve and the origin and the orientation of the current vector relative to a N-S-E-W axis system. The rotational sense of the current direction is indicated by the hour increments before and after each high- and low-water stand.

In the open sea, we can approximate the magnitude of the tidal current from the shallow-water wave theory (see Chapter 16), assuming that the tides behave like simple pregressive waves. The horizontal component of the orbital velocity of a free shallow-water or long wave is

$$u = (A\sigma/kd) \cos (kx - \sigma t), \tag{17.14}$$

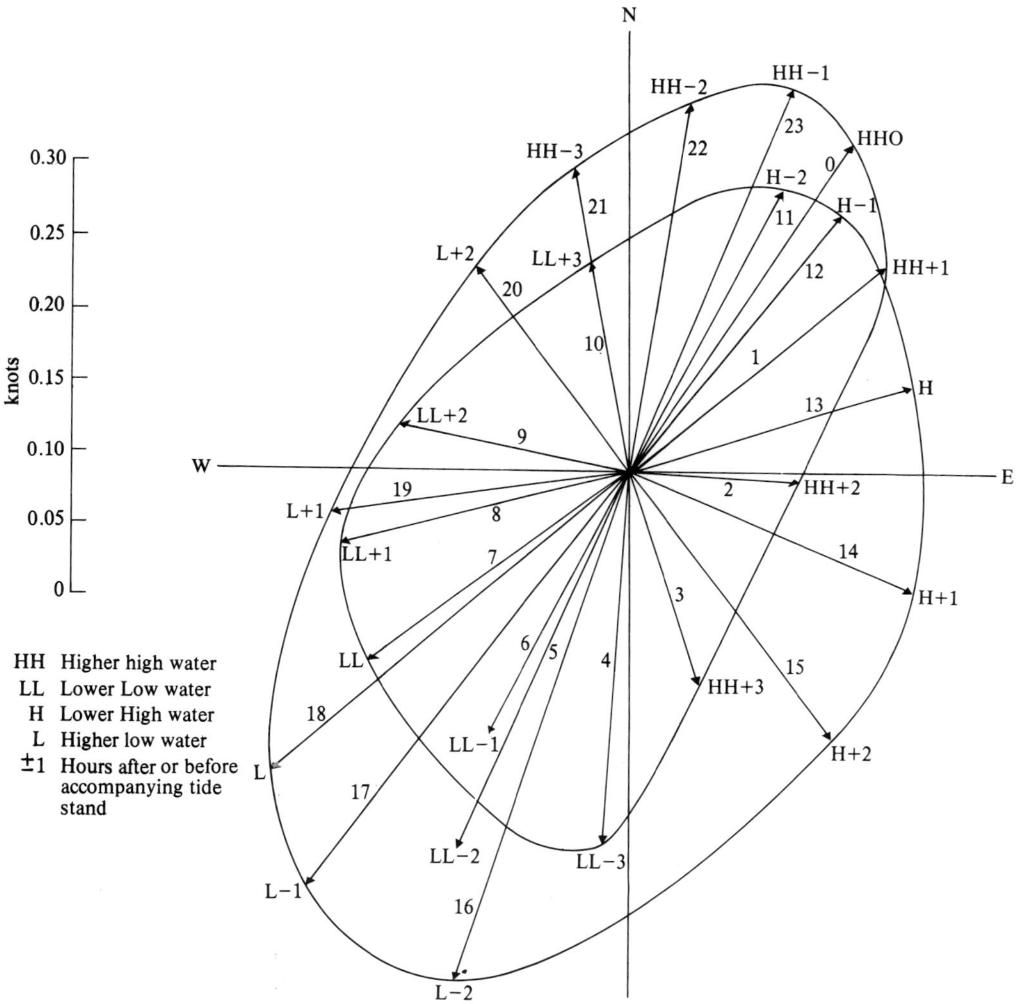

Fig. 17.14 Tidal currents associated with the tide wave are also rotary. The tidal current vectors here depicting the changing direction and strength of the current were obtained at the Columbia River Lightship. Notice the changing direction of the current, which indicates that the current rotates in the direction opposite to the wave (Northern Hemisphere case). Redrawn from H. A. Marmer, *Coastal Currents along the Pacific Coast of the United States*. U.S. Coast and Geodetic Survey, Sp. Pub. 121, Dept. of Commerce, Washington, D.C., 1926.

which approximates the tidal current. Here the maximum current [$\cos(kx - \sigma t) = 1$] occurs at the wave crest (high tide) or at the wave trough (low tide). Since $\sigma = 2\pi/T$ and $k = 2\pi/L$, we can maximize Eq. (17.14) to obtain

$$ud = AC = A\sqrt{gd}. \tag{17.15}$$

where u is the current, d is the average water depth, A is the tidal amplitude, and C is the speed of propagation of the wave.

The maximum current of a simple standing wave occurs out of phase with the crest or trough; it is associated instead with the midtide level (halfway between high and low tides). Its maximum value is not dependent on d, but on the amplitude of the standing wave and other physical factors of the basin.

In the inshore waters of embayments, gulfs, and estuaries, the tidal current is not determined solely by the orbital velocity of the waveform; channel constrictions, river flow, friction, channel curvature and configuration, tide amplitude, and a dozen other factors influence the current. Tides in the appended arms of the ocean are literally maintained by their counterparts in the adjacent large oceanic basins.

Despite the complex nature of tides in interconnecting sea channels and embayments, we can make reasonable estimates of their effect on the current. We can also approximate the average flow at a given cross section in a tidal channel by calculating the volume of water (*tidal prism*) required to raise the inland tide level from low- to high-water. Since this volume must pass through the cross-sectional area of the channel in the time period between the low and high waters, we find that

$$\text{average tidal flow rate} = \text{tidal prism}/(\text{channel area} \times \text{time period between low and high waters}). \tag{17.16}$$

TIDAL BORES

In some embayments and rivers, a miscellaneous tidal phenomenon occurs that can be quite impressive. Under certain conditions, the rising tide may force the tide-wave front to move faster than it can freely propagate into the bay or river (that is, $C > \sqrt{gd}$). When this happens a shock wave is formed called a *bore*, which then moves upriver toward the terminus of the system as a rolling wall of water. A water bore is the counterpart of the sound shock wave ("sonic boom") which occurs when a pressure disturbance is forced to travel at a speed greater than that of sound. Most tidal bores are relatively small, measuring only a foot or so in height; they are often a series of water shock fronts rather than a single front. However, in some cases tide levels can jump 10 ft or more in 25 min when a bore passes. One of the largest known bores, 8 to 11 ft high, occurs in the Chien-Tang River in northern China; that of the Severn River in England is 3 to 5 ft high. The rivers Trent, Amazon, and Seine, and the Bay of Fundy also have famous bores. Most cities with large bores capitalize on them by making them tourist attractions. One can also progress upstream very rapidly by boat in the after-rush of a bore.

Although the tides along populated coastlines have been fairly well described empirically, those in the open sea are only partially understood. Of the oceans, only the Atlantic has been intensively studied for tidal phenomena. Sea-bottom instruments that can accurately determine small changes in sea level in the open ocean have been developed only recently. As oceanographers obtain more data on the tides, our knowledge of dynamic relationships will grow and our use of empirical relationships decrease.

As we have shown, many factors affect the tides: the driving forces, bottom topography, the rotation of the earth, resonance, and friction. Until their interaction is better understood, however, we shall have to rely on tidal measurements and the inadequate equilibrium theory.

STUDY QUESTIONS

1 Show that the vertical component of tide-generating force acting along the line of the earth-moon centers on the side of the earth facing the moon is slightly larger than the force on the side of the earth facing away from the moon.

2 How much weight does a 1-kg mass lose on the surface of the earth if it is on the earth-moon line of centers and the side of the earth facing the moon?

3 Why are diurnal tides also referred to as declinational tides?

4 What causes the fortnightly change in tidal amplitude from neap to spring tides?

5 Calculate the approximate celerity of tides, using the simple progressive water-wave theory, if the water depth is 2.5 km.

6 Why is the Coriolis force considered an important agent in the deflection of water moved by tides but neglected in the case of ordinary wind waves?

7 Given the change in tidal current and height with time, how can you determine whether a coastal estuarian tide is progressive, standing, or a mixture of the two?

8 Why does the crest of a standing tide in a midocean basin rotate in the opposite direction to that of the current associated with the standing wave?

9 What length of an open-ended rectangular basin with a constant depth of 150 m is required to have it resonate with the semidiurnal tide?

10 What is the approximate maximum surface tidal current of a progressive, semidiurnal tide with an amplitude of 0.75 m and a length of 1500 km, if the depth of the water is 4 km?

CHAPTER 18

OCEANOGRAPHIC INSTRUMENTS AND THEIR USES

INTRODUCTION

The instruments used by oceanographers, whether sophisticated or crude, are no more than tools. By themselves, they do not accomplish scientific work nor is their development a scientific achievement. Thus the level of sophistication that instrumentation in the oceanographic field has reached in the past decade is a result of rapid advances in technology and does not necessarily reflect a comparable advance in oceanography as a science.

Oceanographers are, in general, an independent lot. Those that have been working in the field for any length of time have developed measuring methods and tools that suit their particular research interests and personalities. As a result there are as many ways of making measurements with different instruments as there are investigators, if not more. In oceanography, standardization in methods and tools becomes a reality only when the methods and tools are shared by many researchers and the data collected are likely to be of use to the general oceanographic community.

Oceanographic measuring instruments fall into two basic groups: (a) those which are used to make measurements *in situ* and (b) those which are used to make measurements on a sample of water or a specimen that has been removed from the natural setting to a laboratory. The instruments in group (a) have usually been developed specifically for use at sea, while those in group (b) are standard measuring instruments used in any of the basic sciences.

To remove samples from the natural environment for laboratory analyses, we must have at our disposal a variety of collecting equipment. These collecting devices, which usually do not make measurements, are, in general, uniquely designed for use at sea and form a third group.

The task and the available tools dictate the methods that must be employed. We cannot measure currents in the sea by collecting a sample of moving water and bringing it aboard ship to determine its speed and direction of travel. Similarly, we cannot measure the temperature of seawater by removing it from its environment and subjecting it to the laboratory environment. On the other hand, if we want to determine the chemical composition of a sample of water, we can remove it from *in situ* to the laboratory, where precise chemical analyses can be performed. Similarly, studies of microscopic plankton cannot be accomplished in the natural environment. This material must be collected, concentrated, and studied on the ship or ashore. Common sense and an acquaintance with the problems of conducting research at sea will usually enable the researcher to choose the proper tools and methodology for gathering data. If foresight fails in this respect, then nature usually teaches the researcher an expensive and unforgettable lesson.

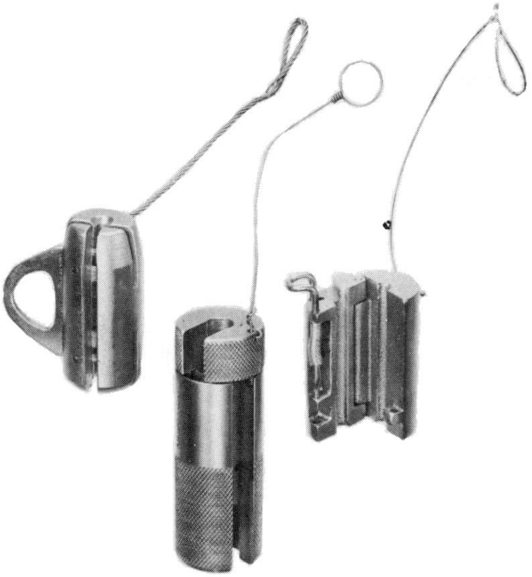

Fig. 18.1 Typical examples of messengers used to activate mechanical devices at depth when suspended from a ship by a hydrographic wire. (Photo courtesy of G. M. Manufacturing and Instrument Corp., N. Y.)

Fig. 18.2 The traditional Nansen reversing water bottle with thermometer holder, or frame. Named after Fridtjof Nansen, 1861–1930, a Norwegian Arctic explorer and scientist. Photo courtesy of Gretchen Motter.

WATER-COLLECTING DEVICES

To remove a water sample from the ocean for laboratory analysis, we use a specially designed water bottle consisting of an open-ended cylinder that can be attached to a weighted wire and lowered to any depth. The bottle is equipped with valves that can be activated by a weight (*messenger*, Fig. 18.1) fastened around the wire. When the weight is released, it slides down to strike the bottle, causing it to close at a predetermined depth. The isolated water sample is then retrieved for analytical work. Water bottles of this type are available in many different materials, shapes, and sizes. The size of a water sample collected in this manner is limited only by the ability of the ship's gear and personnel to hoist it aboard. In some cases, a large sample volume of seawater is required for isotope analyses, determination of trace materials, or organic carbon analyses. At other times, only a minimal amount is needed. As a rule, the smallest size sampler capable of supplying sufficient material is used, because small size means easier handling.

The most common water bottle is the *Nansen* bottle (Fig. 18.2). This bottle is all metal with teflon lining and holds about one liter of water. Because of its

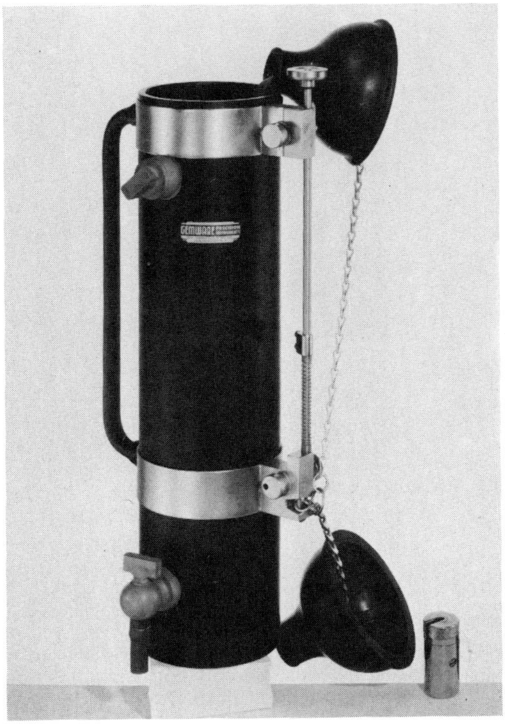

Fig. 18.3 A plastic water bottle shown in cocked position for lowering. This bottle when tripped by messenger will collect a sample at depth that is protected from contamination by metallic ions. (Photo courtesy of G. M. Manufacturing and Instrument Corp., N. Y.)

metallic construction, corrosive salts may adhere to it at the valves, forming a possible contamination source for metallic ions. Chemists therefore prefer a bottle that is made of a relatively inert plastic with few or no metallic parts. Furthermore, chemists and biologists need larger samples and hence bottles of more than one-liter capacity. A typical noncontaminating water bottle holding about two to three liters is shown in Fig. 18.3.

Water samples collected by bottle can be used for any desired chemical or physical property analyses. The particulate matter suspended in the contained sample can be filtered and used by biologists and geologists for their purposes. Concentrations of suspended material can be accurately determined, since the volume of the sample is known. This is an advantage over *in situ* filtering, in which the volume of water filtered by some device is not as accurately known. However, a larger volume of water can be filtered *in situ* than is feasible by using reversing bottles.

DEEP-SEA REVERSING THERMOMETERS

The water bottle acts as a carrier for special thermometers that measure the *in situ* temperature. These deep-sea reversing thermometers (DSRT) are of two types: the protected type, totally encased in a glass shell which protects it from hydrostatic pressure, and the unprotected type, encased in glass but open

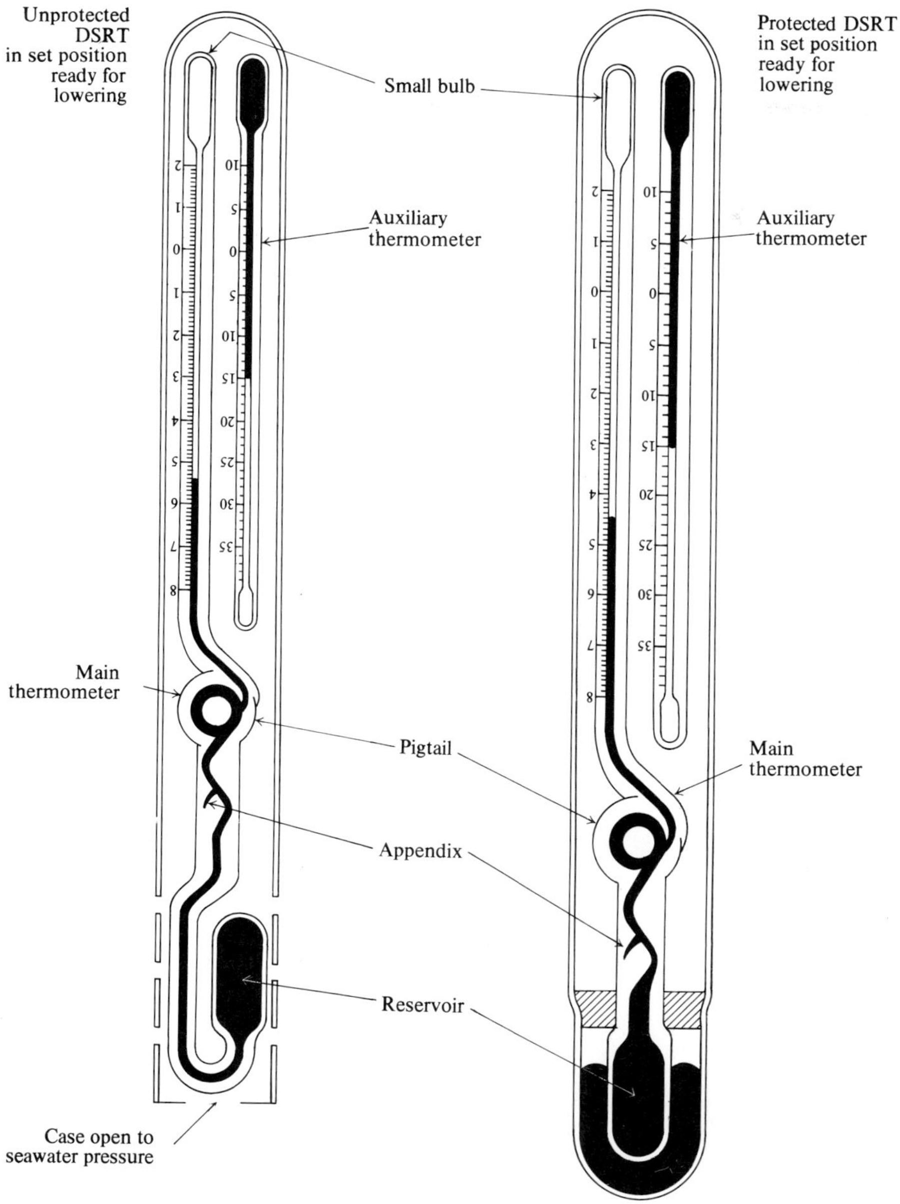

Fig. 18.4a A deep-sea reversing thermometer (DSRT) in set position.

at the bottom to allow water to enter the case and exert hydrostatic pressure on the mercurial bulb (Fig. 18.4a). The DSRT's are placed in a carrier on the water bottle and sent down on the hydrographic wire. The mercury expands or contracts in response to the environmental temperature in the protected DSRT or to temperature and hydrostatic pressure in the unprotected DSRT. When the

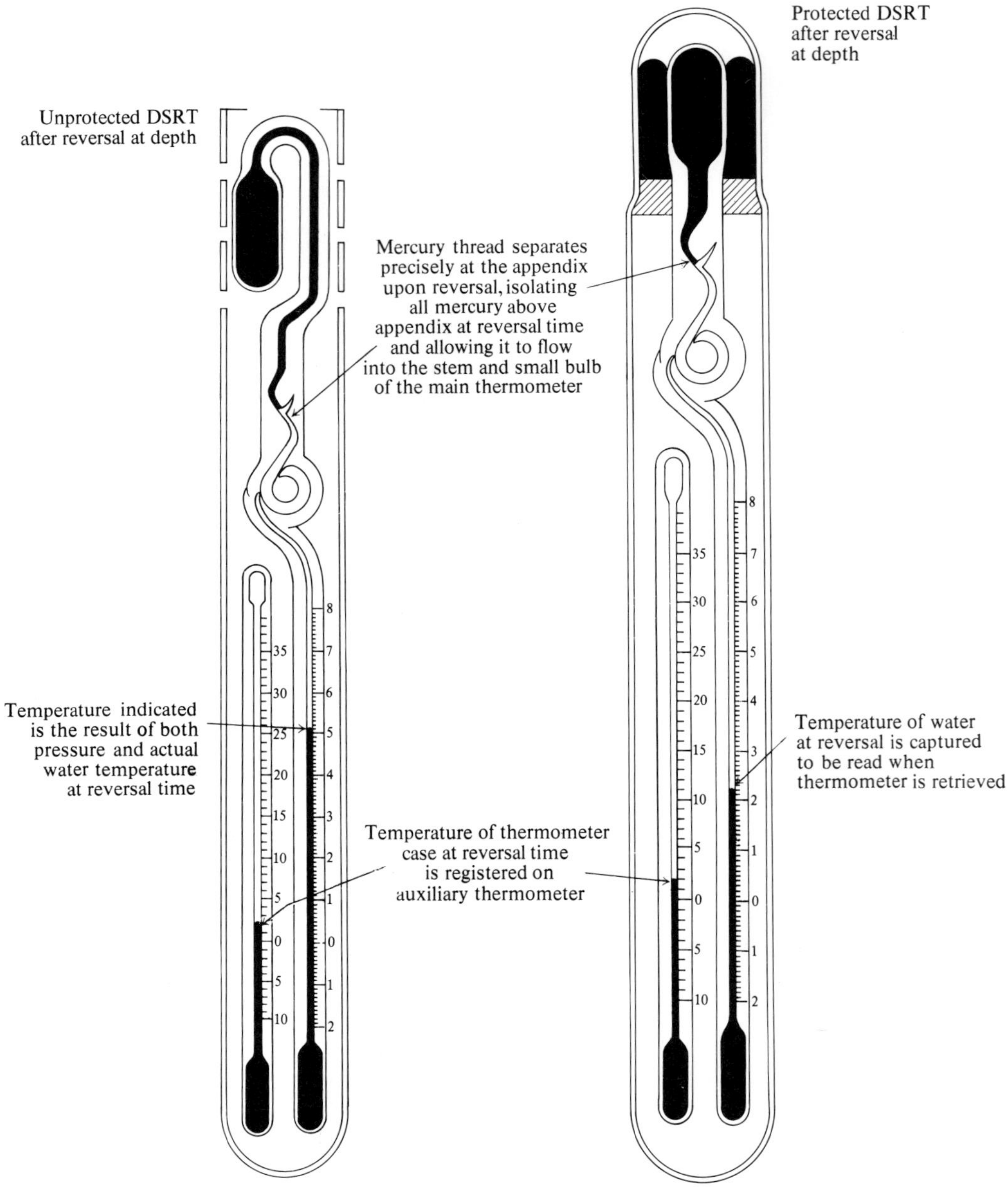

Fig. 18.4b A deep-sea reversing thermometer in reversed position.

messenger slides down and trips the bottle, either the bottle turns upside down or the thermometer holder alone rotates to reverse the thermometers (Fig. 18.4b).

The reversal causes the mercury column above a point in the glass capillary (the *appendix*) to separate from the rest of the mercury in the reservoir and fall

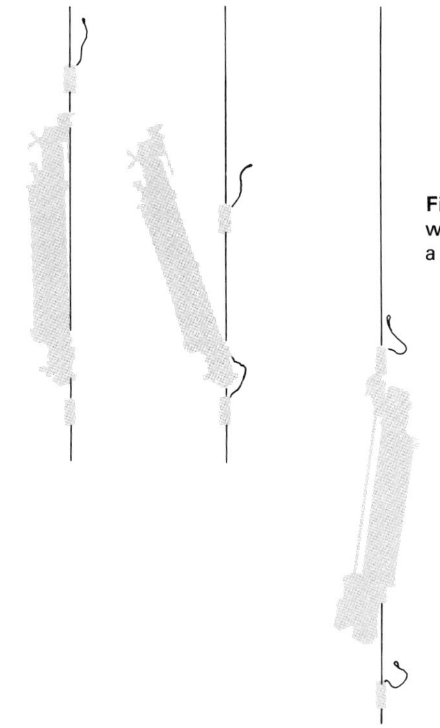

Fig. 18.5 The reversing of a messenger-actuated water bottle equipped with a DSRT, and the release of a second messenger to actuate the next deeper bottle.

to the opposite end, where it now registers the *in situ* temperature or a combined temperature and pressure effect. The thermometer is then retrieved in this position for reading aboard ship. During retrieval and storage on the ship preparatory to reading, the isolated mercury usually expands due to the warmer shipboard environment. An auxiliary thermometer is mounted in the thermometer case, which tells the observer the thermometer temperature at the time of reading. Knowing the case temperature, the volume of mercury that has been isolated, and the coefficient of expansion of mercury in glass, and using the main thermometer reading as an approximation of the water temperature at reversal, he can make a first-order correction for the mercury expansion between the time of reversal and reading. He then adds this correction to the temperature indicated by the DSRT, to obtain the corrected water temperature *in situ* at reversal (Fig. 18.5).

The unprotected thermometer is corrected in the same manner except that the corrected reversal temperature is a resultant of temperature and hydrostatic pressure. If the *in situ* water temperature is determined from a protected thermometer which is paired with an unprotected one, the difference in their corrected readings is due only to the pressure acting on the mercury bulb of the unprotected DSRT. If the pressure constant for the unprotected thermometer is known, i.e., degrees Celsius per meter of hydrostatic pressure head, we can calculate the depth of reversal, the thermometric depth:

$$\text{thermometric depth, meters} = \frac{T_u - T_p}{Q\rho_m},$$

where T_u is the corrected temperature of the unprotected thermometer, T_p is the true water temperature at reversal obtained from the corrected protected thermometer reading, Q is the pressure coefficient, and ρ_m is the mean density of the overlying water column. An unprotected thermometer registers about 1°C above ambient temperature for every 100 m of hydrostatic pressure head or 100 decibars of pressure. The temperature readings on the reversed thermometer in Fig. 18.4(b) indicate the depth of reversal to be ~ 300 m.

Reversing thermometers are very valuable tools, both for measuring the *in situ* temperature and for aiding the researcher to accurately determine the depth at which samples were taken. Careful handling and calibration of the DSRT's make them reliable to ±0.01°C.

DETERMINATION OF SAMPLING DEPTH

Determining accurately the spot from which samples are taken is an important aspect of oceanographic research. Modern navigational aids, radar, loran, and now satellite navigation equipment allow the researcher to obtain latitude and longitude with accuracy. We have already mentioned how the depth at which a sample was taken can be determined by means of DSRT's. However, in general, the depth is determined by measuring the length of the wire between sea surface and the sample; i.e., the wire is passed over a metering wheel (Fig. 18.6). Since the wire seldom hangs straight in the sea due to deflection from the vertical by currents or by the drifting ship, several devices have been developed to ascertain the shape of the distorted cable and the true depth of the sample. Wire angle indicators are hung on the cable at intervals, the wire angle at the surface is measured, depth is measured thermometrically, wire played out is metered, and sound generators are attached to the cable and monitored on depth sounding gear. If samples are taken near the bottom in very deep water, where determination of actual depth from metered cable, etc., is subject to increasing error due to the remoteness of the sample from the sea surface, the depth may be determined relative to the bottom rather than to the sea surface.

Referencing to the bottom is done by attaching a pulsing sound source, called a *pinger*, next to the sampler (Fig. 18.7a). Sound pings are then transmitted directly to the ship and to the bottom to be reflected to the ship. A sonic depth recorder converts the time interval between the arrival of the direct sound beam and the reflected sound beam to the distance between the pinger and the bottom, or the sampler and the bottom (Fig. 18.7b). Excellent depth control can be exercised in this way, and samples can be taken within one or two meters of the bottom even though five or more miles of cable may separate ship and sampler. This same positioning technique can be used to control the

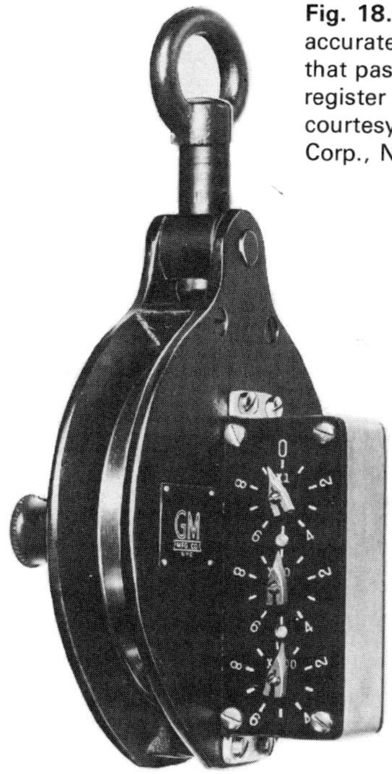

Fig. 18.6 A meter wheel calibrated to register accurately the length of a wire of given diameter that passes over the wheel. The dial counters register in meters times 1, 10, and 100. (Photo courtesy of G. M. Manufacturing and Instrument Corp., N. Y.)

(a)

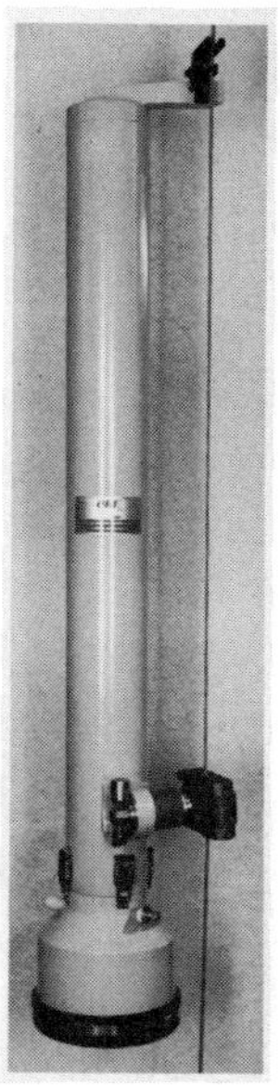

Fig. 18.7 (a) A bottom-finding pinger, sound source, placed on a wire next to an instrument package can be used to maintain a distance between the instrument and the sea floor measured relative to the sea floor. (Photo courtesy of Ocean Research, Inc., Falmouth, Mass.) (b) Facsimile of the recorded trace of a pinger.

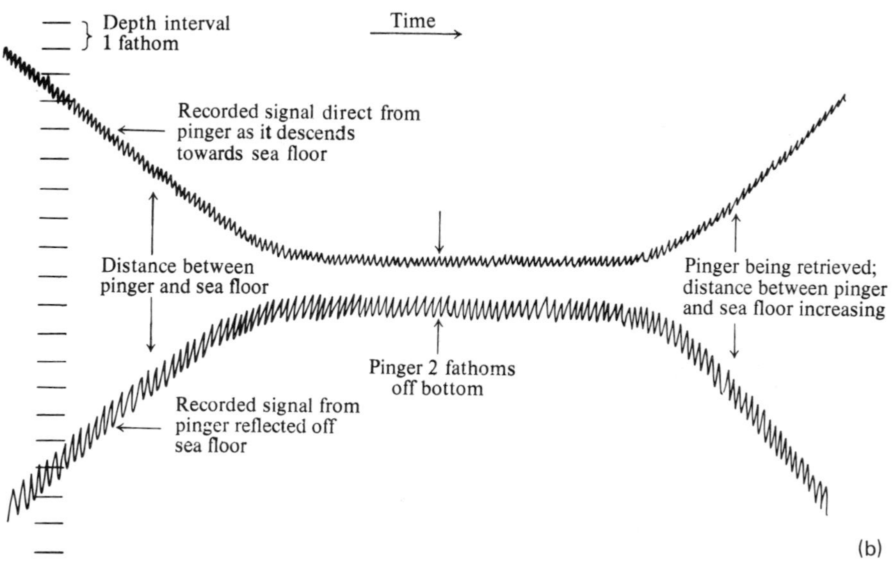

(b)

distance between an underwater camera or TV unit and the bottom, or to monitor depth when it is imperative that a device be set gently on the ocean floor.

DETERMINING SAMPLE TIME

Sample time, as well as position in space, is important. When samples are taken in shallow water, the time at which the messenger is dropped from the ship may suffice as the sample time, but in deep water, the sample is taken some time after messenger release. A messenger slides down the wire at a terminal speed ranging from 150 to 200 m/min, depending on wire angle, messenger size and shape, and cable condition. If the descent rate is about 200 m/min, it takes about 40 min for the messenger to trigger a sampler located at 8000 m below the surface.

Timing is very important. Researchers often invest several hours in getting the sampling gear to depth, drop a messenger, and then wonder how long they should wait before the sampler is tripped and retrieval may start. If retrieval is started too soon, the sampler will be tripped at an unknown shallower depth. If they wait too long, expensive ship time is lost; worse, the ship may drift and, if the depth is shoaling, drag the sampler along the bottom. The longer the sampler remains close to bottom, the higher the loss risk.

Obviously it is very desirable to know exactly when the sample is taken, so retrieval can start immediately. If a sound pinger is being used, it can be rigged to be tripped and shut off by the same messenger that trips the sampler. In this case, the cessation of the sound pings on the depth-sounding equipment signals the sample time. What we have said so far seems quite straightforward and simple, but only the person who has lived through the suspense of taking deep-sea samples under adverse conditions can appreciate the predicaments that can arise, especially when a messenger has been released and there is no feedback indicating that the sample has been successfully taken. Before starting retrieval, the researchers must weigh many factors: how much time was spent in lowering, should they drop another messenger and wait again, how critical is the nature of the sample, what could be the cause of failure? If they decide on retrieval, they'll carefully watch the emerging cable inch by inch to make sure that the messenger has not been stopped in the descent by a wire whisker or a kink in the cable. If this has occurred, the messenger must be spotted and removed from the cable before it passes through the metering block; if this is not done, the cable could be bound and cut and everything lost over the side. These problems, combined with foul weather and a rough sea, can lead to considerable drama.

IN SITU MEASURING DEVICES

The development of electronic gear and new sensors has increased the possibility of *in situ* measurements. One can measure the electrical conductivity and temperature of the sea in place and thus determine the salinity *in situ.*

Strain gauge devices can be used in sensors that measure pressure and thus depth. Even sound velocity can be measured directly rather than being calculated from temperature, salinity, and pressure values. Galvanic electrodes lowered into the sea can determine dissolved oxygen *in situ*, and optical systems, such as immersion photometers, can measure the attenuation of light and turbidity of the water at depth. Instruments that can be used *in situ* can make continuous time and depth measurements as they are lowered, so that the researcher is able to construct vertical profiles of the measured parameters. The data gathered are continuous over depth in contrast to the data obtained from extracted samples taken by water bottles which provide information for discrete depth points only.

The rapid time response, precision, and data-gathering capabilities of these devices make it possible to measure variability of oceanic conditions at fixed spatial points over both short and long time periods. Short-term variations cannot be measured by analyzing extracted samples, since the time increment between collecting samples for analyses is appreciable. Thus *in situ* samplers have opened new areas of research in the sea. At the same time, they have made the interpretation of data, sampling procedures, and seagoing operations much more complicated.

The more sophisticated the sampling tools, the greater the gap between the observer and the system being monitored. Measurements collected at a rate of many times a second must be processed by a computer for later inspection. If there was a malfunction in the sampling apparatus that went undetected at the time of data gathering, the computer could record and generate thousands of erroneous bits of information. Even on later inspection, the observer may not be able to determine whether the data obtained are good or bad unless independent measurements are made by traditional (mechanical or chemical) methods for comparison, reliability checks, and calibration. Data gathered at the slower rates of less sophisticated devices can be analyzed and checked almost as rapidly as they are gathered, giving the observer more confidence in his results.

Reliability of equipment is extremely important in scientific research. This is one of the reasons why some of the mechanical devices, such as water bottles, DSRT's, and the Secchi disk, have survived so long. Simple instruments that are durable and inexpensive, function in an easily understood manner, and can be repaired at sea are a boon to the researcher. Hence these instruments, tested in years of use, have made the older, slower methods the control against which modern techniques are evaluated.

An example of a modern composite *in situ* measuring probe is the salinity–temperature–sound velocity–depth sensor (Fig. 18.8). This array of sensors is lowered into the sea on an electrically conducting coaxial cable. The cable acts as the pathway to supply current to the sensor array and to receive data. Signals from the sensor may be used to generate analog plots of temperature, salinity, and sound velocity versus depth or time. The signals may also be recorded for use in computer programs which give numerical listings of the measured param-

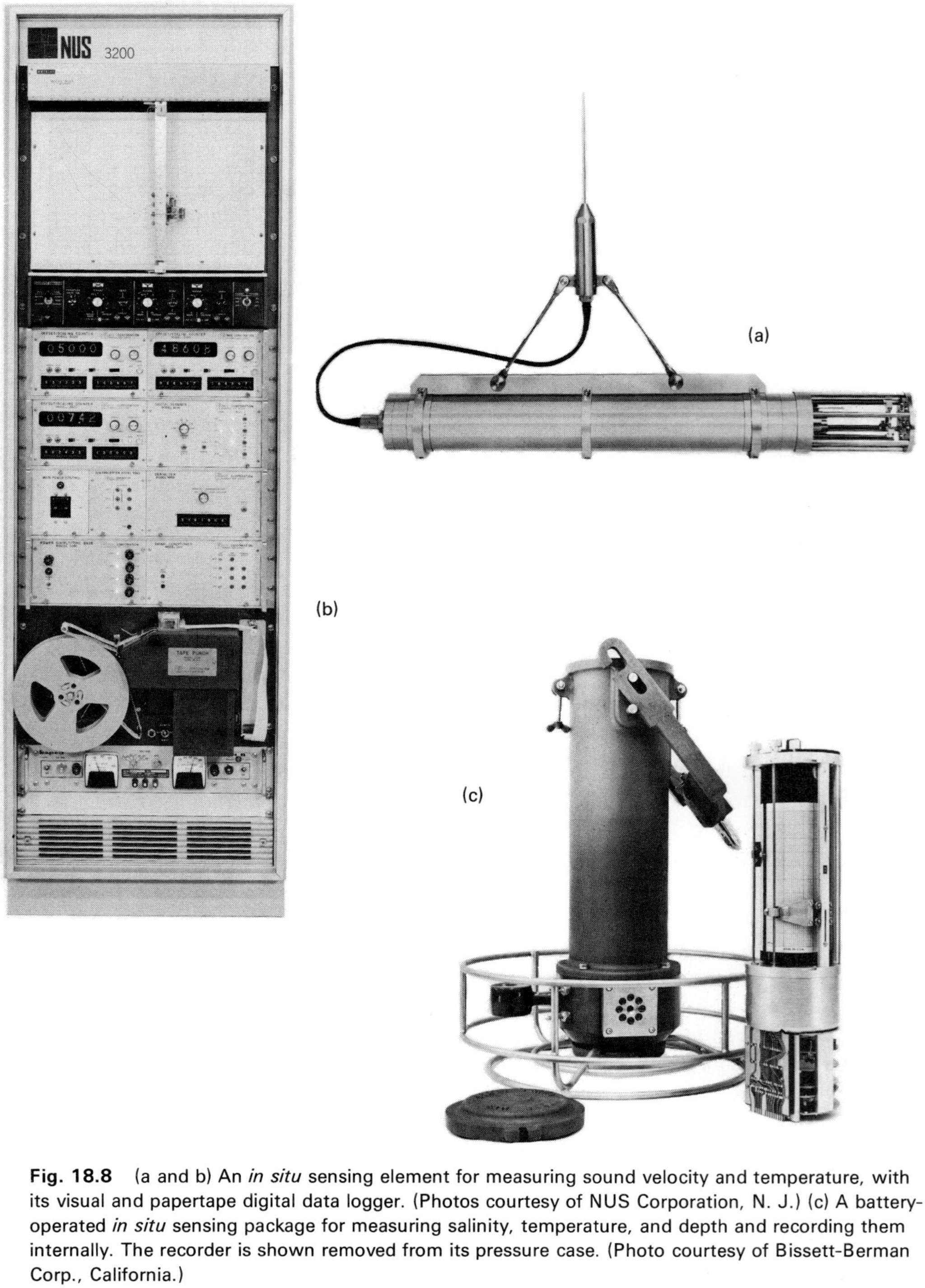

Fig. 18.8 (a and b) An *in situ* sensing element for measuring sound velocity and temperature, with its visual and papertape digital data logger. (Photos courtesy of NUS Corporation, N. J.) (c) A battery-operated *in situ* sensing package for measuring salinity, temperature, and depth and recording them internally. The recorder is shown removed from its pressure case. (Photo courtesy of Bissett-Berman Corp., California.)

eters or other parameters computed from the measured variables. There are also battery-operated instruments that perform the same tasks but record internally. Devices of this sort are capable of rapidly sensing several properties of the water column at the same time. However, the use of these instruments as the sole data sources precludes the gathering of additional data the researcher may be interested in to fully understand the oceanic environment, such as dissolved gas content, trace metals, and biological specimens. Modern instrumentation has its use and is invaluable, but it is not necessarily the ultimate or even desirable alternative for many programs. With the above thoughts in mind, let us consider some of the conventional oceanographic instruments in their older and their modern designs.

TEMPERATURE-MEASURING DEVICES

The temperature field in the sea can be sensed at discrete points by using DSRT's or temperature-sensitive materials such as resistance wires or thermistor beads. Resistance wires and thermistor beads have a rapid time response that allows them to monitor continuous changes in temperature *with depth* as the thermal element is lowered into the sea or *with time* if the element is confined to a single depth. Another temperature-sensing device is the bathythermograph (Fig. 18.9). The mechanical bathythermograph (BT) has a bourdon-tube temperature element that scribes a mark on a coated glass slide. The slide is mounted on spring-loaded bellows which collapse as pressure increases. The two motions, movement of stylus due to temperature change and movement of the slide due to pressure change, cause the stylus to describe a temperature-versus-depth curve on the slide as the instrument is lowered into the sea. Upon retrieval of the BT, the slide is removed and placed against a grid calibrated for the instrument, and the temperature-versus-depth values are determined. Bathythermographs are not very accurate, but they provide a rapid means of determing the thermal structure of the sea to depths of ~ 300 m.

Fig. 18.9 The mechanical bathythermograph, BT, used to record temperature of water versus depth on a coated glass slide.

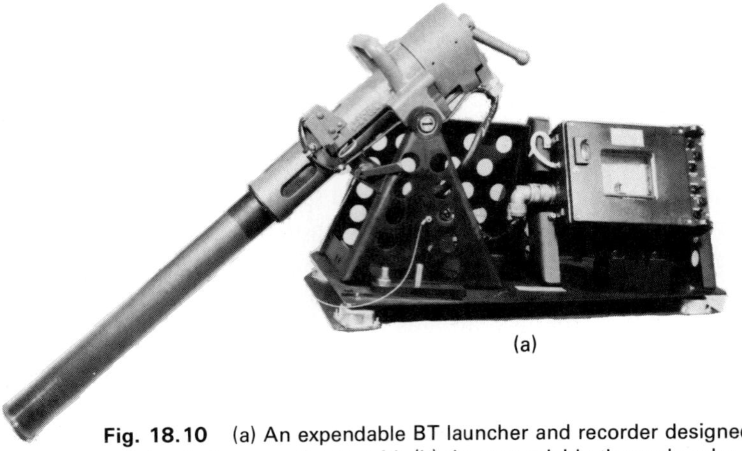
(a)

(b)

Fig. 18.10 (a) An expendable BT launcher and recorder designed to be used from a cruising helicopter, shown with (b) the expendable thermal probe. (Photos courtesy of The Sippican Corp., Marion, Mass.)

An improvement over the mechanical BT is the expendable BT (Fig. 18.10). This is a throwaway thermal probe which is ejected over the ship's side and sinks at a known rate. Thus time defines the depth of the sinking probe. A trailing wire sends back the temperature as an electrical signal, which is recorded on the ship against time and depth. When the probe takes up all the wire it breaks free and is not recovered. Although an expendable BT can be used only once, its cost is more than offset by the fact that it can be used underway and does not require the costly procedure of stopping the ship. The mechanical BT can also be used underway at reduced speeds, but cost per instrument, constant recalibration, and expense in processing and reading the slides make the data gathered by the two systems nearly comparable in cost. Bathythermographs are valuable aids in quickly determining the upper thermal structure, which must be known for underwater sonar work in antisubmarine warfare.

CURRENT-MEASURING DEVICES

The direct measurement of currents in the sea can be accomplished in many ways. The current can be determined by measuring the water moving past a fixed location in space, or it may be observed by following a particular parcel of water as it moves along. The first alternative gives rise to a major problem: it is very difficult, if not impossible, to moor a ship or surface platform in the open sea so that it will not move and by its motion move the current sensor around its principal location. Any motion of the current sensor caused by the ship or buoy, of course, is registered as ocean current and may be of larger magnitude than the water current at the given point. To overcome this problem, one must know the ship or support-buoy motion that is transmitted to the current sensor and correct the data accordingly, or the current sensor must be suspended so

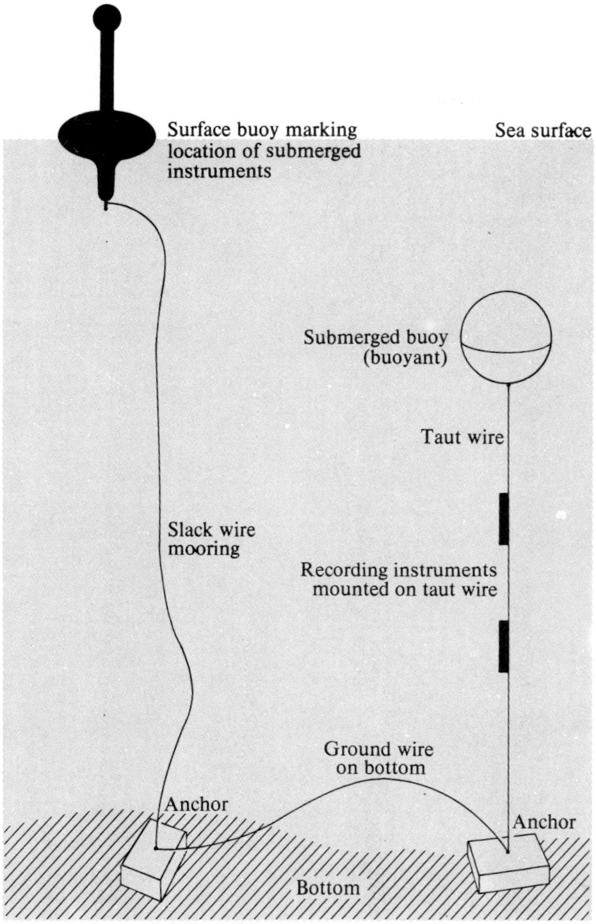

Fig. 18.11 A typical method of mounting instruments using a taut wire mooring to reduce surface platform motion.

that it is relatively independent of surface platform motion and is held fixed in space. Taut wire moors are used at sea that anchor the current sensor to the bottom, that is, an anchor and a wire attached to a submerged float produce a mooring system that is reasonably stationary and can be used to support a current sensor (Fig. 18.11).

If the second alternative is chosen, i.e., the motion of the water parcel is to be followed, platform motion or mooring noise is not a problem. However, in this case the water parcel must be labeled so that it can be followed as it moves through space. This can be done by placing a large current cross or drogue at depth which is compelled to move with the water parcel. A small surface float is secured to the drogue so that its progress through space may be observed and plotted from the surface. If the drag forces exerted on the drogue are much larger than those acting on the surface marker and connecting cable, the system nearly describes the movement of the water that the drogue dis-

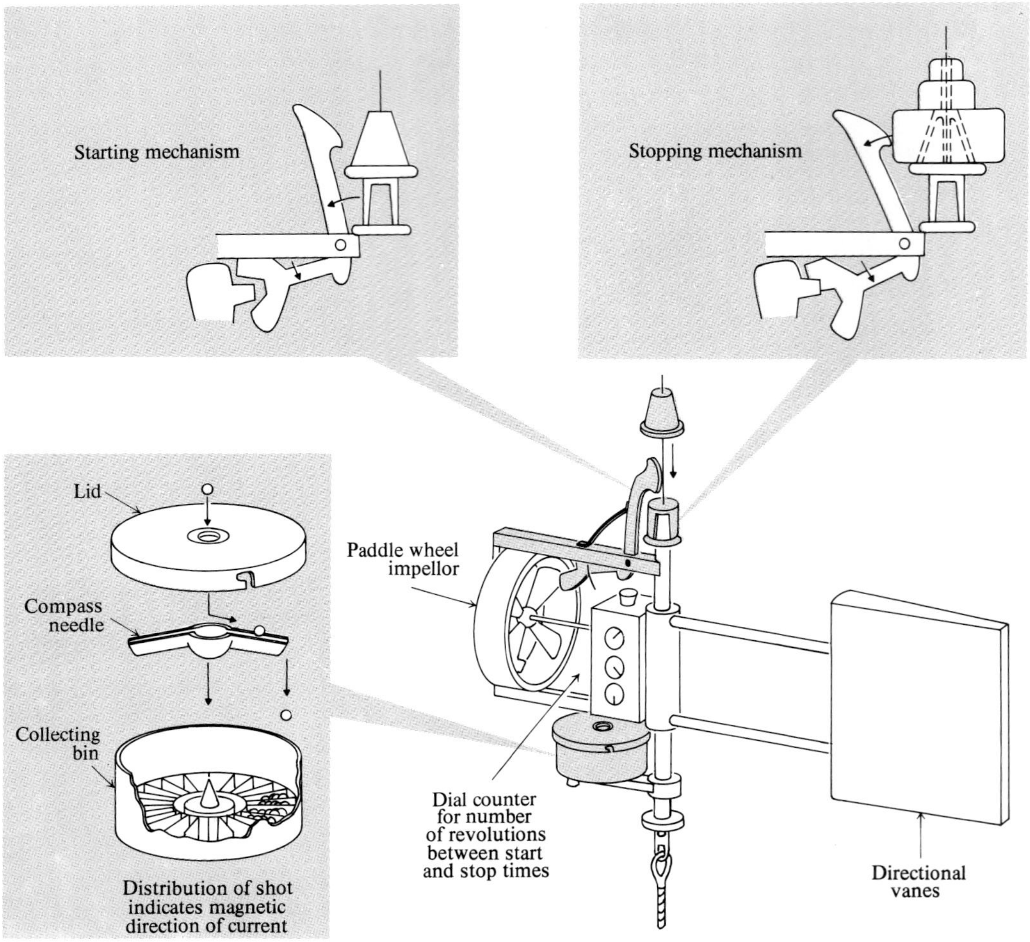

Fig. 18.12 The mechanism of an Ekman-type mechanical current meter. (From A. C. Hardy, *The Open Sea*, Houghton Mifflin, Boston, 1964.)

places. A more advanced technique uses a submersible buoy equipped with a sound pinger whose density can be adjusted to any value found in the sea. Thus a parcel of water of specific density can be followed by adjusting the density of the buoy to that of the parcel and releasing the pinger over the side, where it sinks to the appropriate density level. The neutrally buoyant buoy is then followed along its course by triangulation of its sound ping from known ship positions.

Measurement of current implies measuring both current speed and direction. The measurements may be recorded internally within the meter or telemetered to the ship. They may be integrated, averaged over time, or recorded as nearly instantaneous values of speed and set. They may also be made continuously over time, or the system may be arranged to take spot readings at given time intervals. The older mechanical current meters recorded internally and produced averaged current values. For example, the Ekman current meter (Fig. 18.12) was designed to record revolutions of the current rotor on a system of dials for the period of time the meter was exposed to the current. Approximately every 30 revolutions, a nonmagnetic (bronze) shot was released from a storage bin to fall on top of a magnetic compass needle and roll through a trough toward the south-seeking end of the magnet. The shot then fell into a compartmented bin secured to the meter frame. At the end of the measurement period, i.e., the interval between the time a messenger was released to activate the meter and the time a second messenger was lowered to stop the mechanism, the meter was retrieved. The number of revolutions registered by the dial counters gave the average speed over the measurement period, and the accumulation of shot in the directional bin gave the orientation of the meter relative to magnetic north and thus the current set. Each series of measurements, however, meant resetting the meter, lowering it to the desired depth, starting it, stopping it, and hauling it up for reading. No information on variability of speed and set over the time of the measurement could be obtained.

Today's current meters can be lowered to depth, and the current speed and set are telemetered electrically and recorded continuously. Impellers of many designs may be used in these meters, but all have one thing in common: their rotation in a current is sensed by the closing and breaking of an electrical circuit either magnetically or optically on each revolution. The pulse in the circuit may then be counted directly to yield revolutions per time or it may be integrated to give a direct read-out in speed. Whatever the system used, it is very important to minimize friction on the impeller to ensure that the impeller responds properly to weak and variable flow. A compass, either magnetic or gyroscopic, is housed in the meter body and the orientation of the meter body to the compass or, if the meter body is not directionally controlled by the current, the orientation of a vane is then sensed and transmitted as a signal to the read-out station. Figure 18.13 shows a typical current meter of this type.

If an unattended current meter is to be used at a fixed position for a prolonged period of time, it must be able to internally record data on film, magnetic tape, or some other medium, or to transmit signals to the surface whence they can be radioed to shore or transmitted to an interrogating system, such as a satellite.

There are other crude but effective tools for measuring oceanic drift and current. Plastic cards that float, drift bottles, or sea-bed drifters are released in the sea at known times and locations to be scattered by the surface or bottom currents and recovered on the beaches. The person who finds one of these objects is instructed to send in the following information: location and time

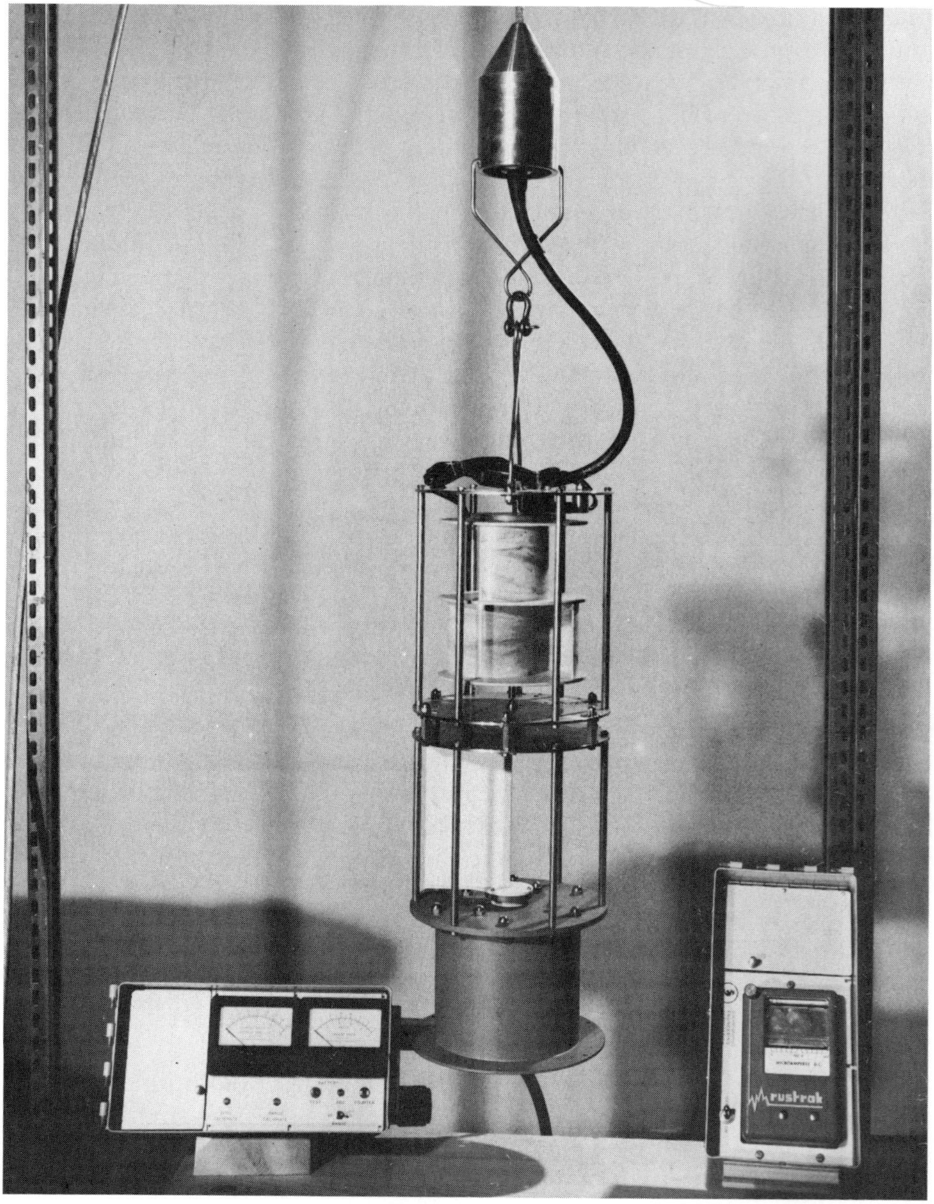

Fig. 18.13 An electrical sensing current meter equipped with a Savonius rotor and a directional vane. This current meter sends electrical signals back to the ship via a wire, to be read directly as current speed and directions on the shown meters or to be recorded. (Photo courtesy of Hydro Products, San Diego, California.)

of recovery and the serial number of the drifter. From this information, it is possible to estimate the direction and rate of travel between the release and recovery points.

FILTERING NETS

The biological populations of the open sea are not usually studied *in situ* but are extracted from the sea for laboratory study. However, since it is essential to know the horizontal and vertical distributions of the populations, as well as their population density, the collecting must be quantitatively controlled. For example, a net may bring up x gm of biological material in one case, and y gm in another. Unless the volume of water passing through the net or, in the case of a dredge, the area of sea floor sampled is known in each instance, it is impossible to say in which of the two cases the material was more abundant.

Nets are available in a wide variety of sizes, mesh types, and designs to provide the researcher who must catch different organisms in different ways and under different conditions with the net best suited for his purposes (Fig. 18.14). A net must pass water and be open at its front to fish effectively. Some nets are designed to open at a given depth and then close to retain the sample and keep it uncontaminated from specimens found at another depth. Towing

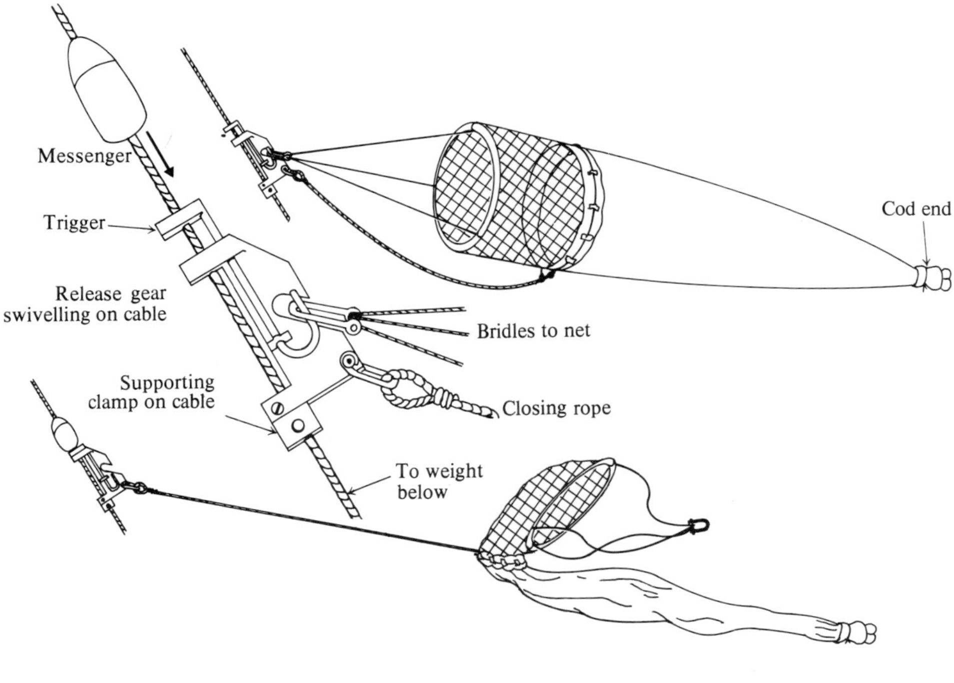

Fig. 18.14 A ring net equipped with a closing device actuated by a messenger.

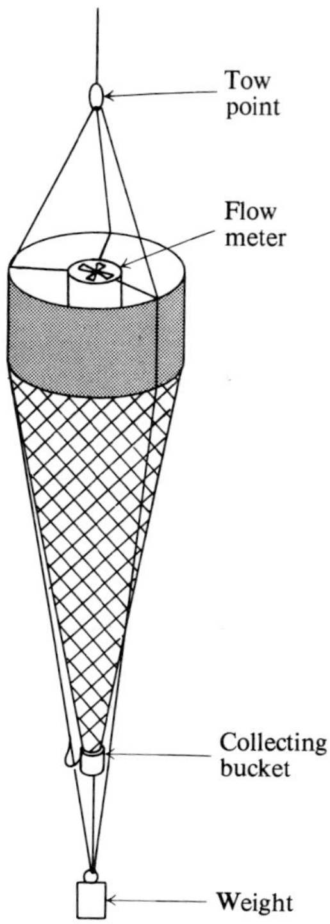

Fig. 18.15 A ring net equipped with a flow meter to register the volume of water filtered by the net.

bridles, tripping mechanisms, choke lines for pursing nets, and net releases come in all sizes, shapes, and forms.

One can estimate the volume of water filtered by the nets by determining the effective area of the net entrance when the net is working and the distance over which it is towed. If towing is always done at the same ship speed, the tow duration can be used to calculate the distance. Then, if the same net is towed in the same way, at the same speed, and for the same period of time, comparable water volumes will be filtered, resulting in a crude quantitative sampling technique. Problems do arise, however, if the net is towed with the current one time and against it another, if the net opening tends to change in area, or if the net clogs up so as not to pass water freely.

To increase the reliability of quantitative measurements, devices have been developed to keep the net entrance area fixed; furthermore, instruments have been placed on the nets that measure distance traveled and depth of net versus time. The net opening can also be provided with a current sensor or flow meter which will record impeller revolutions per time and thus indicate the average speed of the water entering the net (Fig. 18.15). Water velocity in meters per second divided by sensor revolutions per second yields meters per revolution, the calibration constant of the current sensor. This constant × the number of revolutions of the sensor during the net tow × the entrance area gives the

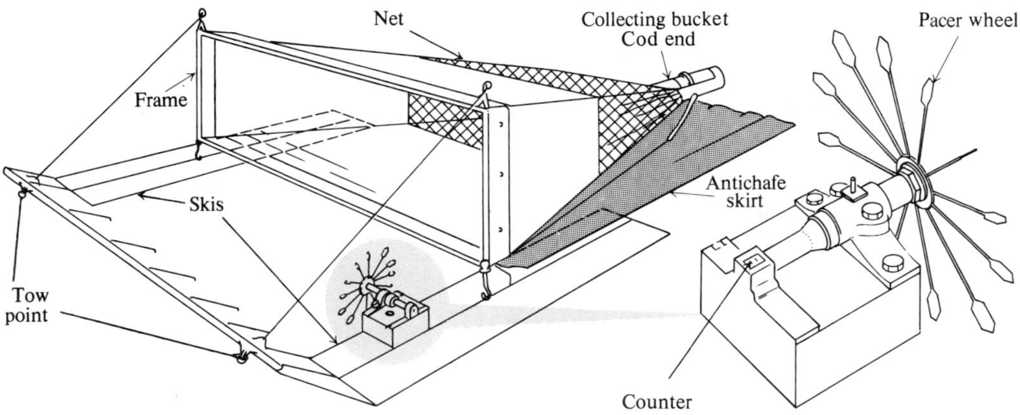

Fig. 18.16 Epibenthic dredge showing a view of the entire ensemble and details of the bottom walker. (From J. W. Hedgpeth, ed., *Treatise on Marine Ecology and Paleoecology*, Ecology, vol. 1. The Geological Society of America, Boulder, Colorado, 1957.)

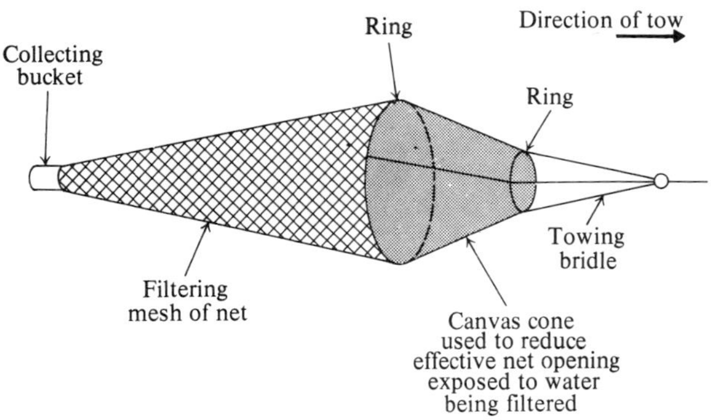

Fig. 18.17 A ring net equipped with a canvas nose cone to reduce speed of water being filtered through filtering mesh.

volume of water filtered. If a net or a dredge is towed across the sea floor to catch epifauna, a pacer wheel incorporated on the net frame will measure the distance the trawl or dredge travels over the bottom (Fig. 18.16). This distance × the width of the net determines the area of the sea floor sampled.

The coarse mesh net or web of the trawl-like nets used for nekton sampling is not particularly subject to clogging so that water freely passes through the mesh. The fine-mesh ring nets used for catching microscopic plankton can be clogged by the trapped organic material, and the fineness of the mesh alone is enough to severely restrict water flow. These nets have to be towed quite slowly for estimates of filtered volume to be at all meaningful. Ring nets come in standard opening sizes, i.e., $\frac{1}{4}$ m, $\frac{1}{2}$ m, and 1 m in diameter, and are equipped with nylon mesh sleeves of varying mesh size (Fig. 18.17). A coarser mesh is used for zooplankton than for phytoplankton. The tow speed can be increased for fine-mesh nets if the net has a large-diameter filtering sleeve and a small entrance. The fast-moving water entering the small opening decreases its speed when it enters the larger area of the filtering sleeve.

Nets can be towed in many ways, three of which have become standard. (1) In a *vertical tow*, the net is lowered to depth and hauled back up in fishing attitude. This type of tow gives a good sample of a vertical water column but yields no information on the vertical distribution of the catch. (2) In an *oblique tow*, the net is lowered to depth alongside the moving ship and hauled back in fishing attitude. This type of tow provides a longer sampling path which covers both the vertical and horizontal extents; hence it is useful when the biopopulation is small. In addition, it evens out patchiness in distribution to yield a more representative biological sample. Like the vertical haul, the oblique tow provides no information on the vertical distribution over the vertical extent of the tow. (3) In the *horizontal tow*, the net is towed at fixed depth, that is, it is opened and closed at depth and thus collects samples from a single depth zone. As a result, this type of tow provides information on the vertical distribution of the organisms caught in the net when tows are made at separate depths.

Even a net that is always in fishing attitude can be used effectively to obtain a relatively pure sample from one depth. If the towing time at the desired depth is very long compared to the time required to lower or retrieve the net, the sample obtained will be representative of the sampled depth. Moreover, one can attach a device that measures depth versus time to the net frame and thus obtain a fairly good record of the fishing effort at a given depth.

DEEP-SEA DREDGES AND BOTTOM SAMPLERS

Let us now consider the problem of sampling the sea floor—a problem of interest to both biologists and geologists. The biologist, of course, wants to obtain quantitative data on the epifauna and infauna as well as information on the sediment type. The geologist is more interested in a sample of sediments, preferably in an undisturbed state, so he can study the depth variation of the sediment. To collect sampling material, dredges dragged over the ocean floor scrape the epifauna off hard surfaces and dig into the softer sediments to separate the infauna and epifauna from the mud. These dredges (Fig. 18.18) are satisfactory for collecting bulk material, but since they may skip and bounce over the sea floor, the resulting sampling is not very quantitative. A geologist may also use a dredge if he is after a bulk quantity of loose material lying scattered on the sea floor.

A more quantitative and undisturbed sample of the sea floor can be obtained by using grab samplers (Fig. 18.19). These devices are lowered to the sea bed and trip on contact, taking a bite of the substrate of (hopefully) fixed surface area and known penetration depth. Depending on the sampler, the

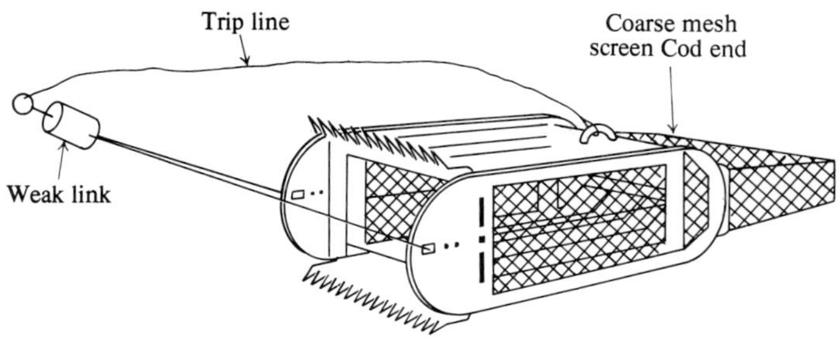

Fig. 18.18 A biological dredge used to collect bottom fauna on a hard sea floor.

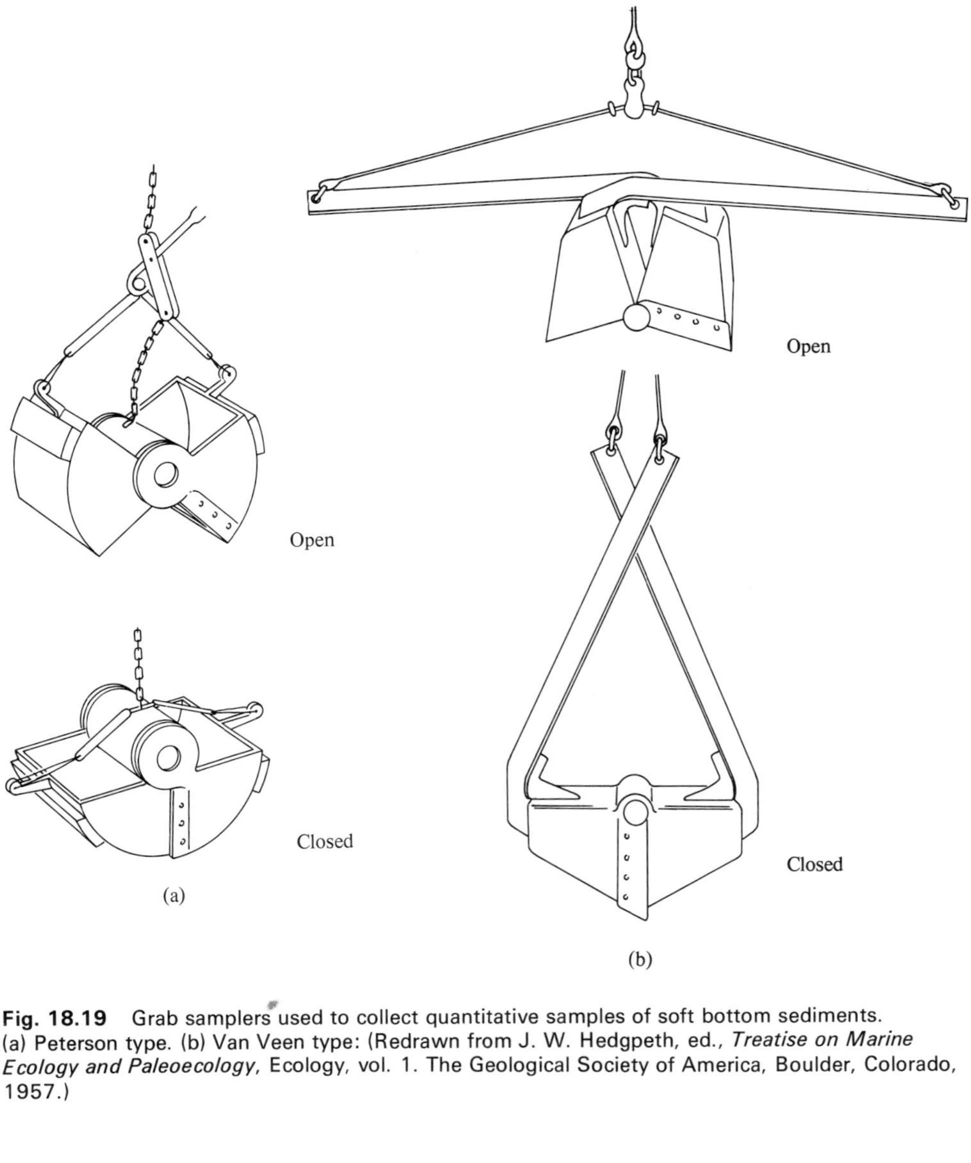

Fig. 18.19 Grab samplers used to collect quantitative samples of soft bottom sediments. (a) Peterson type. (b) Van Veen type: (Redrawn from J. W. Hedgpeth, ed., *Treatise on Marine Ecology and Paleoecology*, Ecology, vol. 1. The Geological Society of America, Boulder, Colorado, 1957.)

sample may be nearly rectangular in shape, a cylinder, or a half-cylinder. Successful manipulation of these samplers is somewhat of an art. Often a small rock or shell may keep the sampler jaws partly open so that the sample is washed out during retrieval. Yet, in many respects, the grab samplers are simple and effective devices.

DEEP-SEA CORING DEVICES

If the geologist needs to penetrate the sediment to a depth greater than that made possible by a bottom grab, he must use a coring device. Corers (Fig. 18.20) consist of a length of pipe equipped with a hard metal nose piece on one end and dead weight on the other. The corer is lowered close to the ocean floor and then the cable is played out rapidly or a tripping mechanism is employed so that the last part of the descent is made in free fall. The core barrel is supposed to penetrate the sediment, cutting out a cylinder of mud that is forced up the center of the barrel (Fig. 18.21). To facilitate the extraction of the sediment core from the barrel, the barrel is equipped with a plastic liner which can be slipped out and which acts as a storage container for the core sample until analysis can be run. An orange-peel core catcher is located between the nose piece and the liner, preventing the core from washing out during retrieval. Several types of coring devices are available: the simple free-fall gravity corers with a barrel length of 1 to 2 m, more complex piston corers with a barrel length of 20+ m, and multiple corers that take several short cores at once.

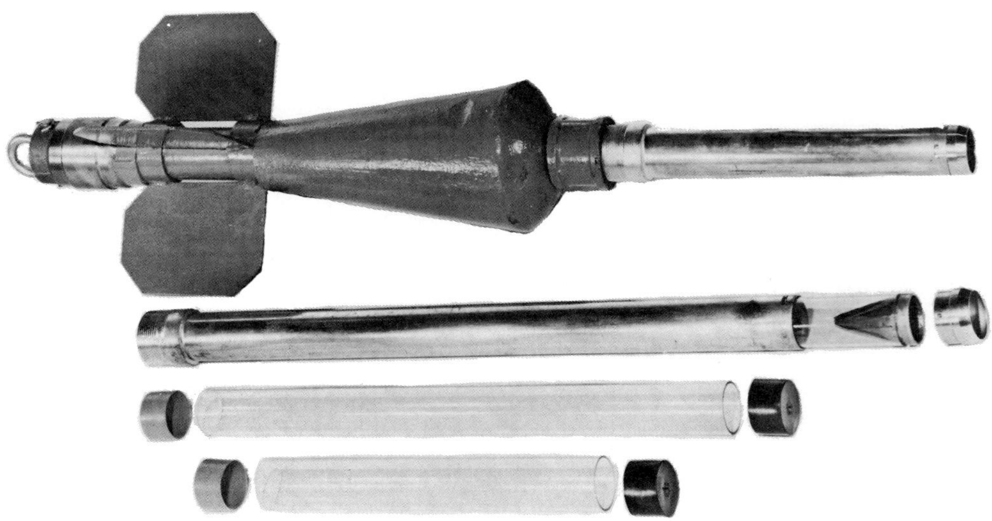

Fig. 18.20 A simple free-fall gravity corer known as a Phleger corer showing assembled corer and disassembled core barrel with plastic core liner, cone catcher, and nose piece. Photo courtesy of G. M. Manufacturing and Instrument Corp., N.Y.

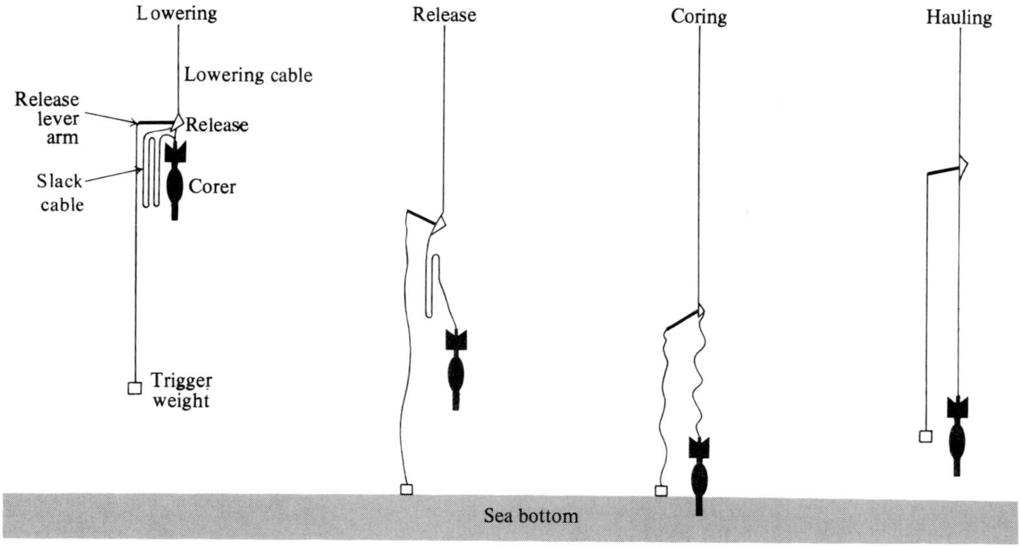

Fig. 18.21 Operating principle of a simple gravity corer.

Several methods have been or are being used to increase the length of the cores. With an older type of corer, the core barrel was shot into the sediment with an explosive charge and then yanked out by a straining cable. Today corers contain a piston inside the liner near the nose piece. These devices, called piston corers, are suspended from an auxiliary hanger on the lowering cable (Fig. 18.22). As soon as a tripping weight hanging a known distance below the corer senses the ocean floor, it trips the corer so that it is released from its hanger and free-falls a known distance. During its fall, the corer simultaneously takes up the loose slack in the suspending cable, touches bottom for penetration, and puts a strain on the piston which is fastened to the end of the cable. The piston moves up the liner as the corer penetrates the sediments, permitting the hydrostatic pressure head of the water column to aid the weighted corer in penetrating the sediments and removing the water from the barrel to reduce the resistance to the entering core. By means of piston corers, cores in excess of 20 m have been obtained from soft sediments. This coring technique has provided the longest cores from the ocean floor except for those obtained with drilling apparatus.

A recent innovation in coring is the boomerang throw-away corer (Fig. 18.23). This gravity corer is tossed over the ship's side unattached to a cable. It falls in a penetrating attitude and takes a core. As soon as penetration has occurred, two glass floats break loose from the upper end of the corer and pull out the core liner from the top of the corer. The floats and the liner containing the core rise to the surface where they are recovered. A bright flashing xeon

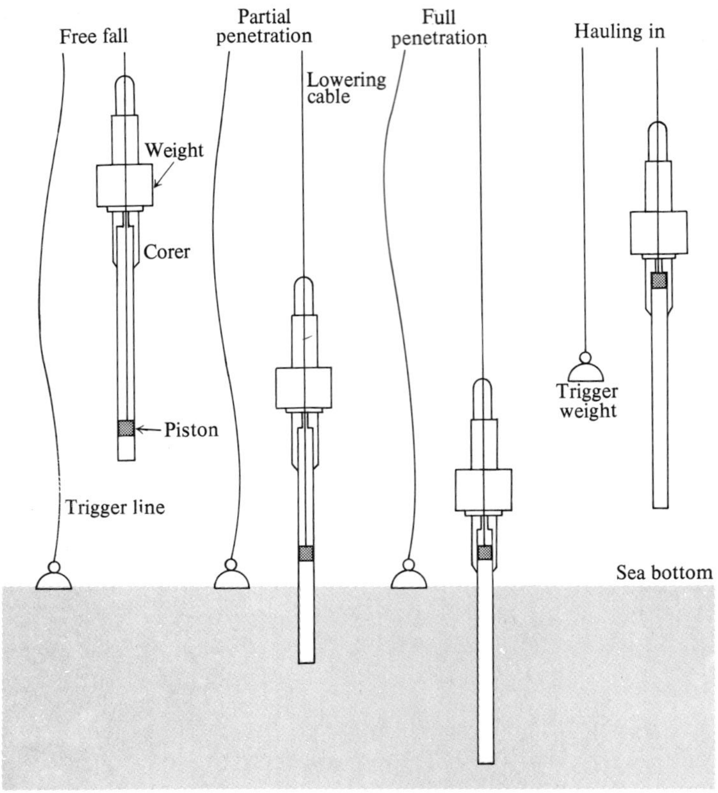

Fig. 18.22 A diagramatic sketch of a piston corer. The penetration of this corer into the sediments is enhanced by the moving piston in the core barrel. This corer is tripped and allowed to fall free-fall, the same as the simple gravity corer.

Fig. 18.23 Boomerang corer being readied for dropping over the side of a research vessel. (Photo courtesy of Benthos, Inc., North Falmouth, Mass.)

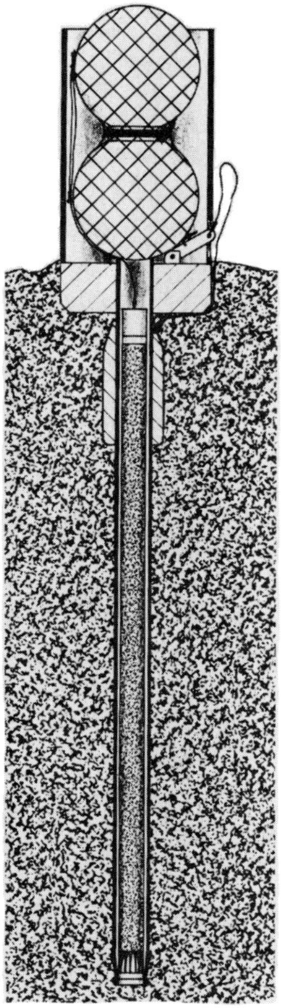

Fig. 18.24 A diagrammatic sketch showing a boomerang corer penetrating the sediments.

light in one of the floats aids in recovery. The core barrel and weighted cannister are lost in the operation. Only relatively short cores can be obtained in this manner (Fig. 18.24).

The use of corers with long barrels is restricted to soft sediments. If, during free fall, a piston corer strikes rock or even compacted semisolidified materials, penetration is stopped quickly and the corer will topple over. Bent or broken core barrels are not at all uncommon.

To reduce the friction between the barrel and the sediments and allow rapid penetration of the core barrel into the sediments, a weighted mechanical vibrator can be mounted on the top of the barrel. When the core barrel is brought into contact with the bottom, the vibrator makes the barrel shimmy; as a result friction is reduced.

INSTRUMENTS USED IN SUB-BOTTOM INVESTIGATIONS

The structure of the earth's crust below the surface sediments is sampled using the geophysical techniques discussed earlier. High-intensity energy sources generate pulses that penetrate the sediments, where they are reflected and refracted by the crustal layers. *Hydrophones* (underwater listening devices) receive the returning signals which are recorded and converted to graphical representations of the reflecting layers in the sediments, thus revealing features of the earth's crust hidden by its blanket of sediments. Temperature probes are driven into the sediments to measure the temperature gradient. At the same time, a small core of the sediment is retrieved for measurement of its thermal conductivity. Knowledge of the thermal conductivity and the *in situ* temperature gradient of the sediment allows the geophysicist to measure heat flux through the sea floor.

INSTRUMENTS THAT INCREASE MAN'S ABILITY TO SEE FEATURES OF THE OCEAN ENVIRONMENT

The instruments we have discussed so far are examples of tools used in oceanographic field survey work. They either collect material for analyses or measure parameters *in situ*. There are a few tools that do not actually measure or collect but instead serve as field aids; that is, they enable the researcher to descend to the depths of the sea to directly view the oceanic environment or to view the depths from a remote vantage point. (Self-contained underwater breathing apparatus (SCUBA) can, of course, be used effectively in the warm and protected waters along the shallow edges of the sea; however, from the point of view of oceanographic research, this region is a relatively insignificant portion of the oceans.)

Instead of trying to descend directly into the hazardous open-sea environment to observe its features, the researcher may use underwater still and television cameras to make visual records. The instruments are lowered by cable or mounted on remote-controlled underwater vehicles that travel along the sea floor (Fig. 18.25). While cameras take stereo pictures for later analysis, TV images can be stored on tape and viewed simultaneously, serving as the underwater eyes of the researcher. Thus a remote-controlled vehicle can be directed in accordance with what the TV camera sees in its field of vision.

Visual records of the seas are important because they give the researcher a much better understanding of the surface structure of the sea floor and the distribution of its populations than do remote sampling techniques. Let us consider the following example. Imagine an investigator in an aircraft at an altitude of 10,000 m. If he were to drop a grab sampler and pick up a random sample from the earth's surface without being able to see the earth at all, what would he surmise about its surface features? Contrast his conclusions with those he would be able to arrive at if he suddenly had a clear view of the earth. This example is directly applicable to the seas, where much of the research is

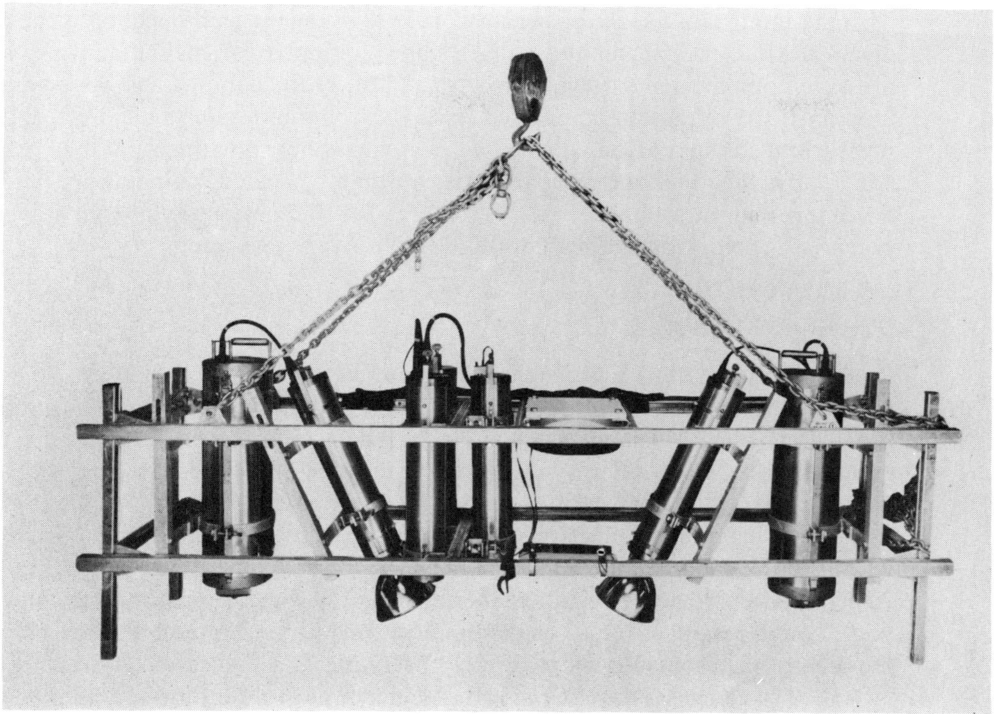

Fig. 18.25 An underwater stereo camera and strobe light array equipped with its own pinger so that it can be positioned at a precise distance above the sea floor. An assembly like this was used to find and photograph the remains of the nuclear submarine *Scorpion*, lost in the Atlantic Ocean. (Photo courtesy of E G & G International Geodyne Division, Waltham, Mass.)

conducted by men groping about rather blindly with an unfeeling hand suspended on a very long arm.

There are, of course, the deep-sea submersibles that are now so much in the news. Although these vehicles serve for the most part only as containers which carry man to depth in the sea with some degree of comfort and safety and provide him with a means of doing simple mechanical manipulation while there, they do contribute greatly to oceanographic research by allowing an observer to view this hidden environment directly, make decisions on the spot, and respond immediately—within limits—to what is around him. No complex of machinery or computers can compete with a human on location in the sea.

For the time being, the deep-sea submersibles are not satisfactory sampling devices except in cases of selective sampling carried out at the discretion of a human at the controls. The limitations imposed on their operations by sea state, their slowness and limited submersion time, and the prohibitive cost of construction and operation would make them a very expensive sampler indeed for routine use. For some tasks, they are irreplaceable, but they are not the ultimate tool for much of the research carried on at sea.

One important aspect of the most recent vintage of submersibles is their diver lockout compartments. Submersibles equipped for lockout have two pressure compartments. One compartment houses the conning and observing functions, the other can be entered from the conning region, sealed off, and pressurized so that personnel may exit from this sphere into the water for work and return. The lockout compartment can then be sealed, pumped dry, and decompressed at required rates. This facility increases the submersible's potential for work on the sea floor and for recovery operations.

OTHER SYSTEMS

In addition to the instruments discussed so far, there exists a wide variety of equipment and gear which we cannot discuss here in detail. Instruments and other tools range from old-fashioned to very advanced, the basic requirement being that they work reliably, accurately, and with ease. Modern research ships are equipped with laboratories whose facilities equal those of installations on land. Even complex electronic computers go to sea on modern ships. There is only one fundamental difference between the two environments: motion. Accelerations and deccelerations experienced on a rolling, yawing, and pitching vessel make it impossible to use equipment sensitive to motion; at times they handicap the researcher more directly by giving him a dose of sea-sickness.

The last ten years have witnessed a tremendous proliferation of equipment for oceanographic research. Industry offers off-the-shelf hardware and electronic packages in many designs. Some of these have been found to be satisfactory, others have not. The competition among manufacturers trying to corner the market for tools for specific research tasks has produced a healthy climate, and designs of gear are steadily improving with a concomitant increase in the sophistication level of ocean research and the quality of instrumentation. It is to be hoped that this trend will continue so that design requirements will keep pace with the needs of the researcher while research in turn will be stimulated by the advances in technology.

The national effort in oceanographic research has already increased the researcher's ability to monitor the oceans over larger areas of space and longer periods of time. Worldwide networks of self-contained instrument buoys which can measure a variety of oceanic parameters at their open-sea moored positions are contemplated. These buoy stations will radio in data which can then be used to forecast oceanographic conditions in much the same way that weather station data are presently used for meteorological predictions.

Nearly synoptic coverage of the surface of the oceans is possible from orbiting satellites. However, the instruments necessary for remote sensing of the sea surface are still in the experimental stage. Surface roughness, waves, surface temperature, and color can be perceived by remote sensors, but some of the sensors used in these monitoring programs do not have all-weather capability; they are subject to interference by atmospheric conditions, clouds, and moisture content. Furthermore, coverage and the resolution of the ocean features are additional problems at satellite altitudes. If full coverage of the earth is desired every 24 hours, definition suffers. If the coverage is decreased to obtain more detail, measurements become discontinuous. It is hoped that sensors and sensing techniques will be developed that allow all-weather surveillance and direct measurements of the 58% of sea surface normally hidden by clouds. Then oceanographers will be able to use world-coverage oceanographic data of the sea surface in the same way that meteorologists use Tiros weather data to improve their forecasts and understanding of the rates at which the character of the ocean-atmospheric system changes.

INDEX

INDEX

Italic page numbers indicate figures.